INTERNATIONAL UNION OF PURE AND APPLIED CHEMISTRY

ANALYTICAL CHEMISTRY DIVISION
COMMISSION ON SOLUBILITY DATA

SOLUBILITY DATA SERIES

Volume 25

METALS IN MERCURY

SOLUBILITY DATA SERIES

Volume 1	H. L. Clever, *Helium and Neon*
Volume 2	H. L. Clever, *Krypton, Xenon and Radon*
Volume 3	M. Salomon, *Silver Azide, Cyanide, Cyanamides, Cyanate, Selenocyanate and Thiocyanate*
Volume 4	H. L. Clever, *Argon*
Volume 5/6	C. L. Young, *Hydrogen and Deuterium*
Volume 7	R. Battino, *Oxygen and Ozone*
Volume 8	C. L. Young, *Oxides of Nitrogen*
Volume 9	W. Hayduk, *Ethane*
Volume 10	R. Battino, *Nitrogen and Air*
Volume 11	B. Scrosati and C. A. Vincent, *Alkali Metal, Alkaline Earth Metal and Ammonium Halides. Amide Solvents*
Volume 12	C. L. Young, *Sulfur Dioxide, Chlorine, Fluorine and Chlorine Oxides*
Volume 13	S. Siekierski, T. Mioduski and M. Salomon, *Scandium, Yttrium, Lanthanum and Lanthanide Nitrates*
Volume 14	H. Miyamoto, M. Salomon and H. L. Clever, *Alkaline Earth Metal Halates*
Volume 15	A. F. M. Barton, *Alcohols with Water*
Volume 16/17	E. Tomlinson and A. Regosz, *Antibiotics: I. β-Lactam Antibiotics*
Volume 18	O. Popovych, *Tetraphenylborates*
Volume 19	C. L. Young, *Cumulative Index: Volumes 1–18*
Volume 20	A. L. Horvath and F. W. Getzen, *Halogenated Benzenes, Toluenes and Phenols with Water*
Volume 21	C. L. Young and P. G. T. Fogg, *Ammonia, Amines, Phosphine, Arsine, Stibine, Silane, Germane and Stannane in Organic Solvents*
Volume 22	T. Mioduski and M. Salomon, *Scandium, Yttrium, Lanthanum and Lanthanide Halides in Nonaqueous Solvents*
Volume 23	T. P. Dirkse, *Copper, Silver, Gold, and Zinc, Cadmium, Mercury Oxides and Hydroxides*
Volume 24	W. Hayduk, *Propane, Butane and 2-Methylpropane*
Volume 25	C. Hirayama, Z. Galus and C. Guminski, *Metals in Mercury*
Volume 26	M. R. Masson, H. D. Lutz and B. Engelen, *Sulfites, Selenites and Tellurites*

Selected Volumes in Preparation

E. Tomlinson, *Antibiotics: II. Peptide Antibiotics*

H. L. Clever and C. L. Young, *Methane*

H. Miyamoto, *Copper and Silver Halates*

J. W. Lorimer, *Beryllium, Strontium, Barium and Radium Sulfates*

H. L. Clever and C. L. Young, *Carbon Dioxide*

NOTICE TO READERS

Dear Reader

If your library is not already a standing-order customer or subscriber to the Solubility Data Series, may we recommend that you place a standing order or subscription order to receive immediately upon publication all new volumes published in this valuable series. Should you find that these volumes no longer serve your needs, your order can be cancelled at any time without notice.

Robert Maxwell
Publisher at Pergamon Press

SOLUBILITY DATA SERIES

Editor-in-Chief
A. S. KERTES

Volume 25

METALS IN MERCURY

Volume Editors

C. HIRAYAMA
Westinghouse Corporation
Pittsburgh, PA, USA

Z. GALUS
University of Warsaw
Warsaw, Poland

C. GUMINSKI
University of Warsaw
Warsaw, Poland

Contributors

Z. GALUS
University of Warsaw
Warsaw, Poland

C. GUMINSKI
University of Warsaw
Warsaw, Poland

J. BALEJ
Czechoslovak Academy of Science
Prague, Czechoclovakia

M. SALOMON
Solubility Data Project
PO Box 254, Fair Haven, NJ, USA

PERGAMON PRESS
OXFORD · NEW YORK · BEIJING · FRANKFURT
SÃO PAULO · SYDNEY · TOKYO · TORONTO

U.K.	Pergamon Press, Headington Hill Hall, Oxford OX3 0BW, England
U.S.A.	Pergamon Press, Maxwell House, Fairview Park, Elmsford, New York 10523, U.S.A.
PEOPLE'S REPUBLIC OF CHINA	Pergamon Press, Qianmen Hotel, Beijing, People's Republic of China
FEDERAL REPUBLIC OF GERMANY	Pergamon Press, Hammerweg 6, D-6242 Kronberg, Federal Republic of Germany
BRAZIL	Pergamon Editora, Rua Eça de Queiros, 346, CEP 04011, São Paulo, Brazil
AUSTRALIA	Pergamon Press Australia, P.O. Box 544, Potts Point, N.S.W. 2011, Australia
JAPAN	Pergamon Press, 8th Floor, Matsuoka Central Building, 1-7-1 Nishishinjuku, Shinjuku-ku, Tokyo 160, Japan
CANADA	Pergamon Press Canada, Suite 104, 150 Consumers Road, Willowdale, Ontario M2J 1P9, Canada

Copyright © 1986 International Union of Pure and Applied Chemistry

All Rights Reserved. No part of this publication may be reproduced, stored in a retrieval system or transmitted in any form or by any means: electronic, electrostatic, magnetic tape, mechanical, photocopying, recording or otherwise, without permission in writing from the copyright holders.

First edition 1986

Library of Congress Cataloging in Publication Data

Solubility data series.—Vol. 1 —Oxford; New York:
Pergamon, c 1979-
v.; 28 cm.
Separately cataloged and classified in LC before no. 18.
ISSN 0191-5622 = Solubility data series.
1. Solubility—Tables—Collected works.
QD543.S6629   541.3'42'05-dc19   85-641351
AACR 2 MARC-S

British Library Cataloguing in Publication Data

Metals in mercury.—(Solubility data
series; v. 25)
1. Mercury compounds—Solubility
I. Hirayama, C. II. Galus, Z. III. Guminski, C.
IV. Balej, J. V. Salomon, M. VI. Series
546'.6632   QD181.H6
ISBN 0-08-023921-8

Printed in Great Britain by A. Wheaton & Co. Ltd., Exeter

CONTENTS

Foreword   vii

Preface   x

Introduction: The Solubility of Solids in Liquids   xiii

1	Lithium, Sodium, Potassium, Rubidium, Cesium	1
2	Beryllium, Magnesium, Calcium, Strontium, Barium	55
3	Boron, Aluminum, Gallium, Indium, Thallium	83
4	Carbon, Silicon, Germanium, Tin, Lead	134
5	Arsenic, Antimony, Bismuth	172
6	Tellurium	194
7	Scandium, Yttrium, Lanthanum, Lanthanides	206
	7.1 Scandium, Yttrium, Lanthanum	206
	7.2 Cerium, Praseodymium, Neodymium, Samarium, Europium, Gadolinium, Terbium, Dysprosium, Holmium, Erbium, Thulium, Ytterbium, Lutetium	214
8	Titanium, Zirconium, Hafnium	258
9	Vanadium, Niobium, Tantalum	268
10	Chromium, Molybdenum, Tungsten	277
11	Manganese, Rhenium	285
12	Iron, Ruthenium, Osmium	301
13	Cobalt, Rhodium, Iridium	310
14	Nickel, Palladium, Platinum	317
15	Copper, Silver, Gold	335
16	Zinc, Cadmium	385
17	Radioactive Elements	421
	17.1 Technetium, Promethium, Polonium, Francium, Radium, Actinium	421
	17.2 Thorium, Protactinium, Uranium, Plutonium	422

System Index   437

Registry Number Index   439

Author Index   441

SOLUBILITY DATA SERIES

Editor-in-Chief

A. S. KERTES
The Hebrew University
Jerusalem, Israel

EDITORIAL BOARD

H. Akaiwa (Japan)
A. F. M. Barton (Australia)
R. Battino (USA)
Kathryn R. Bullock (USA)
H. L. Clever (USA)
R. Cohen-Adad (France)
T. P. Dirkse (USA)
P. Franzosini* (Italy)
J. Fu (China)
L. H. Gevantman (USA)
J. Hala (Czechoslovakia)
G. T. Hefter (Australia)
C. Kalidas (India)
Irma Lambert (France)
J. W. Lorimer (Canada)
J. D. Navratil (USA)
M. Salomon (USA)
D. G. Shaw (USA)
R. P. T. Tomkins (USA)
V. M. Valyashko (USSR)
C. L. Young (Australia)

*Deceased January 1986

Publication Coordinator
P. D. GUJRAL
IUPAC Secretariat, Oxford, UK

INTERNATIONAL UNION OF PURE AND APPLIED CHEMISTRY
IUPAC Secretariat: Bank Court Chambers, 2–3 Pound Way,
Cowley Centre, Oxford OX4 3YF, UK

FOREWORD

*If the knowledge is
undigested or simply wrong,
more is not better*

How to communicate and disseminate numerical data effectively in chemical science and technology has been a problem of serious and growing concern to IUPAC, the International Union of Pure and Applied Chemistry, for the last two decades. The steadily expanding volume of numerical information, the formulation of new interdisciplinary areas in which chemistry is a partner, and the links between these and existing traditional subdisciplines in chemistry, along with an increasing number of users, have been considered as urgent aspects of the information problem in general, and of the numerical data problem in particular.

Among the several numerical data projects initiated and operated by various IUPAC commissions, the *Solubility Data Project* is probably one of the most ambitious ones. It is concerned with preparing a comprehensive critical compilation of data on solubilities in all physical systems, of gases, liquids and solids. Both the basic and applied branches of almost all scientific disciplines require a knowledge of solubilities as a function of solvent, temperature and pressure. Solubility data are basic to the fundamental understanding of processes relevant to agronomy, biology, chemistry, geology and oceanography, medicine and pharmacology, and metallurgy and materials science. Knowledge of solubility is very frequently of great importance to such diverse practical applications as drug dosage and drug solubility in biological fluids, anesthesiology, corrosion by dissolution of metals, properties of glasses, ceramics, concretes and coatings, phase relations in the formation of minerals and alloys, the deposits of minerals and radioactive fission products from ocean waters, the composition of ground waters, and the requirements of oxygen and other gases in life support systems.

The widespread relevance of solubility data to many branches and disciplines of science, medicine, technology and engineering, and the difficulty of recovering solubility data from the literature, lead to the proliferation of published data in an ever increasing number of scientific and technical primary sources. The sheer volume of data has overcome the capacity of the classical secondary and tertiary services to respond effectively.

While the proportion of secondary services of the review article type is generally increasing due to the rapid growth of all forms of primary literature, the review articles become more limited in scope, more specialized. The disturbing phenomenon is that in some disciplines, certainly in chemistry, authors are reluctant to treat even those limited-in-scope reviews exhaustively. There is a trend to preselect the literature, sometimes under the pretext of reducing it to manageable size. The crucial problem with such preselection - as far as numerical data are concerned - is that there is no indication as to whether the material was excluded by design or by a less than thorough literature search. We are equally concerned that most current secondary sources, critical in character as they may be, give scant attention to numerical data.

On the other hand, tertiary sources - handbooks, reference books and other tabulated and graphical compilations - as they exist today are comprehensive but, as a rule, uncritical. They usually attempt to cover whole disciplines, and thus obviously are superficial in treatment. Since they command a wide market, we believe that their service to the advancement of science is at least questionable. Additionally, the change which is taking place in the generation of new and diversified numerical data, and the rate at which this is done, is not reflected in an increased third-level service. The emergence of new tertiary literature sources does not parallel the shift that has occurred in the primary literature.

With the status of current secondary and tertiary services being as briefly stated above, the innovative approach of the *Solubility Data Project* is that its compilation and critical evaluation work involve consolidation and reprocessing services when both activities are based on intellectual and scholarly reworking of information from primary sources. It comprises compact compilation, rationalization and simplification, and the fitting of isolated numerical data into a critically evaluated general framework.

The *Solubility Data Project* has developed a mechanism which involves a number of innovations in exploiting the literature fully, and which contains new elements of a more imaginative approach for transfer of reliable information from primary to secondary/tertiary sources. *The fundamental trend of the Solubility Data Project is toward integration of secondary and tertiary services with the objective of producing in-depth critical analysis and evaluation which are characteristic to secondary services, in a scope as broad as conventional tertiary services.*

Fundamental to the philosophy of the project is the recognition that the basic element of strength is the active participation of career scientists in it. Consolidating primary data, producing a truly critically-evaluated set of numerical data, and synthesizing data in a meaningful relationship are demands considered worthy of the efforts of top scientists. Career scientists, who themselves contribute to science by their involvement in active scientific research, are the backbone of the project. The scholarly work is commissioned to recognized authorities, involving a process of careful selection in the best tradition of IUPAC. This selection in turn is the key to the quality of the output. These top experts are expected to view their specific topics dispassionately, paying equal attention to their own contributions and to those of their peers. They digest literature data into a coherent story by weeding out what is wrong from what is believed to be right. To fulfill this task, the evaluator must cover *all* relevant open literature. No reference is excluded by design and every effort is made to detect every bit of relevant primary source. Poor quality or wrong data are mentioned and explicitly disqualified as such. In fact, it is only when the reliable data are presented alongside the unreliable data that proper justice can be done. The user is bound to have incomparably more confidence in a succinct evaluative commentary and a comprehensive review with a complete bibliography to both good and poor data.

It is the standard practice that the treatment of any given solute-solvent system consists of two essential parts: I. Critical Evaluation and Recommended Values, and II. Compiled Data Sheets.

The Critical Evaluation part gives the following information:

(i) a verbal text of evaluation which discusses the numerical solubility information appearing in the primary sources located in the literature. The evaluation text concerns primarily the quality of data after consideration of the purity of the materials and their characterization, the experimental method employed and the uncertainties in control of physical parameters, the reproducibility of the data, the agreement of the worker's results on accepted test systems with standard values, and finally, the fitting of data, with suitable statistical tests, to mathematical functions;

(ii) a set of recommended numerical data. Whenever possible, the set of recommended data includes weighted average and standard deviations, and a set of smoothing equations derived from the experimental data endorsed by the evaluator;

(iii) a graphical plot of recommended data.

The Compilation part consists of data sheets of the best experimental data in the primary literature. Generally speaking, such independent data sheets are given only to the best and endorsed data covering the known range of experimental parameters. Data sheets based on primary sources where the data are of a lower precision are given only when no better data are available. Experimental data with a precision poorer than considered acceptable are reproduced in the form of data sheets when they are the only known data for a particular system. Such data are considered to be still suitable for some applications, and their presence in the compilation should alert researchers to areas that need more work.

The typical data sheet carries the following information:

(i) components - definition of the system - their names, formulas and Chemical Abstracts registry numbers;
(ii) reference to the primary source where the numerical information is reported. In cases when the primary source is a less common periodical or a report document, published though of limited availability, abstract references are also given;
(iii) experimental variables;
(iv) identification of the compiler;
(v) experimental values as they appear in the primary source. Whenever available, the data may be given both in tabular and graphical form. If auxiliary information is available, the experimental data are converted also to SI units by the compiler.

Under the general heading of Auxiliary Information, the essential experimental details are summarized:

(vi) experimental method used for the generation of data;
(vii) type of apparatus and procedure employed;
(viii) source and purity of materials;
(ix) estimated error;
(x) references relevant to the generation of experimental data as cited in the primary source.

This new approach to numerical data presentation, formulated at the initiation of the project and perfected as experience has accumulated, has been strongly influenced by the diversity of background of those whom we are supposed to serve. We thus deemed it right to preface the evaluation/compilation sheets in each volume with a detailed discussion of the principles of the accurate determination of relevant solubility data and related thermodynamic information.

Finally, the role of education is more than corollary to the efforts we are seeking. The scientific standards advocated here are necessary to strengthen science and technology, and should be regarded as a major effort in the training and formation of the next generation of scientists and engineers. Specifically, we believe that there is going to be an impact of our project on scientific-communication practices. The quality of consolidation adopted by this program offers down-to-earth guidelines, concrete examples which are bound to make primary publication services more responsive than ever before to the needs of users. The self-regulatory message to scientists of the early 1970s to refrain from unnecessary publication has not achieved much. A good fraction of the literature is still cluttered with poor-quality articles. The Weinberg report (in 'Reader in Science Information', ed. J. Sherrod and A. Hodina, Microcard Editions Books, Indian Head, Inc., 1973, p. 292) states that 'admonition to authors to restrain themselves from premature, unnecessary publication can have little effect unless the climate of the entire technical and scholarly community encourages restraint...' We think that projects of this kind translate the climate into operational terms by exerting pressure on authors to avoid submitting low-grade material. The type of our output, we hope, will encourage attention to quality as authors will increasingly realize that their work will not be suited for permanent retrievability unless it meets the standards adopted in this project. It should help to dispel confusion in the minds of many authors of what represents a permanently useful bit of information of an archival value, and what does not.

If we succeed in that aim, even partially, we have then done our share in protecting the scientific community from unwanted and irrelevant, wrong numerical information.

A. S. Kertes

PREFACE

This volume is concerned with the solubility of metals in mercury, and includes all of the metals and the metalloids carbon, silicon and boron. The solubility only in the seventy-six binary amalgams is considered here. The compilation of the solubility data for these binary systems includes numerous reports, such as those published by the U.S. Atomic Energy Commission from its various laboratories. The literature coverage for this volume extends through 1983.

The solubility of a metal in mercury at a given temperature is represented by the concentration of the saturated solution which is in equilibrium with the solid phase. The solid phase may be the pure metal, the metal saturated with mercury, or an intermetallic compound with mercury. This concentration also is represented by the liquidus point at the given temperature on the binary phase diagram. Clearly, the solubility also is represented by the crystallization temperature of the liquid amalgam.

Only those parts of the complete metal-mercury systems are included in which the solid metal, or a metal amalgam, appear as solid phases. In those systems where a phase diagram has been accurately determined, the equilibrium solid phase is clearly defined; the published phase diagrams for these systems are included in the Critical Evaluation, and should correctly aid the reader in assigning the solid-liquid equilibrium. However, there are some systems where there is disagreement on the equilibrium solid phase so that the solid-liquid equilibrium for these systems cannot be accurately defined. There are certain phase diagrams which have been constructed from precise data, but the liquidus data may be somewhat questionable because equilibrium may not have been attained during the short equilibration times employed. Instances of possible supersaturation in the determination of the liquidus from cooling curves are noted by the evaluators. In this volume, the emphasis is on accurate, evaluated solubility data rather than phase relations in the various systems.

Concentrations in the metal-mercury systems are mostly reported in atomic percent, at %, rather than in mole percent. The rationale for the non-SI unit is that each system is represented by the equilibrium of two atomic species, and much of the literature data on binary metallic systems are reported so.

The solubility of a number of metals in mercury, especially the refractory metals, is very low, and often below the experimental detection limit. For such systems only a selected number of data sheets were compiled for those reports which gave the highest solubility limit as determined by a well defined method. However, the solubility in these systems may be estimated by the semiempirical equations of Kozin. The first equation (1) is given by

$$\ln (100x_1) = -0.4 \left[1 + \frac{\Delta H_m (T_m - T)}{RT_m T} \right]^2 \left[\frac{\Delta H_m (T_m - T)}{RT_m T} \right]^{0.001} \quad [1]$$

where the atomic percent solubility of the metal, $100x_1$, is a function of its enthalpy of fusion, ΔH_m, and its melting point, T_m. Kozin subsequently reported (2) a second solubility equation,

$$\ln (100x_1) = -\left[\frac{\Delta H_m (T_m - T)}{RT_m T} \right]^{1.39} \quad [2]$$

Equation [2] was derived from the Schroeder relation in which the exponent is unity for ideal solutions. The exponent, 1.39, in eq. [2] results from fitting known values of solubilities in the binary amalgams to ΔH_m and T_m. It was reported by Kozin (2) that the mean standard scatter of points for systems of known solubilities is ±0.028 at a 95% confidence level in eq. [2]. Estimates from this equation for the solubility at 298 K for some of the binary systems are near the experimental values, but there also are systems where the estimates are at great variance from experimental values. For systems of very low solubility, where experimental data are not available, eq. [2] may be applied only as a first approximation.

For some of the metal-Hg systems the data were reported only graphically; some of the liquidus covered an extensive composition range, others only a narrow composition range. Because the numerical data are of interest to many workers, the data points from these graphical presentations were visually read from the curves and are compiled on the data sheets. Admittedly, the error in abstracting such data from the curves may be large, depending on the size of the original figure.

For every system where experimental solubility data were reported, all of the data were plotted on a semilogarithmic paper (of 60 x 20 cm dimensions) as log $(100x_1)$ vs. $(T/K)^{-1}$. The data were then evaluated by visually fitting the best curve. Evaluated solubility data are tabulated at the end of the Critical Evaluation. When at least two independent works agreed within experimental error, the solubility values were assigned to the recommended category. Values were assigned as tentative when only one reliable work was reported, or when the mean value from two or more reliable works was outside of the error limits. In this tabulation, three, two, or one significant figures is assigned for respective precisions that are better than ±1 and ±10% and worse than ±10%. There were no data that agreed to within ±0.1%.

In a number of papers the temperature of the measurement was reported as "room temperature"; in plotting these data on the solubility curve, the temperature was arbitrarily assigned as 293 K.

Data for concentrated solutions which were reported in mol atom dm^{-3} without specifying the density were not useful for this compilation; solubilities in atom percent could not be assigned to these data.

Because of the large number of binary systems in this volume, the presentation is grouped according to the Periodic Table. The non-transition metals are given first in sequence starting from the alkali metals, followed by the transition metals in similar order. The actinides and the unstable radioactive elements are presented at the end of the volume.

Some previous compilations dealing with solubilities in selected amalgam systems (3-10) are considered incomplete, and the data in some of these references erroneous.

The editors acknowledge the encouragement of IUPAC Commission V.5 under whose authorization this work was initiated. The Editor also acknowledges the helpful advice and suggestions made by Dr. Mark Salomon during the course of editing this volume. Acknowledgment also is made to the Westinghouse Electric Corporation for providing the Editor with library and stenographic services during this project. It is also a pleasure to acknowledge the aid of Mrs. Joyce Walsh for the complete typing of this volume.

Acknowledgment is made to the following for permission to reproduce various phase diagrams directly from their publications: The American Society of Metals; McGraw-Hill Book Company; Elsevier Science Publishers; R. Oldenbourg Verlag; Der Deutschen Gesellschaft Fur Metallkunde; Academic Press Inc., Acta Metallurgica Inc.; and VAAP, the Copyright Agency of the USSR.

REFERENCES

1. Kozin, L.F. *Tr. Inst. Khim. Nauk Akad. Nauk Kaz. SSR* 1962, *9*, 101.
2. Kozin. L.F. *Fiziko-Khimicheskie Osnovy Amalgamnoi Metallurgii*, Nauka, Alma-Ata, 1964.
3. Hansen, M.; Anderko, K. *Constitution of Binary Alloys*, McGraw-Hill, New York, NY, 1958.
4. Kozlovsky, M.T.; Zebreva, A.I.; Gladyshev, V.P. *Amalgamy i ikh Primenenie*, Nauka, Alma-Ata, 1971.
5. Kozlovsky, M.T.; Zebreva, A.I. *Progress in Polarography*, Vol. III, Zuman, P. and Meites, L., Eds., Wiley-Interscience, New York, NY, 1972.
6. Gavze, M.N. *Koroziya i Smachivaemost Metallov Rtutei*, Nauka, Moskva, 1969.
7. Kozin, L.F.; Nigmetova, R.Sh.; Dergacheva, M.B. *Termodinamika Binarnykh Amalgamnykh Sistem*, Nauka, Alma-Ata, 1977.
8. Jangg, G. *Metall* 1965, *19*, 442.
9. Vol, A.E. *Stroenie i Svoistva Dvoinykh Metallicheskikh Sistem*, Fizmatgiz, Moskva, Vol. I 1959; Vol. II 1962; Vol. III (with Kagan, I.K.) 1976; Vol. IV 1979.
10. Seidel, A. *Solubilities of Inorganic and Organic Compounds*, Vol. I, 3rd ed., D. Van Nostrand Company, Inc., New York, NY 1940; also Supplements to this volume.

C. Guminski and Z. Galus
Department of Chemistry
University of Warsaw
Warsaw, Poland

C. Hirayama
Westinghouse Electric Corporation
Research and Development Center
Pittsburgh, PA, USA

November, 1985

INTRODUCTION: THE SOLUBILITY OF SOLIDS IN LIQUIDS

Nature of the Project

The Solubility Data Project (SDP) has as its aim a comprehensive search of the literature for solubilities of gases, liquids, and solids in liquids or solids. Data of suitable precision are compiled on data sheets in a uniform format. The data for each system are evaluated, and where data from different sources agree sufficiently, recommended values are proposed. The evaluation sheets, recommended values, and compiled data sheets are published on consecutive pages.

This series of volumes includes solubilities of solids of all types in liquids of all types.

Definitions

A *mixture* (1,2) describes a gaseous, liquid, or solid phase containing more than one substance, when the substances are all treated in the same way.

A *solution* (1,2) describes a liquid or solid phase containing more than one substance, when for convenience one of the substances, which is called the *solvent* and may itself be a mixture, is treated differently than the other substances, which are called *solutes*. If the sum of the mole fractions of the solutes is small compared to unity, the solution is called a *dilute solution*.

The *solubility* of a substance B is the relative proportion of B (or a substance related chemically to B) in a mixture which is saturated with respect to solid B at a specified temperature and pressure. *Saturated* implies the existence of equilibrium with respect to the processes of dissolution and precipitation; the equilibrium may be stable or metastable. The solubility of a metastable substance is usually greater than that of the corresponding stable substance. (Strictly speaking, it is the activity of the metastable substance that is greater.) Care must be taken to distinguish true metastability from supersaturation, where equilibrium does not exist.

Either point of view, mixture or solution, may be taken in describing solubility. The two points of view find their expression in the quantities used as measures of solubility and in the reference states used for definition of activities and activity coefficients.

The qualifying phrase "substance related chemically to B" requires comment. The composition of the saturated mixture (or solution) can be described in terms of any suitable set of thermodynamic components. Thus, the solubility of a salt hydrate in water is usually given as the relative proportion of anhydrous salt in solution, rather than the relative proportions of hydrated salt and water.

Quantities Used as Measures of Solubility

1. *Mole fraction* of substance B, x_B:

$$x_B = n_B / \sum_{i=1}^{c} n_i \qquad (1)$$

where n_i is the amount of substance of substance i, and c is the number of distinct substances present (often the number of thermodynamic components in the system). *Mole per cent* of B is $100 \, x_B$.

2. *Mass fraction* of substance B, w_B:

$$w_B = m'_B / \sum_{i=1}^{c} m'_i \qquad (2)$$

where m'_i is the mass of substance i. *Mass per cent* of B is $100 \, w_B$. The equivalent terms weight fraction and weight per cent are not used.

3. *Solute mole (mass) fraction* of solute B (3,4):

$$x_{S,B} = n_B / \sum_{i=1}^{c'} n_i = x_B / \sum_{i=1}^{c'} x_i \qquad (3)$$

where the summation is over the solutes only. For the solvent A, $x_{S,A} = x_A$. These quantities are called *Jänecke mole (mass) fractions* in many papers.

4. *Molality* of solute B (1,2) in a solvent A:

$$m_B = n_B/n_A M_A \qquad \text{SI base units: mol kg}^{-1} \qquad (4)$$

where M_A is the molar mass of the solvent.

5. *Concentration* of solute B (1,2) in a solution of volume V:

$$c_B = [B] = n_B/V \qquad \text{SI base units: mol m}^{-3} \qquad (5)$$

The terms molarity and molar are not used.

    Mole and mass fractions are appropriate to either the mixture or the solution points of view. The other quantities are appropriate to the solution point of view only. In addition of these quantities, the following are useful in conversions between concentrations and other quantities.

6. *Density*: $\rho = m/V$ SI base units: kg m^{-3} (6)

7. *Relative density*: d; the ratio of the density of a mixture to the density of a reference substance under conditions which must be specified for both (1). The symbol $d_t^{t'}$ will be used for the density of a mixture at t°C, 1 atm divided by the density of water at t'°C, 1 atm.

    Other quantities will be defined in the prefaces to individual volumes or on specific data sheets.

Thermodynamics of Solubility

    The principal aims of the Solubility Data Project are the tabulation and evaluation of: (a) solubilities as defined above; (b) the nature of the saturating solid phase. Thermodynamic analysis of solubility phenomena has two aims: (a) to provide a rational basis for the construction of functions to represent solubility data; (b) to enable thermodynamic quantities to be extracted from solubility data. Both these aims are difficult to achieve in many cases because of a lack of experimental or theoretical information concerning activity coefficients. Where thermodynamic quantities can be found, they are not evaluated critically, since this task would involve critical evaluation of a large body of data that is not directly relevant to solubility. The following discussion is an outline of the principal thermodynamic relations encountered in discussions of solubility. For more extensive discussions and references, see books on thermodynamics, e.g., (5-10).

Activity Coefficients (1)

    (a) *Mixtures*. The activity coefficient f_B of a substance B is given by

$$RT \ln(f_B x_B) = \mu_B - \mu_B^* \qquad (7)$$

where μ_B is the chemical potential, and μ_B^* is the chemical potential of pure B at the same temperature and pressure. For any substance B in the mixture,

$$\lim_{x_B \to 1} f_B = 1 \qquad (8)$$

    (b) *Solutions*.

        (i) *Solute substance, B*. The molal activity coefficient γ_B is given by

$$RT \ln(\gamma_B m_B) = \mu_B - (\mu_B - RT \ln m_B)^\infty \qquad (9)$$

where the superscript ∞ indicates an infinitely dilute solution. For any solute B,

$$\gamma_B^\infty = 1 \qquad (10)$$

Activity coefficients y_B connected with concentration c_B, and $f_{x,B}$ (called the *rational activity coefficient*) connected with mole fraction x_B are defined in analogous ways. The relations among them are (1,9):

$$\gamma_B = x_A f_{x,B} = V_A^*(1 - \sum_s c_s) y_B \qquad (11)$$

or

$$f_{x,B} = (1 + M_A \sum_s m_s)\gamma_B = V_A^* y_B/V_m \qquad (12)$$

or

$$y_B = (V_A + M_A \sum_s m_s V_s)\gamma_B/V_A^* = V_m f_{x,B}/V_A^* \qquad (13)$$

where the summations are over all solutes, V_A^* is the molar volume of the pure solvent, V_i is the partial molar volume of substance i, and V_m is the molar volume of the solution.

For an electrolyte solute $B \equiv C_{\nu+}A_{\nu-}$, the molal activity is replaced by (9)

$$\gamma_B m_B = \gamma_{\pm}^{\nu} m_B^{\nu} Q^{\nu} \qquad (14)$$

where $\nu = \nu_+ + \nu_-$, $Q = (\nu_+^{\nu_+}\nu_-^{\nu_-})^{1/\nu}$, and $\gamma_{\pm}$ is the mean ionic molal activity coefficient. A similar relation holds for the concentration activity $y_B c_B$. For the mol fractional activity,

$$f_{x,B} x_B = \nu_+^{\nu_+}\nu_-^{\nu_-} f_{\pm}^{\nu} x_{\pm}^{\nu} \qquad (15)$$

The quantities x_+ and x_- are the ionic mole fractions (9), which for a single solute are

$$x_+ = \nu_+ x_B/[1+(\nu-1)x_B]; \qquad x_- = \nu_- x_B/[1+(\nu-1)x_B] \qquad (16)$$

(ii) *Solvent*, A:

The *osmotic coefficient*, ϕ, of a solvent substance A is defined as (1):

$$\phi = (\mu_A^* - \mu_A)/RT\, M_A \sum_s m_s \qquad (17)$$

where μ_A^* is the chemical potential of the pure solvent.

The *rational osmotic coefficient*, ϕ_x, is defined as (1):

$$\phi_x = (\mu_A - \mu_A^*)/RT \ln x_A = \phi M_A \sum_s m_s / \ln(1 + M_A \sum_s m_s) \qquad (18)$$

The activity, a_A, or the activity coefficient f_A is often used for the solvent rather than the osmotic coefficient. The activity coefficient is defined relative to pure A, just as for a mixture.

The Liquid Phase

A general thermodynamic differential equation which gives solubility as a function of temperature, pressure and composition can be derived. The approach is that of Kirkwood and Oppenheim (7). Consider a solid mixture containing c' thermodynamic components i. The Gibbs-Duhem equation for this mixture is:

$$\sum_{i=1}^{c'} x_i'(S_i' dT - V_i' dp + d\mu_i) = 0 \qquad (19)$$

A liquid mixture in equilibrium with this solid phase contains c thermodynamic components i, where, usually, $c \geq c'$. The Gibbs-Duhem equation for the liquid mixture is:

$$\sum_{i=1}^{c'} x_i(S_i dT - V_i dp + d\mu_i) + \sum_{i=c'+1}^{c} x_i(S_i dT - V_i dp + d\mu_i) = 0 \qquad (20)$$

Eliminate $d\mu_1$ by multiplying (19) by x_1 and (20) x_1'. After some algebra, and use of:

$$d\mu_i = \sum_{j=2}^{c} G_{ij} dx_j - S_i dT + V_i dp \qquad (21)$$

where (7)

$$G_{ij} = (\partial \mu_i/\partial x_j)_{T,P,x_i \neq x_j} \qquad (22)$$

it is found that

$$\sum_{i=2}^{c'}\sum_{j=2}^{c}(x_i' - x_i x_1'/x_1)G_{ij}dx_j - (x_1'/x_1)\sum_{i=c'+1}^{c}\sum_{j=2}^{c} x_i G_{ij}dx_j$$

$$= \sum_{i=1}^{c'} x_i'(H_i - H_i')dT/T - \sum_{i=1}^{c'} x_i'(V_i - V_i')dp \qquad (23)$$

where

$$H_i - H_i' = T(S_i - S_i') \tag{24}$$

is the enthalpy of transfer of component i from the solid to the liquid phase, at a given temperature, pressure and composition, and H_i, S_i, V_i are the partial molar enthalpy, entropy, and volume of component i. Several special cases (all with pressure held constant) will be considered. Other cases will appear in individual evaluations.

(a) Solubility as a function of temperature.
Consider a binary solid compound A_nB in a single solvent A. There is no fundamental thermodynamic distinction between a binary compound of A and B which dissociates completely or partially on melting and a solid mixture of A and B; the binary compound can be regarded as a solid mixture of constant composition. Thus, with $c = 2$, $c' = 1$, $x_A' = n/(n+1)$, $x_B' = 1/(n+1)$, eqn (23) becomes

$$(1/x_B - n/x_A)\left\{1 + \left(\frac{\partial \ln f_B}{\partial \ln x_B}\right)_{T,P}\right\} dx_B = (nH_A + H_B - H_{AB}^*) dT/RT^2 \tag{25}$$

where the mole fractional activity coefficient has been introduced. If the mixture is a non-electrolyte, and the activity coefficients are given by the expression for a simple mixture (6):

$$RT \ln f_B = w x_A^2 \tag{26}$$

then it can be shown that, if w is independent of temperature, eqn (25) can be integrated (cf. (5), Chap. XXIII, sect. 5). The enthalpy term becomes

$$nH_A + H_B - H_{AB}^* = \Delta H_{AB} + n(H_A - H_A^*) + (H_B - H_B^*)$$

$$= \Delta H_{AB} + w(nx_B^2 + x_A^2) \tag{27}$$

where ΔH_{AB} is the enthalpy of melting and dissociation of one mole of pure solid A_nB, and H_A^*, H_B^* are the molar enthalpies of pure liquid A and B. The differential equation becomes

$$R \, d \ln\{x_B(1-x_B)^n\} = -\Delta H_{AB} \, d\left(\frac{1}{T}\right) - w \, d\left(\frac{x_A^2 + nx_B^2}{T}\right) \tag{28}$$

Integration from x_B, T to $x_B = 1/(1+n)$, $T = T^*$, the melting point of the pure binary compound, gives:

$$\ln\{x_B(1-x_B)^n\} \simeq \ln\left\{\frac{n^n}{(1+n)^{n+1}}\right\} - \left\{\frac{\Delta H_{AB}^* - T^* \Delta C_p^*}{R}\right\}\left(\frac{1}{T} - \frac{1}{T^*}\right)$$

$$+ \frac{\Delta C_p^*}{R} \ln\left(\frac{T}{T^*}\right) - \frac{w}{R}\left\{\frac{x_A^2 + nx_B^2}{T} - \frac{n}{(n+1)T^*}\right\} \tag{29}$$

where ΔC_p^* is the change in molar heat capacity accompanying fusion plus decomposition of the compound at temperature T^*, (assumed here to be independent of temperature and composition), and ΔH_{AB}^* is the corresponding change in enthalpy at $T = T^*$. Equation (29) has the general form

$$\ln\{x_B(1-x_B)^n\} = A_1 + A_2/T + A_3 \ln T + A_4(x_A^2 + nx_B^2)/T \tag{30}$$

If the solid contains only component B, $n = 0$ in eqn (29) and (30).
If the infinite dilution standard state is used in eqn (25), eqn (26) becomes

$$RT \ln f_{x,B} = w(x_A^2 - 1) \tag{31}$$

and (27) becomes

$$nH_A + H_B - H_{AB} = (nH_A^* + H_B^\infty - H_{AB}^*) + n(H_A - H_A^*) + (H_B - H_B^\infty) = \Delta H_{AB}^\infty + w(nx_B^2 + x_A^2 - 1) \tag{32}$$

where the first term, ΔH_{AB}^∞, is the enthalpy of melting and dissociation of solid compound A_nB to the infinitely dilute state of solute B in solvent A; H_B^∞ is the partial molar enthalpy of the solute at infinite dilution. Clearly, the integral of eqn (25) will have the same form as eqn (29), with $\Delta H_{AB}^\infty(T^*)$, $\Delta C_p^\infty(T^*)$ replacing ΔH_{AB}^* and ΔC_p^* and $x_A^2 - 1$ replacing x_A^2 in the last term.

If the liquid phase is an aqueous electrolyte solution, and the solid is a salt hydrate, the above treatment needs slight modification. Using rational mean activity coefficients, eqn (25) becomes

$$R\nu(1/x_B - n/x_A)\{1+(\partial \ln f_\pm/\partial \ln x_\pm)_{T,P}\}dx_B/\{1+(\nu-1)x_B\}$$
$$= \{\Delta H_{AB}^\infty + n(H_A - H_A^*) + (H_B - H_B^\infty)\}d(1/T) \tag{33}$$

If the terms involving activity coefficients and partial molar enthalpies are negligible, then integration gives (cf. (11)):

$$\ln\left\{\frac{x_B^\nu(1-x_B)^n}{1+(\nu-1)x_B}^{n+\nu}\right\} = \ln\left\{\frac{n^n}{(n+\nu)^{n+\nu}}\right\} - \left\{\frac{\Delta H_{AB}^\infty(T^*) - T^*\Delta C_p^*}{R}\right\}\left(\frac{1}{T} - \frac{1}{T^*}\right) + \frac{\Delta C_p^*}{R}\ln(T/T^*) \tag{34}$$

A similar equation (with $\nu=2$ and without the heat capacity terms) has been used to fit solubility data for some MOH=H_2O systems, where M is an alkali metal; the enthalpy values obtained agreed well with known values (11). In many cases, data on activity coefficients (9) and partial molal enthalpies (8,10) in concentrated solution indicate that the terms involving these quantities are not negligible, although they may remain roughly constant along the solubility temperature curve.

The above analysis shows clearly that a rational thermodynamic basis exists for functional representation of solubility-temperature curves in two-component systems, but may be difficult to apply because of lack of experimental or theoretical knowledge of activity coefficients and partial molar enthalpies. Other phenomena which are related ultimately to the stoichiometric activity coefficients and which complicate interpretation include ion pairing, formation of complex ions, and hydrolysis. Similar considerations hold for the variation of solubility with pressure, except that the effects are relatively smaller at the pressures used in many investigations of solubility (5).

(b) *Solubility as a function of composition.*
At constant temperature and pressure, the chemical potential of a saturating solid phase is constant:

$$\mu_{A_nB}^* = \mu_{A_nB}(sln) = n\mu_A + \mu_B \tag{35}$$
$$= (n\mu_A^* + \nu_+\mu_+^\infty + \nu_-\mu_-^\infty) + nRT \ln f_A x_A$$
$$+ \nu RT \ln \gamma_\pm m_\pm Q_\pm \tag{36}$$

for a salt hydrate A_nB which dissociates to water, (A), and a salt, B, one mole of which ionizes to give ν_+ cations and ν_- anions in a solution in which other substances (ionized or not) may be present. If the saturated solution is sufficiently dilute, $f_A = x_A = 1$, and the quantity K_{S0}^0 in

$$\Delta G^\infty \equiv (\nu_+\mu_+^\infty + \nu_-\mu_-^\infty + n\mu_A^* - \mu_{AB}^*)$$
$$= -RT \ln K_{S0}^0$$
$$= -RT \ln Q^\nu \gamma_\pm^\nu m_+^{\nu_+} m_-^{\nu_-} \tag{37}$$

is called the *solubility product* of the salt. (It should be noted that it is not customary to extend this definition to hydrated salts, but there is no reason why they should be excluded.) Values of the solubility product are often given on mole fraction or concentration scales. In dilute solutions, the theoretical behaviour of the activity coefficients as a function of ionic strength is often sufficiently well known that reliable extrapolations to infinite dilution can be made, and values of K_{S0}^0 can be determined. In more concentrated solutions, the same problems with activity coefficients that were outlined in the section on variation of solubility with temperature still occur. If these complications do not arise, the solubility of a hydrate salt $C_{\nu_+}A_{\nu_-} \cdot nH_2O$ in the presence of other solutes is given by eqn (36) as

$$\nu \ln\{m_B/m_B(0)\} = -\nu\ln\{\gamma_\pm/\gamma_\pm(0)\} - n \ln(a_{H_2O}/a_{H_2O}(0)) \tag{38}$$

where a_{H_2O} is the activity of water in the saturated solution, m_B is the molality of the salt in the saturated solution, and (0) indicates absence of other solutes. Similar considerations hold for non-electrolytes.

The Solid Phase

The definition of solubility permits the occurrence of a single solid phase which may be a pure anhydrous compound, a salt hydrate, a non-stoichiometric compound, or a solid mixture (or solid solution, or "mixed crystals"), and may be stable or metastable. As well, any number of solid phases consistent with the requirements of the phase rule may be present. Metastable solid phases are of widespread occurrence, and may appear as polymorphic (or allotropic) forms or crystal solvates whose rate of transition to more stable forms is very slow. Surface heterogeneity may also give rise to metastability, either when one solid precipitates on the surface of another, or if the size of the solid particles is sufficiently small that surface effects become important. In either case, the solid is not in stable equilibrium with the solution. The stability of a solid may also be affected by the atmosphere in which the system is equilibrated.

Many of these phenomena require very careful, and often prolonged, equilibration for their investigation and elimination. A very general analytical method, the "wet residues" method of Schreinemakers (12) (see a text on physical chemistry) is usually used to investigate the composition of solid phases in equilibrium with salt solutions. In principle, the same method can be used with systems of other types. Many other techniques for examination of solids, in particular X-ray, optical, and thermal analysis methods, are used in conjunction with chemical analyses (including the wet residues method).

COMPILATIONS AND EVALUATIONS

The formats for the compilations and critical evaluations have been standardized for all volumes. A brief description of the data sheets has been given in the FOREWORD; additional explanation is given below.

Guide to the Compilations

The format used for the compilations is, for the most part, self-explanatory. The details presented below are those which are not found in the FOREWORD or which are not self-evident.

Components. Each component is listed according to IUPAC name, formula, and Chemical Abstracts (CA) Registry Number. The formula is given either in terms of the IUPAC or Hill (13) system and the choice of formula is governed by what is usual for most current users: i.e. IUPAC for inorganic compounds, and Hill system for organic compounds. Components are ordered according to:
  (a) saturating components;
  (b) non-saturating components in alphanumerical order;
  (c) solvents in alphanumerical order.

The saturating components are arranged in order according to a 18-column, 2-row periodic table:
  Columns 1,2: H, groups IA, IIA;
       3,12: transition elements (groups IIIB to VIIB, group VIII, groups IB, IIB);
       13-18: groups IIIA-VIIA, noble gases.
  Row 1: Ce to Lu;
  Row 2: Th to the end of the known elements, in order of atomic number.
Salt hydrates are generally not considered to be saturating components since most solubilities are expressed in terms of the anhydrous salt. The existence of hydrates or solvates is carefully noted in the texts, and CA Registry Numbers are given where available, usually in the critical evaluation. Mineralogical names are also quoted, along with their CA Registry Numbers, again usually in the critical evaluation.

Original Measurements. References are abbreviated in the forms given by *Chemical Abstracts Service Source Index (CASSI)*. Names originally in other than Roman alphabets are given as transliterated by *Chemical Abstracts*.

Experimental Values. Data are reported in the units used in the original publication, with the exception that modern *names* for units and quantities are used; e.g., mass per cent for weight per cent; mol dm^{-3} for molar; etc. Both mass and molar values are given. Usually, only one type of value (e.g., mass per cent) is found in the original paper, and the compiler has added the other type of value (e.g., mole per cent) from computer calculations based on 1976 atomic weights (14). Errors in calculations and fitting equations in original papers have been noted and corrected, by computer calculations where necessary.

Method. Source and Purity of Materials. Abbreviations used in *Chemical Abstracts* are often used here to save space.

Estimated Error. If these data were omitted by the original authors, and if relevant information is available, the compilers have attempted to estimate errors from the internal consistency of data and type of apparatus used. Methods used by the compilers for estimating and reporting errors are based on the papers by Ku and Eisenhart (15).

Comments and/or Additional Data. Many compilations include this section which provides short comments relevant to the general nature of the work or additional experimental and thermodynamic data which are judged by the compiler to be of value to the reader.

References. See the above description for Original Measurements.

Guide to the Evaluations

The evaluator's task is to check whether the compiled data are correct, to assess the reliability and quality of the data, to estimate errors where necessary, and to recommend "best" values. The evaluation takes the form of a summary in which all the data supplied by the compiler have been critically reviewed. A brief description of the evaluation sheets is given below.

Components. See the description for the Compilations.

Evaluator. Name and date up to which the literature was checked.

Critical Evaluation

(a) Critical text. The evaluator produces text evaluating *all* the published data for each given system. Thus, in this section the evaluator review the merits or shortcomings of the various data. Only published data are considered; even published data can be considered only if the experimental data permit an assessment of reliability.

(b) Fitting equations. If the use of a smoothing equation is justifiable, the evaluator may provide an equation representing the solubility as a function of the variables reported on all the compilation sheets.

(c) Graphical summary. In addition to (b) above, graphical summaries are often given.

(d) Recommended values. Data are *recommended* if the results of at least two independent groups are available and they are in good agreement, and if the evaluator has no doubt as to the adequacy and reliability of the applied experimental and computational procedures. Data are reported as *tentative* if only one set of measurements is available, or if the evaluator considers some aspect of the computational or experimental method as mildly undesirable but estimates that it should cause only minor errors. Data are considered as *doubtful* if the evaluator considers some aspect of the computational or experimental method as undesirable but still considers the data to have some value in those instances where the order of magnitude of the solubility is needed. Data determined by an inadequate method or under ill-defined conditions are *rejected*. However references to these data are included in the evaluation together with a comment by the evaluator as to the reason for their rejection.

(e) References. All pertinent references are given here. References to those data which, by virtue of their poor precision, have been rejected and not compiled are also listed in this section.

(f) Units. While the original data may be reported in the units used by the investigators, the final recommended values are reported in S.I. units (1,16) when the data can be accurately converted.

References

1. Whiffen, D. H., ed., *Manual of Symbols and Terminology for Physicochemical Quantities and Units. Pure Applied Chem.* 1979, 51, No. 1.
2. McGlashan, M.L. *Physicochemical Quantities and Units.* 2nd ed. Royal Institute of Chemistry. London. 1971.
3. Jänecke, E. *Z. Anorg. Chem.* 1906, 51, 132.
4. Friedman, H.L. *J. Chem. Phys.* 1960, 32, 1351.
5. Prigogine, I.; Defay, R. *Chemical Thermodynamics.* D.H. Everett, transl. Longmans, Green. London, New York, Toronto. 1954.
6. Guggenheim, E.A. *Thermodynamics.* North-Holland. Amsterdam. 1959. 4th ed.
7. Kirkwood, J.G.; Oppenheim, I. *Chemical Thermodynamics.* McGraw-Hill, New York, Toronto, London. 1961.
8. Lewis, G.N.; Randall, M. (rev. Pitzer, K.S.; Brewer, L.). *Thermodynamics.* McGraw Hill. New York, Toronto, London. 1961. 2nd ed.
9. Robinson, R.A.; Stokes, R.H. *Electrolyte Solutions.* Butterworths. London. 1959, 2nd ed.
10. Harned, H.S.; Owen, B.B. *The Physical Chemistry of Electrolytic Solutions.* Reinhold. New York. 1958. 3rd ed.
11. Cohen-Adad, R.; Saugier, M.T.; Said, J. *Rev. Chim. Miner.* 1973, 10, 631.
12. Schreinemakers, F.A.H. *Z. Phys. Chem., stoechiom. Verwandschaftsl.* 1893, 11, 75.
13. Hill, E.A. *J. Am. Chem. Soc.* 1900, 22, 478.
14. IUPAC Commission on Atomic Weights. *Pure Appl. Chem.,* 1976, 47, 75.

15. Ku, H.H., p. 73; Eisenhart, C., p. 69; in Ku, H.H., ed. *Precision Measurement and Calibration*. NBS Special Publication 300. Vol. 1. Washington. 1969.
16. *The International System of Units*. Engl. transl. approved by the BIPM of *Le Système International d'Unités*. H.M.S.O. London. 1970.

R. Cohen-Adad, Villeurbanne, France
J.W. Lorimer, London, Canada
M. Salomon, Fair Haven, New Jersey, U.S.A.

COMPONENTS:	EVALUATOR:
(1) Lithium; Li; [7439-93-2] (2) Mercury; Hg; [7439-97-6]	C. Guminski; Z. Galus Department of Chemistry University of Warsaw Warsaw, Poland July, 1985

CRITICAL EVALUATION:

Maey (1) was the first to report the solubility of lithium in mercury at room temperature by determining the specific volume of the amalgam, but the solubility of 0.9 at % is too low and is rejected. Kerp and coworkers (2) determined the solubility by the analyses of the samples after filtration of the equilibrated mixture of Li and Hg. These authors determined the lithium solubilities at four temperatures between 273 and 373 K, with values ranging from 1.1 to 3.6 at %, respectively. Smith and Bennett (3) determined a solubility of 1.34 at % at 295 K by a method similar to that of Kerp et al. Richards and Garrod-Thomas (4) reported a solubility of 1.05 at % at room temperature, but this value is too low and is rejected. Zukovsky (5) reported the first extensive determination of the solubility curve over the complete composition range by thermal analysis; it was found that the concentration of Li in the saturated amalgam was 0.9 at % at the eutectic temperature of 231 K, and that the concentration increased to 49.6 at % at 872 K. Above the latter temperature the liquids were completely miscible. Grube and Wolf (6) also determined the solubility curve over the complete concentration range by thermal analysis, and the results of these authors agreed with those of Zukovsky in the concentration range of 20-85 at % Li. Also, Grube and Wolf confirmed the eutectic temperature of 231 K, but at 0.6 at % Li. However, there was a wide discrepancy between the solubility curve of Zukovsky and of Grube and Wolf at lithium concentrations above 85 at %. Strachan and Harris (7) reported a room temperature solubility of 0.66 at % that is too low and is rejected. Kozin (8) estimated a solubility of 66.49 at % at 298 K, but this solubility is inconsistent with experimental data because the author neglected the strong interactions of lithium and mercury.

Gladyshev and coworkers (9) determined a consistent lithium solubility of 1.37 and 2.1 at % at 293 and 313 K, respectively, by a potentiometric method. Cogley and Butler (10) determined the EMF of concentration cells with a non-aqueous electrolyte, and also obtained a consistent solubility of 1.33 at % at 299 K; however, their earlier result of 2.0 at % at 298 K (11) was overstated and is rejected. Korshunov et al. (12) reported a solubility of 1.1 at % at 293 K, but no experimental details were given by these authors. Dean (16) reported a 298 K solubility of 1.25 at % which is consistent with accepted values; the amalgam was prepared by electrolysis from LiOH, but no experimental details were described by this author. Onstott and coworkers (17,18) performed careful determinations at 295.4 K and obtained a solubility of 1.27 at %. A value of 1.3 at % at 296 may be suggested from potentiometric measurements of Horner and Schmitt (19). Based on calorimetric titration, Filippova and coworkers (13-15) reported that the saturated Li amalgam contains 1.20 at % Li at 298 K.

In summary, there is good agreement among the results of (10, 16-19), whereas the thermoanalytical data of (5,6) are significantly overstated at temperatures below 473 K.

Figure 1 shows the phase diagram reported by Hultgren et al. (20); this phase diagram is based mainly on the data of (2), (5) and (6). The intermetallic compounds which have been verified are Hg_3Li, Hg_2Li, $HgLi$, $HgLi_2$, $HgLi_3$ and $HgLi_6$.

Recommended (r) and tentative values of Li solubility in Hg:

(Continued next page)

COMPONENTS:

(1) Lithium; Li; [7439-93-2]
(2) Mercury; Hg; [7439-97-6]

EVALUATOR:

C. Guminski; Z. Galus
Department of Chemistry
University of Warsaw
Warsaw, Poland

July, 1985

CRITICAL EVALUATION: (continued)

Recommended (r) and tentative values of Li solubility in Hg:

T/K	Soly/at %	Reference
231	0.6	[6]
293	1.2[a]	[3,9,12,17,18]
298	1.3 (r)	[10,16-19]
323	2.2[b]	[5,9]
373	5[b]	[5]
473	13	[5]
573	25	[5]
673	33	[5,6]
773	39[b]	[5,6]
873	50.0	[5]

[a] Mean value from data of cited references.
[b] Interpolated value from data of cited references.

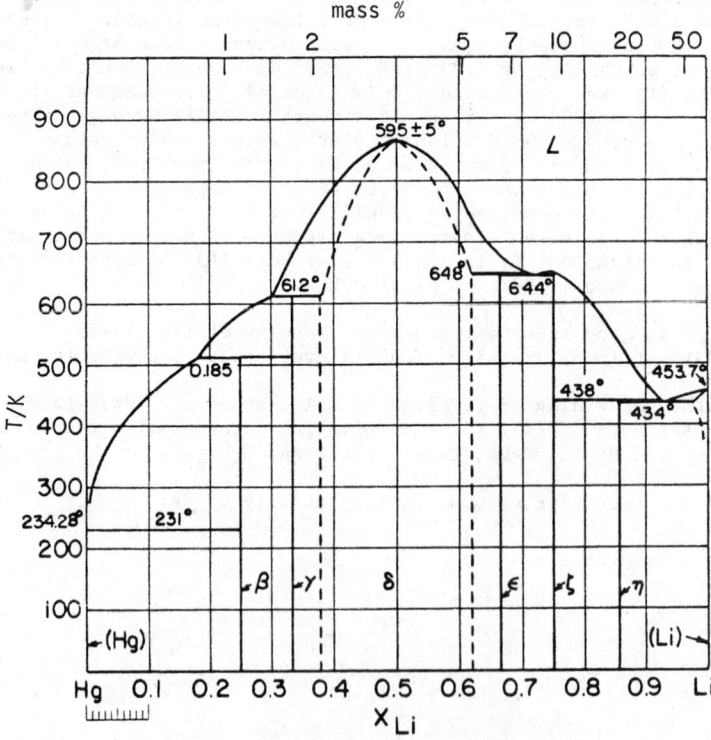

Fig. 1. The Li-Hg phase diagram (20).

(Continued next page)

COMPONENTS:	EVALUATOR:
(1) Lithium; Li; [7439-93-2] (2) Mercury; Hg; [7439-97-6]	C. Guminski; Z. Galus Department of Chemistry University of Warsaw Warsaw, Poland July, 1985

CRITICAL EVALUATION: (Continued)

References

1. Maey, E. *Z. Phys. Chem.* 1899, *29*, 119.
2. Kerp, W.; Böttger, W.; Winter, H. *Z. Anorg. Chem.* 1900, *25*, 1.
3. Smith, G.McP.; Bennett, H.C. *J. Am. Chem. Soc.* 1909, *31*, 799; 1910, *32*, 622.
4. Richards, T.W.; Garrod-Thomas, R.N. *Z. Phys. Chem.* 1910, *72*, 165.
5. Zukovsky, G.J. *Z. Anorg. Chem.* 1911, *71*, 403.
6. Grube, G.; Wolf, W. *Z. Elektrochem.* 1935, *41*, 675.
7. Strachan, J.F.; Harris, N.L. *J. Inst. Metals* 1956-57, *85*, 17.
8. Kozin, L.F. *Fiziko-Khimicheskie Osnovy Amalgamnoi Metallurgii*, Nauka, Alma-Ata, 1964.
9. Gladyshev, V.P.; Ruban, L.M.; Kuleshov, V.A. *Tr. Inst. Khim. Nauk Akad. Nauk Kaz. SSR* 1969, *24*, 111.
10. Cogley, D.R.; Butler, J.N. *J. Phys. Chem.* 1968, *72*, 1017.
11. Same authors. *J. Electrochem. Soc.* 1966, *113*, 1074.
12. Korshunov, V.N.; Kuznetsova, N.K.; Gradkih, I.P.; Volkov, A.G. *Elektrokhimiya* 1971, *7*, 1501.
13. Filippova, L.M.; Zhumakanov, V.Z.; Zebreva, A.I. *Izv. Vyssh. Ucheb. Zaved., Khim. Khim. Tekhnol.* 1980, *23*, 204.
14. Filippova, L.M.; Zebreva, A.I.; Zhumakanov, V.Z. *Ukr. Khim. Zh.* 1981, *47*, 473.
15. Same authors. *Izv. Vyssh. Ucheb. Zaved., Khim. Khim. Tekhnol.* 1982, *25*, 827.
16. Dean, O.C. *U.S. At. Ener. Comm. Rep.*, CF-58-11, 1958, p. 23.
17. Onstott, E.I.; Goddard, J.B. *U.S. At. Ener. Comm. Rep.*, LA-DC-7013, 1964.
18. Goddard, J.B.; Campbell, J.M.; Onstott, E.I. *U.S. At. Ener. Comm. Rep.*, LA-DC-8393, 1965.
19. Horner, L.; Schmitt, R.E. *Z. Naturforsch.*, B 1982, *37*, 1163.
20. Hultgren, R.; Desai, P.D.; Hawkins, D.T.; Gleiser, M.; Kelley, K.K. *Selected Values of the Thermodynamic Properties of Binary Alloys*, Am. Soc. Metals, Metals Park, OH 1973, p. 964.

COMPONENTS:	ORIGINAL MEASUREMENTS:
(1) Lithium; Li; [7439-93-2] (2) Mercury; Hg; [7439-97-6]	Kerp, W.; Böttger, W.; Winter, H. Z. Anorg. Chem. 1900. 25, 1-71.
VARIABLES: Temperature: 0-100°C	PREPARED BY: C. Guminski; Z. Galus

EXPERIMENTAL VALUES:

Solubility of lithium in mercury:

t/°C	Soly/mass %	Soly/at %[a]
0	0.04	1.1
64.5	0.10	2.8
81	0.11	3.1
99.8	0.13	3.6

[a] by compilers.

The experimental procedure may give results that are too low, especially at the higher temperatures (compilers).

Analysis of the solid phase resulted in the formula $LiHg_5$.

AUXILIARY INFORMATION

METHOD/APPARATUS/PROCEDURE:	SOURCE AND PURITY OF MATERIALS:
The amalgam was prepared by electrolysis of a saturated aqueous LiCl solution with Hg as the cathode. Subsequent experimental operations with the amalgam were performed in a dry hydrogen atmosphere. After separation with a leder plate in a Gooch crucible, the content of Li in the amalgam was determined by back-titration of an acidified solution with a standard baryta water solution.	Nothing specified.
	ESTIMATED ERROR: Soly: nothing specified; precision no better than ± 10% (compilers). Temp: nothing specified.
	REFERENCES:

COMPONENTS:	ORIGINAL MEASUREMENTS:
(1) Lithium; Li; [7439-93-2] (2) Mercury; Hg; [7439-97-6]	Smith, G.McP.; Bennett, H.C. *J. Am. Chem. Soc.* 1909, *31*, 799-806. *J. Am. Chem. Soc.* 1910, *32*, 622-26.
VARIABLES:	PREPARED BY:
One temperature: 22°C	C. Guminski; Z. Galus

EXPERIMENTAL VALUES:

The solubility of lithium in mercury was reported to be 4.7×10^{-2} mass %.

The corresponding atomic % solubility calculated by the compilers is 1.34 at %.

Analysis of the solid phase corresponded to the compound $LiHg_4$.

AUXILIARY INFORMATION

METHOD/APPARATUS/PROCEDURE:	SOURCE AND PURITY OF MATERIALS:
Lithium amalgam was prepared electrolytically from saturated LiCl solution; 250 g of Hg was used as the cathode. The amalgam, after preparation, was washed, dried, allowed to stand 2 days, and finally filtered at 22°C. The filtrate was treated with standard HCl and back-titrated with standard NaOH.	"Very pure" salts from Kahlbaum were used. Mercury purity not specified.
	ESTIMATED ERROR: Soly: nothing specified; precision probably no better than several percent (compilers). Temp: nothing specified.
	REFERENCES:

COMPONENTS:	ORIGINAL MEASUREMENTS:
(1) Lithium; Li; [7439-93-2] (2) Mercury; Hg; [7439-97-6]	Zukovsky, G.J. Z. Anorg. Chem. 1911, 71, 403-18.
VARIABLES: Temperature: (-30)-600°C	PREPARED BY: C. Guminski; Z. Galus

EXPERIMENTAL VALUES:

Freezing points of Li-Hg alloys were reported; the solubilities corresponding to the liquidus concentrations are as follows:

t/°C	Soly/at %	t/°C	Soly/at %	t/°C	Soly/at %	t/°C	Soly/at %
-30	0.97	256	19.9	584	48.4	453	62.2
11	2.5	261	20.5	593	49.0	440	63.1
110	5.8	270	21.3	593	49.4	406	65.3
128	7.2	276	21.6	597	49.5	379	68.9
132	7.5	298	23.6	599	49.6	379	75.1
140	7.7	305	25.4	600.5	50.0	376	75.2
160	9.1	320	27.1	600.3	50.1	369	76.4
173	10.7	325	27.5	597	50.3	364	76.9
184	11.2	332	29.0	595	50.6	355	78.1
203	13.1	338	29.4	578.7	50.8	348	78.8
216	14.2	338	30.1	578.7	51.7	315	82.0
228	16.4	360	30.9	579.5	52.6	275	83.6
229	17.2	358	31.2	580	52.9	272	86.0
232	17.6	378	32.2	568	54.4	270	86.3
234	18.4	297	33.0	564	65.0	265	87.3
238	18.7	415	35.4	534	57.7	260	90.0
242	19.1	448	36.5	496	60.1	253	90.7
246	19.3	476	38.2	490	60.5	250	91.2
247	19.4	580	47.6	478	61.2	232	92.7
249	19.8	585	48.3	464	61.8	226	93.3
						207	95.4
						162	97.6

AUXILIARY INFORMATION

METHOD/APPARATUS/PROCEDURE:	SOURCE AND PURITY OF MATERIALS:
Freezing points were determined by determination of the temperature of primary crystallization. Porcelain and steel containers were used with the same results. The alloys were prevented from oxidation by covering with paraffin at lower temperatures, and with melted chlorides of Li, Rb, and K at temperatures higher than 312°C.	Purest lithium and mercury from Kahlbaum were used. Only traces of sodium were found in the lithium by spectroscopic analysis.

Additional Data:

The saturated amalgams were in equilibrium with the solid phases, $LiHg_3$, $LiHg_2$, $LiHg$ and Li_3Hg.

ESTIMATED ERROR:
Soly: nothing specified. Temp: precision ± 0.6 K.

REFERENCES:

COMPONENTS:	ORIGINAL MEASUREMENTS:
(1) Lithium; Li; [7439-93-2] (2) Mercury; Hg; [7439-97-6]	Grube, G.; Wolf, W. *Z. Elektrochem.* 1935, *41*, 675-79.
VARIABLES:	PREPARED BY:
Temperature: (-42)-585°C	C. Guminski; Z. Galus

EXPERIMENTAL VALUES:

Freezing points of Li-Hg alloys were reported; the solubilities corresponding to the liquidus concentrations are as follows:

$t/°C$	Soly/at %	$t/°C$	Soly/at %	$t/°C$	Soly/at %
-42	0.6	474	37.1	344	79.6
24	3.6	510	39.6	322	81.7
42	5.2	556	43.6	290	83.4
48	5.7	584	49.0	270	84.7
79	6.7	585	50.4	257	85.1
85	7.3	580	51.5	254	85.6
87	8.0	561	55.0	240	86.4
148	11.5	520	59.4	227	86.9
162	13.1	434	63.8	220	87.2
205	15.1	412	65.0	200	88.7
213	16.9	402	66.2	180	89.2
219	17.5	397	66.9	165	90.6
260	19.3	388	68.0	162	91.5
293	23.0	384	69.0	161	91.9
314	25.0	382	69.5	161	92.8
335	27.7	375	70.7	163	93.8
341	29.2	372	72.4	164	94.9
361	30.3	373	73.5	165	95.8
382	31.5	375	74.4	171	96.8
397	32.3	371	75.8	176	98.0
435	34.4	366	76.7	178	99.0
467	36.1	347	79.3		

AUXILIARY INFORMATION

METHOD/APPARATUS/PROCEDURE:	SOURCE AND PURITY OF MATERIALS:
The temperatures of the primary crystallization of the alloys were determined by thermoanalysis in a furnace of high-carbon steel. After the measurement the alloys were decomposed with water and analyzed for lithium content by acid-base titration with HCl. Hg content was determined gravimetrically by weighing the Hg after it was washed with water and dried.	Freshly distilled mercury and lithium from Metallgeselshaft A.G., Frankfurt, were employed.
Additional Data: The following solid phases were reported: $LiHg_3$, $LiHg_2$, $LiHg$, Li_2Hg, Li_3Hg, and Li_6Hg.	ESTIMATED ERROR: Nothing specified.
Comments: The values seem to be reliable at temperatures higher than 200°C. Except for the -42°C eutectic, the lower temperature values are too high (compilers).	REFERENCES:

COMPONENTS:	ORIGINAL MEASUREMENTS:
(1) Lithium; Li; [7439-93-2] (2) Mercury; Hg; [7439-97-6]	Onstott, E.I.; Goddard, J.B. *U.S. At. Ener. Comm. Rep., LA-DC-7013,* <u>1964</u>. Goddard, J.B.; Campbell, J.M.; Onstott, E.I. *U.S. At. Ener. Comm. Rep., LA-DC-8393,* <u>1965</u>.
VARIABLES: One temperature: 22°C	PREPARED BY: C. Guminski; Z. Galus

EXPERIMENTAL VALUES:

The solubility of lithium in mercury at 22.0°C was reported to be 0.0440 mass %. The corresponding atomic % solubility calculated by the compilers is 1.27 at %.

The solubility value was based on six separate determinations.

AUXILIARY INFORMATION

METHOD/APPARATUS/PROCEDURE:	SOURCE AND PURITY OF MATERIALS:
The amalgams were prepared by electrolysis of saturated LiOH solution on a Hg pool cathode; a carbon bar served as the anode. The amalgam was drained, sometimes through cotton gauze, and stored under mineral oil until used. Composition of the amalgam was determined by reacting with known amount of 1 mol dm^{-3} HCl, then adding excess of 0.1 mol dm^{-3} NaOH, followed by titration of the excess NaOH with standard 0.1 mol dm^{-3} HCl.	Purified Hg was used. Purity of LiOH not specified.
	ESTIMATED ERROR: Soly: standard deviation 0.9% (compilers). Temp: precision ± 0.2 K.
	REFERENCES:

COMPONENTS:	ORIGINAL MEASUREMENTS:
(1) Lithium; Li; [7439-93-2] (2) Mercury; Hg; [7439-97-6]	Cogley, D.R.; Butler, J.N. *J. Phys. Chem.* 1968, 72, 1017-20.
VARIABLES: One temperature: 26°C	PREPARED BY: C. Guminski; Z. Galus

EXPERIMENTAL VALUES:

Solubility of lithium in mercury at 26.0°C was reported to be 1.33 at %.

AUXILIARY INFORMATION

METHOD/APPARATUS/PROCEDURE:	SOURCE AND PURITY OF MATERIALS:
Amalgams were prepared by combining weighed quantities of the metals. Amalgams were analyzed by decomposition with acid, followed by determination of Li in the resulting solution by flame photometry. Electrolytes were prepared from anhydrous LiCl or LiClO$_4$ and dimethyl sulfoxide. Employing a high-impedance differential voltmeter, the potentials of the following cell were determined as a function of the amalgam concentration: Li(s)|LiCl or LiClO$_4$ in DMSO|Li(Hg). All manipulations were carried out in an argon atmosphere containing less than 1×10^{-6} mol/mol of H$_2$O, O$_2$ or N$_2$.	Mercury was triple-distilled material from Doe-Ingalls; it was freed from oxygen by passing through a porous frit in argon atmosphere. Lithium was 99.97% pure from Foote Mineral Co. LiCl and LiClO$_4$ were ultrapure from Anderson Physics Labs. Chromatographic grade DMSO from Matheson, Coleman, Bell; water content was less than 0.001%.
	ESTIMATED ERROR: Soly: nothing specified; precision ± 1% (compilers). Temp: precision ± 0.2 K.
	REFERENCES:

COMPONENTS:	ORIGINAL MEASUREMENTS:
(1) Lithium; Li; [7439-93-2] (2) Mercury; Hg; [7439-97-6]	Gladyshev, V.P.; Ruban, L.M.; Kuleshov, V.A. *Tr. Inst. Khim. Nauk Akad. Nauk Kaz. SSR* 1969, 24, 111-19.
VARIABLES:	PREPARED BY:
Temperature: 20-40°C	C. Guminski; Z. Galus

EXPERIMENTAL VALUES:

Solubility of lithium in mercury at 20°C was reported to be $(4.8 \pm 0.5) \times 10^{-2}$ mass %. The corresponding atomic % solubility calculated by the compilers is 1.37 at %.

On the basis of reported potentials at 40°C, the compilers calculated a solubility of $(7.5 \pm 1.0) \times 10^{-2}$ mass %, corresponding to 2.1 at %.

AUXILIARY INFORMATION

METHOD/APPARATUS/PROCEDURE:	SOURCE AND PURITY OF MATERIALS:
Potentials were measured by the compensation method for the cell, $Li(Hg) \mid 2 \text{ mol dm}^{-3} \text{ LiCl} \mid Li(Hg)_x$. Concentration of Li in one half-cell was kept constant at 1.7×10^{-2} mol dm^{-3} while that in the other half-cell was varied. Lithium amalgam was obtained electrolytically. Measurements were carried out in an atmosphere of hydrogen in a constant temperature system. With the cell employed, the effect of corrosion on the measurements should be minimal.	Chemically pure compounds of lithium were used. Mercury was purified electrolytically, then distilled.
	ESTIMATED ERROR:
	Soly: precision approximately $\pm$ 10%. Temp: nothing specified.
	REFERENCES:

COMPONENTS:	ORIGINAL MEASUREMENTS:
(1) Lithium; Li; [7439-93-2] (2) Mercury; Hg; [7439-97-6]	Filippova, L.M.; Zhumakanov, V.Z.; Zebreva, A.I. *Izv. Vyssh. Ucheb. Zaved., Khim. Khim. Technol.* 1980, *23*, 204-7.
VARIABLES:	PREPARED BY:
One temperature: 298 K	C. Guminski; Z. Galus

EXPERIMENTAL VALUES:

The 298 K solubility of lithium in mercury was reported to be 1.20 ± 0.05 at %, or 0.83 ± 0.03 mol dm^{-3}.

This solubility is also reported in (1) and (2).

AUXILIARY INFORMATION

METHOD/APPARATUS/PROCEDURE:	SOURCE AND PURITY OF MATERIALS:
The amalgam was prepared by an electrolytic method. Sample of the amalgam was analyzed for the lithium content by acid titration. The homogeneous and heterogeneous amalgams were titrated with mercury and the thermal effects were determined. A bend on the plot of the thermal effect versus concentration corresponds to concentration of the saturated amalgam. All experiments were performed under argon atmosphere.	Nothing specified.
	ESTIMATED ERROR: Soly: precision about 4%. Temp: nothing specified.
	REFERENCES: 1. Filippova, L.M.; Zebreva, A.I.; Zhumakanov, V.Z. *Ukr. Khim. Zh.* 1981, *47*, 473. 2. Same authors. *Izv. Vyssh. Ucheb. Zaved., Khim. Khim. Tekhnol.* 1982, *25*, 827.

COMPONENTS:	ORIGINAL MEASUREMENTS:
(1) Lithium; Li; [7439-93-2] (2) Mercury; Hg; [7439-97-6]	Horner, L.; Schmitt, R.E. *Z. Naturforsch.*, B **1982**, *37*, 1163. 1163-8.
VARIABLES: One temperature: 23°C	PREPARED BY: C. Guminski; Z. Galus

EXPERIMENTAL VALUES:

On the basis of potentiometric measurements reported by the authors, the compilers obtained a 23°C solubility of 1.3 at %.

The following EMF data were reported for the cell at 23°C:

at % Li	−E/V
17.6	2.104
13.6	2.085
1.9	2.018
1.2	2.005
0.46	1.998
0.21	1.990
0.075	1.992

AUXILIARY INFORMATION

METHOD/APPARATUS/PROCEDURE:	SOURCE AND PURITY OF MATERIALS:
Lithium amalgam was obtained by potentiostatic or galvanostatic electrolysis on Hg cathode from 2 mol dm^{-3} LiClO$_4$ in THF, AN or DMF solutions; a carbon cylinder was used as anode. More dilute amalgams were prepared by adding defined amounts of Hg to the solid amalgams obtained; the resulting amalgam was homogenized by heating. Lithium content in the amalgams was determined by addition of 0.1 mol dm^{-3} HCl and back-titration with 0.1 mol dm^{-3} NaOH with Phenolphthalein indicator. Potentials of the following cell were determined: Li(Hg)$_x$ | 0.1 mol dm^{-3} LiClO$_4$ in AN | | KCl$_{aq}$ | Hg$_2$Cl$_2$, Hg.	Nothing specified.
	ESTIMATED ERROR: Soly: precision ± 10%. Temp: precision ± 1 K.
A plot of potential vs. logarithm of Li content was constructed by the compilers; the breakpoint in the curve corresponds to the saturation concentration of Li in Hg.	REFERENCES:

COMPONENTS:	EVALUATOR:
(1) Sodium; Na; [7440-23-5] (2) Mercury; Hg; [7439-97-6]	J. Balej Institute of Inorganic Chemistry Czechoslovak Academy of Sciences Prague, Czechoslovakia C. Guminski; Z. Galus Department of Chemistry University of Warsaw Warsaw, Poland July, 1985

CRITICAL EVALUATION:

The existence of various intermetallic compounds in the Na-Hg system is clearly evident from the phase diagram. Because of the formation of these compounds the solubility of sodium in mercury, and vice versa, must be considered in relation to the crystallization region of the phase diagram. Many compounds have been proposed for this system, but the existence of most of these has never been proved. Some have been invoked in attempts to explain the observed properties of the liquid amalgams (1-3), while others have been proposed on the basis of analyses of the crystal phases which were separated from saturated liquid amalgams (4-8,10). Based on all published data for this system, the following may be considered as proved at the present time: $NaHg_4$, $NaHg_2$, Na_7Hg_8, $NaHg$, Na_3Hg_2, Na_5Hg_2, and Na_3Hg. With the exception of Na_7Hg_8, the existence of the compounds has been confirmed by independent measurements of concentration cells of the type $Na|Na^+|Na(Hg)$ (14).

The complete phase diagram for the Na-Hg system has been investigated by Kurnakov (9), Schüller (11), Vanstone (12) and Jänecke (13). In all of these works the classical thermal analysis of cooling curves was employed, and temperatures of primary (9-13) and secondary (11,12) crystallization were determined. The coexisting solid compounds were identified by measuring the molar volumes of liquid and solid amalgams, and by microscopic examination (12). Only Kurnakov and Vanstone presented their results in both numerical and graphical forms; Schüller listed the compositions and the corresponding primary crystallization temperatures for the characteristic points only, and the results for about 100 other samples have been presented in the form of a phase diagram. Jänecke (13) presented his results in a graphical form only. The early results by Merz and Weith (32) are rejected because of poor accuracy.

Only one congruently melting compound, $NaHg_2$, was found in the Na-Hg system (9, 11, 12).

A summary of chracteristic data of the phase diagram for the Na-Hg system is presented in Table I.

Hansen and Anderko (15) presented the Na-Hg phase diagram which has been generally accepted. In the present evaluation, a revised phase diagram is presented in Fig. 1. This phase diagram was constructed by graphical smoothing of all the reliable data on solubility of sodium in mercury. Figure 1 shows good agreement in the 0-17 at % Na range with that of (15). However, the latter shows a slightly lower liquidus temperature of 421 K at 17.1 at % Na. It appears that more reliable data are needed in the range of the peritectic at 18 at % Na, as well as for the other peritectic points. The solubility of sodium in mercury, and vice versa, for various crystallization regions have been presented in (3,4,7,8,10,16-26). Graphically smoothed solubilities of sodium in the Hg-rich region are presented in Table II.

For the crystallization region of the very dilute amalgams the results by Tammann (16), on the melting point depression of pure mercury by small additions of sodium, agree very well with the latest data of Balej and Biros (25); the latter authors utilized differential scanning microcalorimetry with maximum possible suppression of undercooling. There is satisfactory agreement between these data (16,25) and those of (11,12) for the given crystallization region. However, the results of (3) for this region are not consistent with thermodynamic analysis (27).

Most reports have dealt with the solubility in the crystallization region of $NaHg_4$. The solubility data in this region were obtained by classical thermal analysis (3,9, 11-13); by the chemical analyses of the saturated liquid amalgams at various temperatures after separating the crystal of coexisting solid phases (4,7,8,21); and by less common methods, such as the measurement of the anodic limiting currents of sodium dissolution as a function of its concentration at various temperatures (22), and by the calorimetric titration of one- and two-phase amalgams with mercury (23,24,31). In general, the most reliable results are those obtained by chemical analysis of the saturated amalgams after separation of the coexisting solid phase (7,8,21), and by EMF measurements of

(continued next page)

COMPONENTS:	EVALUATOR:
(1) Sodium; Na; [7440-23-5] (2) Mercury; Hg; [7439-97-6]	J. Balej Institute of Inorganic Chemistry Czechoslovak Academy of Sciences Prague, Czechoslovakia C. Guminski; Z. Galus Department of Chemistry University of Warsaw Warsaw, Poland July, 1985
CRITICAL EVALUATION: (continued)	

Na|Na$^+$|Na(Hg) concentration cells (18,26,28,29). These methods allow the determination of true equilibrium data, whereas those obtained by thermal analysis often are in error because of undercooling and supersaturation. Nevertheless, good agreement between the results of the equilibrium methods has been found only for temperatures up to 313 K. At higher temperatures the equilibrium data of Kerp et al. (7,8) are several percent lower than the recent data of Balej (26); the latter data are in good agreement with those of (12) which were obtained by thermal analysis. The solubility reported by Strachan and Harris (19), of 0.88 at % at room temperature, is obviously in error since it is nearly an order of magnitude lower than other more reliable data. The data of Lange et al. (21,22) show satisfactory agreement only for 293 K and 313 K; at 333 K there is an appreciable deviation caused probably by fluctuations of the anodic limiting currents in their experiments. For the solubility in the region between 18 and 85 at % Na the data of (9,11,12) agree in the overall shape of the phase diagram. In the region of NaHg$_2$, however, (9) obtained primary crystallization temperatures that were consistently lower than those recorded by (11,12). The differences were ascribed to the possibility of uncertain thermometer stem corrections and to the effect of oxidation. It should be indicated, however, that some differences exist even between the first and second series of Vanstone's (9) measurements in the more concentrated region above 47.26 at % Na; this finding was ascribed by the author to amalgam oxidation during the measurements. The mutual agreement of data by the above authors is shown in Table I. The controversial views with regard to the composition of some coexisting compounds have been discussed above. In our opinion, some discrepancies may arise also from different degrees of purity of the metallic sodium used by the various authors; an indication of this is suggested by the variation of the melting points, shown in Table I, as compared to the most recent value of 370.98 (30).

From the data in Table II, the liquidus curve in the Hg-rich region, where the sodium concentration is less than 2.8 at %, may be expressed by

$$\log x(\text{Na}) = 0.27786 - 65.235/(T/\text{K}) \qquad [1]$$

The solubility calculated from eq. [1] shows a mean relative deviation of 0.1% from the data in Table II.

For the crystallization region of the Na-rich region, at concentrations above 85.2 at % Na, the agreement between the various authors (9,11,12,16,17) is excellent, especially when compared with the data of various authors for the other crystallization regions. Some discrepancies have been ascribed to partial oxidation of sodium in the amalgams during measurements, and also to errors in the temperature determinations due to probable uncertainties of the thermometer stem corrections. There have been reports of significant effect of pressure on the composition of coexisting compounds and on the solubility (3,4,7,8), but this effect has not yet been investigated quantitatively, and the reported qualitative observations are rather inconsistent.

For the region above 85.2 at % Na the solubility may be expressed by

$$\log x(\text{Na}) = 0.26618 - 99.631/(T/\text{K}) \qquad [2]$$

with a mean relative deviation of 0.43%.

(continued next page)

COMPONENTS:	EVALUATOR:
(1) Sodium; Na; [7440-23-5]	J. Balej
(2) Mercury; Hg; [7439-97-6]	C. Guminski; Z. Galus

CRITICAL EVALUATION: (continued)

TABLE I

Characteristic Data of the Phase Diagram for the Na-Hg System

	Reference			
	9	11	12	15
Hg, m.p., T/K	--	--	234.6	234.3
Eutectic, Hg-NaHg$_4$				
T/K	--	225	226.4	225.2
at % Na	--	2.8	2.7	2.8
Peritectic, NaHg$_4$-NaHg$_2$				
T/K	428.2[a]	432.2	429.4	430.2
at % Na	17.95	18.1	17.9	18
NaHg$_2$, m.p., T/K	619.2	633.2	627.2	626.2
Peritectic, NaHg$_2$-Na$_7$Hg$_8$ (ref. 12), or Na$_{12}$Hg$_{13}$ (ref. 11)[b]				
T/K	491.2	500.2	494.8 (495.2)[c]	496±5
at % Na	47.6	48.1	47.5 (47.6)[c]	48
Peritectic, Na$_7$Hg$_8$ (or Na$_{12}$Hg$_{13}$)-NaHg				
T/K	483.2	492.2	485.4 (485.9)[c]	~488
at % Na	50.6	50.9	51.5 (51.0)[c]	51
Peritectic, NaHg-Na$_3$Hg$_2$				
T/K	(392.2)	396.2	391.7 (393.2)[c]	394.2
at % Na	63	61.9	63.3 (62.5)[c]	62
Peritectic, Na$_3$Hg$_2$-Na$_5$Hg$_2$				
T/K	340.2	339.2	338.9 (338.7)[c]	339.2
at % Na	71.9	71.8	71.7 (73.5)[c]	71.8
Peritectic, Na$_5$Hg$_2$-Na$_3$Hg[d]				
T/K	--	307.2	307.6[e]	307.2
at % Na	--	84.1	83.4 (83.7)[c]	84.1
Eutectic, Na$_3$Hg-Na				
T/K	294.4	294.6	294.6	294.6
at % Na	85.09	85.2	85.2	85.2
Na, m.p., T/K	369.60	--	370.65 (370.75)[c]	370.65

(a) For T/K <428.2 Kurnakov (9) considered the composition of the coexisting solid phase to be NaHg$_5$ or NaHg$_6$.
(b) Kurnakov (9) specified the composition as NaHg$_n$ only, for (2 > n > 1).
(c) According to Vanstone's second series of measurements (12).
(d) Vanstone (12) assigned this peritectic to Na$_3$Hg$_2$-Na$_3$Hg.
(e) Taken as the temperature of polymorphic transformation of Na$_3$Hg (12). Moreover, Na$_{12}$Hg$_{13}$ undergoes (11) a polymorphic transformation at 453.2 K. Similar polymorphic transformations of Na$_5$Hg$_2$ were observed by Schüller (11) at 333.2 and 322.2 K, respectively, whereas Vanstone (12) ascribed these transformations (at 333.2 and 325.2 K, respectively) to Na$_3$Hg$_2$.

COMPONENTS:

(1) Sodium; Na; [7440-23-5]

(2) Mercury; Hg; [7439-97-6]

EVALUATOR:

J. Balej
Institute of Inorganic Chemistry
Czechoslovak Academy of Sciences
Prague, Czechoslovakia

C. Guminski; Z. Galus
Department of Chemistry
University of Warsaw
Warsaw, Poland

July, 1985

CRITICAL EVALUATION: (continued)

TABLE II

Recommended Smoothed Solubility of Sodium in the Hg-Rich Region

T/K	Soly/at %	Solid Phase	Remark
234.28	0.00	Hg	m.p.
232.0	0.843	Hg	
229.8	1.48	Hg	
225.3	2.56	Hg + $NaHg_4$	eutectic
248.2	3.39	$NaHg_4$	
273.2	4.25	$NaHg_4$	
293.2	5.10	$NaHg_4$	
298.2	5.40	$NaHg_4$	
313.2	6.15	$NaHg_4$	
333.2	7.33	$NaHg_4$	
353.2	8.67	$NaHg_4$	
373.2	10.2	$NaHg_4$	
423.2	16.0	$NaHg_4$	
430.2	18.0	$NaHg_4$ + $NaHg_2$	peritectic
498.2	20.3	$NaHg_2$	
523.2	21.5	$NaHg_2$	
573.2	24.9	$NaHg_2$	
623.2	31.4	$NaHg_2$	
626.2	33.3	$NaHg_2$	m.p.

Sodium

COMPONENTS:

(1) Sodium; Na; [7440-23-5]
(2) Mercury; Hg; [7439-97-6]

EVALUATOR:

J. Balej
Institute of Inorganic Chemistry
Czechoslovak Academy of Science
Prague, Czechoslovakia

C. Guminski; Z. Galus
Department of Chemistry
University of Warsaw
Warsaw, Poland

July, 1985

CRITICAL EVALUATION:

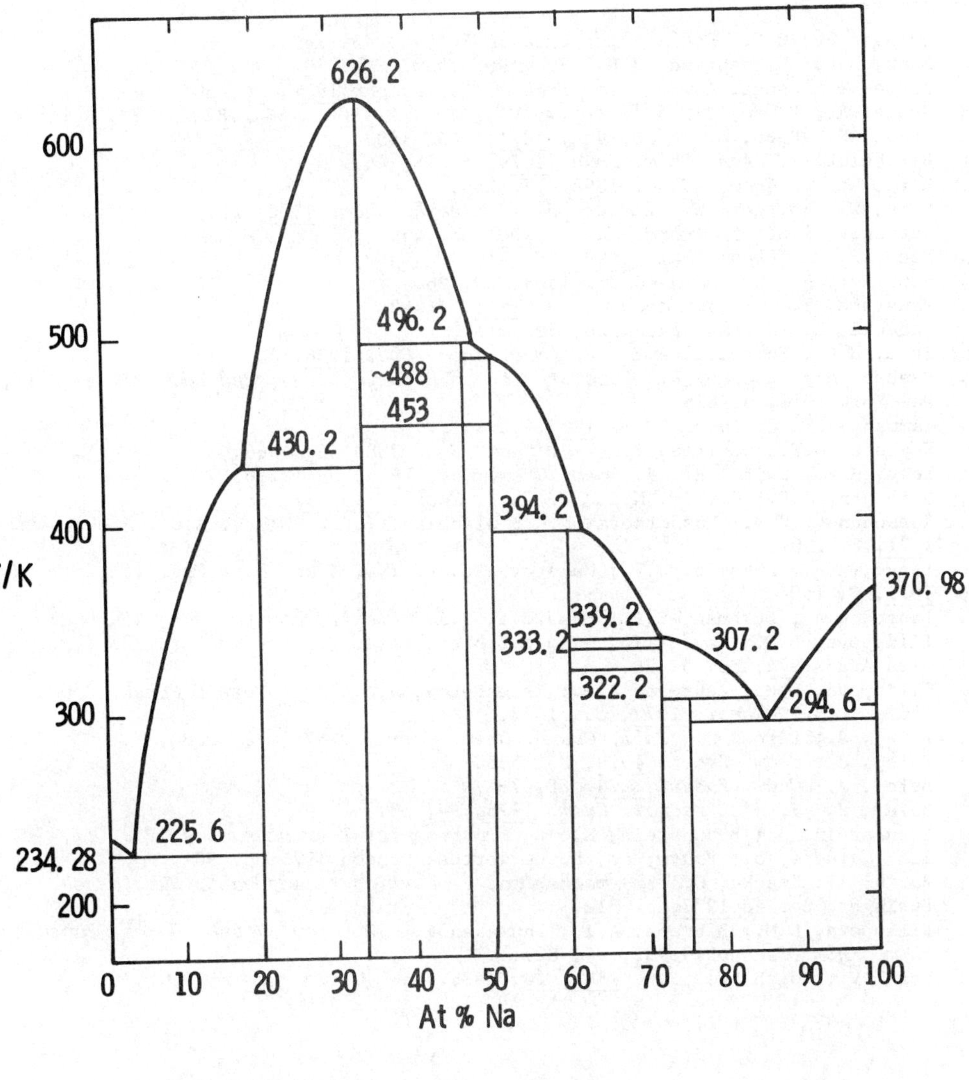

Fig. 1. The Na-Hg phase diagram. Eutectics at 2.6 and 85.2 at % Na.

COMPONENTS:	EVALUATOR:
(1) Sodium; Na; [7440-23-5] (2) Mercury; Hg; [7439-97-6]	J. Balej Institute of Inorganic Chemistry Czechoslovak Academy of Sciences Prague, Czechoslovakia C. Guminski; Z. Galus Department of Chemistry University of Warsaw Warsaw, Poland July, 1985
CRITICAL EVALUATION: (continued)	

References

1. Zvjagincev, O.E. *Dokl. Akad. Nauk SSSR* 1944, *43*, 163.
2. Bent, H.E.; Hildebrand, J.H. *J. Amer. Chem. Soc.* 1927, *49*, 3011.
3. Inoue, Y.; Osugi, A. *J. Electrochem. Soc. Japan* 1952, *20*, 502.
4. Guntz, A.; Férée, J. *C.R. Acad. Sci., Ser. 2* 1900, *131*, 182.
5. Kraut, K.; Popp, O. *Lieb. Ann.* 1871, *159*, 188.
6. Berthelot, M. *Ann. Chim. Phys.* 1879, *5*, 18, 442.
7. Kerp, W. *Z. Anorg. Chem.* 1898, *17*, 284.
8. Kerp, W.; Böttger, W.; Winter, H. *Z. Anorg. Chem.* 1900, *25*, 1.
9. Kurnakov, N.S. *Z. Anorg. Chem.* 1900, *23*, 439.
10. Maey, E. *Z. Phys. Chem.* 1899, *29*, 119.
11. Schüller, A. *Z. Anorg. Chem.* 1904, *40*, 26.
12. Vanstone, E. *Trans. Faraday Soc.* 1911, *7*, 42.
13. Jänecke, E. *Z. Metallk.* 1928, *20*, 113.
14. Bent, H.E.; Forziati, A.E. *J. Amer. Chem. Soc.* 1936, *58*, 2220.
15. Hansen, M.; Anderko, K. *Constitution of Binary Alloys*, 2nd Ed., McGraw-Hill, New York 1958, p. 825.
16. Tammann, G. *Z. Phys. Chem.* 1889, *3*, 441.
17. Heycock, C.T.; Neville, F.H. *J. Chem. Soc.* 1889, *55*, 666.
18. Bent, H.E.; Swift, E. *J. Amer. Chem. Soc.* 1936, *58*, 2216.
19. Strachan, J.F.; Harris, N.L. *J. Inst. Metals* 1956-57, *85*, 17.
20. Korschunov, V.N.; Kuznetsova, N.K.; Gladkikh, I.P.; Volkov, A.G. *Elektrokhimiya* 1971, *7*, 1501.
21. Lange, A.A.; Bukhman, S.P.; Makarova, J.A. *Izv. Akad. Nauk Kaz. SSR, Ser. Khim.* 1971, *7*, 1501.
22. Lange, A.A.; Bukhman, S.P.; Makarova, J.A. *Elektrokhimiya* 1979, *15*, 618.
23. Filippova, L.M.; Gayfullin, A.Sh.; Zebreva, A.I. *Prikl. Teoret. Khimiya*, Alma-Ata 1974, No. 5, 76.
24. Filippova, L.M.; Zebreva, A.I.; Espenbetov, A.A. *Izv. Vyssh. Ucheb. Zaved., Khim. Khim. Tekhnol.* 1977, *20*, 1468.
25. Balej, J.; Biros, J. *Coll. Czech. Chem. Commun.* 1978, *43*, 2834.
26. Balej, J. *Chem. Zvesti* 1979, *33*, 585.
27. Balej, J. *Chem. Zvesti* 1978, *32*, 767.
28. Balej, J. *J. Electroanal. Chem.* 1978, *94*, 13.
29. Braunstein, J.; Braunstein, H. in *Experimental Thermodynamics*, Vol. II, Eds. Neindre, B.; Vodar, B., Butterworths, London 1975, p. 901.
30. Barin, I.; Knacke, O. *Thermochemical Properties of Inorganic Substances*, Springer, Berlin 1973, p. 512.
31. Filippova, L.M.; Zebreva, A.I.; Zhumakanov, V.Z. *Izv. Vyssh. Ucheb. Zaved., Khim. Khim. Tekhnol.* 1982, *25*, 827.
32. Merz, V.; Weith, W. *Ber.* 1881, *14*, 1438.

COMPONENTS:	ORIGINAL MEASUREMENTS:
(1) Sodium; Na; [7440-23-5] (2) Mercury; Hg; [7439-97-6]	Maey, E. Z. Phys. Chem. 1899, 29, 119-38.
VARIABLES: Room temperature measurement	PREPARED BY: C. Guminski; Z. Galus

EXPERIMENTAL VALUES:

Solubility of sodium at room temperature was reported to be 0.62 mass %. The corresponding atomic % solubility calculated by the compilers is 5.2 at %. The temperature was probably 293 K (compilers).

The following phases were reportedly found: $NaHg_5$, $NaHg_2$, $NaHg$, Na_3Hg.

AUXILIARY INFORMATION

METHOD/APPARATUS/PROCEDURE:	SOURCE AND PURITY OF MATERIALS:
The amalgams were prepared by dissolution of sodium in mercury under petroleum. The specific volume of the amalgams was determined with a pycnometer. The specific volumes were plotted as a function of Na concentration, and the solubility was determined from the breakpoint of the curve. The concentration of the amalgams was determined by decomposition with water with subsequent titration with standard sulfuric acid to obtain Na content; the residual mercury was washed and weighed for gravimetric determination.	Nothing specified.
	ESTIMATED ERROR: Soly: nothing specified; precision ± 1% (compilers). Temp: nothing specified.
	REFERENCES:

COMPONENTS:	ORIGINAL MEASUREMENTS:
(1) Sodium; Na; [7440-23-5] (2) Mercury; Hg; [7439-97-6]	Heycock, C.T.; Neville, F.H. *J. Chem. Soc.* 1889, 666-76.
VARIABLES:	PREPARED BY:
Temperature: 83-97°C	C. Guminski; Z. Galus

EXPERIMENTAL VALUES:

The melting points of Na-Hg alloys were reported; the composition of the melting point, or liquidus temperature, corresponds to the solubility of sodium:

$t/°C$	Soly/at %
96.6	0.1982
95.95	0.333
95.38	0.4588
94.46	0.6599
93.64	0.840
92.25	1.172
90.93	1.454
83.35	3.127

AUXILIARY INFORMATION

METHOD/APPARATUS/PROCEDURE:	SOURCE AND PURITY OF MATERIALS:
Melting points were determined with mercury thermometers. The amalgams were protected from oxidation by immersion under paraffin.	Nothing specified.
	ESTIMATED ERROR: Soly: nothing specified. Temp: precision $\pm$ 0.05 K.
	REFERENCES:

COMPONENTS:	ORIGINAL MEASUREMENTS:
(1) Sodium; Na; [7440-23-5] (2) Mercury; Hg; [7439-97-6]	Tammann, G. *Z. Phys. Chem.* 1889, *3*, 441-9.
VARIABLES: Temperature	PREPARED BY: C. Guminski; Z. Galus

EXPERIMENTAL VALUES:

The melting point depression, $-\Delta T/\text{K}$, when sodium was added to pure mercury and when mercury was added to pure sodium:

Sodium Content			Mercury Content		
$-\Delta T/\text{K}$	mass %	at %[a]	$-\Delta T/\text{K}$	mass %	at %[a]
0.39	0.022	0.19	0.01	0.11	0.013
0.72	0.043	0.37	0.11	0.33	0.038
2.23	0.112	0.96	0.27	0.65	0.074
			0.99	2.22	0.260
			1.59	3.39	0.401
			2.40	4.39	0.523
			3.83	7.34	0.900
			7.09	12.76	1.649

[a] by compilers

The melting point of Hg was reported to be 244 instead of 234 K, but it is the opinion of the compilers that the former value was a typographical error in the original publication.

AUXILIARY INFORMATION

METHOD/APPARATUS/PROCEDURE:	SOURCE AND PURITY OF MATERIALS:
The melting temperatures were determined with thermometers. Although no experimental details were given, this work presents a set of precise data which were confirmed in other works.	Nothing specified.
	ESTIMATED ERROR: Soly: nothing specified. Temp: precision $\pm$ 0.05 K.
	REFERENCES:

COMPONENTS:	ORIGINAL MEASUREMENTS:
(1) Sodium; Na; [7440-23-5] (2) Mercury; Hg; [7439-97-6]	Guntz, A.; Férée, J. *C.R. Acad. Sci., Ser. 2*, <u>1900</u>, *131*, 182-4.
VARIABLES: Room temperature	PREPARED BY: C. Guminski; Z. Galus

EXPERIMENTAL VALUES:

Solubility of sodium in mercury at room temperature was reported to be 0.57 mass %. The corresponding atomic % solubility calculated by the compilers is 4.8 at %.

Crystals of the following formulae were reported to exist in the solid phases: $NaHg_8$, $NaHg_6$, $NaHg_5$, and $NaHg_4$. Solid $NaHg_6$ was reported to decompose under pressure to yield solid $NaHg_4$ and liquid amalgam containing 0.57 mass % Na.

AUXILIARY INFORMATION

METHOD/APPARATUS/PROCEDURE:	SOURCE AND PURITY OF MATERIALS:
The amalgams were prepared by dissolution of sodium in mercury; The solids were separated by filtration of the saturated amalgam through chamois leather.	Nothing specified.
	ESTIMATED ERROR: Soly: nothing specified; precision probably better than few percent (compilers). Temp: nothing specified.
	REFERENCES:

COMPONENTS:	ORIGINAL MEASUREMENTS:
(1) Sodium; Na; [7440-23-5] (2) Mercury; Hg; [7439-97-6]	Kerp, W.; Böttger, W.; Winter, H. Z. Anorg. Chem. 1900, 25, 1-71.
VARIABLES: Temperature: 0-161°C	PREPARED BY: C. Guminski; Z. Galus

EXPERIMENTAL VALUES:

The solubility of sodium in mercury:

$t/°C$	mass %	at %[a]	$t/°C$	mass %	at %[a]	$t/°C$	mass %	at %[a]
0	0.53	4.44	40.5	0.72±0.01	5.95	90.4	0.98±0.02	7.95
25	0.65±0.01	5.40	42	0.72±0.01	5.95	99.8	1.10±0.03	8.84
30	0.67	5.56	50	0.74	6.11	124	1.47±0.03	11.52
35	0.70±0.01	5.79	56.7	0.79±0.01	6.50	139	1.69±0.03	13.04
37.7	0.71±0.01	5.87	64.9	0.85±0.02	6.96	161	2.01±0.05	15.18
39.9	0.72±0.01	5.95	81	0.92±0.01	7.49			

---

[a] by compilers

The authors made some systematic errors in their experiments; the result at 0°C is 5% too high and those at temperatures higher than 30°C are too low; the error is as high as 20% at 161°C (compilers). Part of the data and the method were previously reported in (1).

The analysis of crystals yielded $NaHg_6$, at 0 to 40°C, and $NaHg$, at 42 to 100°C.

AUXILIARY INFORMATION

METHOD/APPARATUS/PROCEDURE:	SOURCE AND PURITY OF MATERIALS:
A solid Na-Hg alloy was first prepared in a closed container. Samples were prepared by diluting the alloy with Hg accompanied by heating. The investigated amalgams were transferred to small vessels and thermostated. The equilibrated samples were filtered through a leder plate placed inside of Gooch crucible; filtration in hydrogen atmosphere above 100°C. The filtrates and crystals were analyzed by addition of excess standard HCl and back-titrating with standard baryta water.	Mercury was purified with HNO_3, dried and filtered. Sodium purity not specified.
	ESTIMATED ERROR: Soly: precision better than ± 3%. Temp: nothing specified.
	REFERENCES: 1. Kerp, W. Z. Anorg. Chem. 1898, 17, 284.

COMPONENTS:	ORIGINAL MEASUREMENTS:
(1) Sodium; Na; [7440-23-5] (2) Mercury; Hg; [7439-97-6]	Kurnakov, N.S. *Z. Anorg. Chem.* 1900, *23*, 439-62. *Zh. Russ. Fiz. Khim. Obshch. Ser Khim.* 1899, *31*, 921-48.
VARIABLES:	PREPARED BY:
Temperature: 16-346°C	C. Guminski; Z. Galus

EXPERIMENTAL VALUES:

The freezing points of sodium alloys over the complete composition range were reported; the liquidus composition corresponds to the solubility:

t/°C	Soly at %	t/°C	Soly at %	t/°C	Soly at %	t/°C	Soly at %	t/°C	Soly at %
16.4	4.97	150.5	17.12	341.0	35.91	209.0	50.92	66.3	72.31
33.0	6.22	151.8	17.27	324.0	38.93	207.4	51.78	66.0	73.06
37.0	6.33	155.0	17.95	302.0	41.94	204.8	52.59	65.5	73.52
~46.0	7.25	160.0	18.45	276.5	43.76	201.2	53.43	65.0	74.06
61.0	8.65	163.5	18.76	269.0	44.25	198.5	54.14	62.6	75.70
69.0	9.00	172.5	19.38	238.0	46.31	194.4	54.93	59.3	77.13
91.0	11.66	237.0	21.38	229.9	46.86	169.7	58.09	53.5	78.73
118.4	13.00	281.0	26.01	221.0	47.38	152.2	60.80	47.0	80.46
120.5	13.18	320.5	29.15	218.0	47.60	129.9	61.68	33.65	82.80
123.3	13.50	328.0	30.11	217.5	47.92	114.6	64.43	30.0	83.77
126.4	13.80	330.5	30.41	216.2	48.50	105.5	66.54	25.15	84.43
137.2	15.05	339.5	31.29	215.0	49.07	92.1	68.80	21.25	85.05
145.9	16.24	345.8	32.43	212.7	49.64	85.8	69.95	23.4	85.54
148.9	16.68	345.9	32.79	210.8	50.23	75.2	71.10	32.4	87.34
149.4	16.95	346.0	33.26	209.7	50.60	67.0	71.90	44.9	89.30
								87.65	98.11
								91.95	99.27

The results in the following composition ranges are too high (compilers):
6-14, 26-30, 39, 45-50 and 99 at %.

AUXILIARY INFORMATION

METHOD/APPARATUS/PROCEDURE:	SOURCE AND PURITY OF MATERIALS:
The amalgams were prepared by dissolving sodium into mercury in a hydrogen atmosphere. The amalgams were covered with paraffin and heated, then the freezing points were determined with a mercury thermometer.	Melting point of Na indicates either some impurity or some error in temperature measurement. Mercury purity not specified.
	ESTIMATED ERROR: Soly: nothing specified. Temp: precision ± 0.2 K below 473 K; ± 1 K above 473 K.
	REFERENCES:

COMPONENTS:	ORIGINAL MEASUREMENTS:
(1) Sodium; Na; [7440-23-5] (2) Mercury; Hg; [7439-97-6]	Schüller, A. Z. Anorg. Chem. 1904, 40, 385-99.
VARIABLES:	PREPARED BY:
Temperature: (-48)-360°C	C. Guminski; Z. Galus

EXPERIMENTAL VALUES:

Solidification temperatures of sodium amalgams:

$t/°C$	at % Na
-48.2	2.8
159	18.1
360	33.3
227	48.1
219	50.9
123	61.9
66.2	71.8
33.9	84.1
21.4	85.2

The complete phase diagram was presented and the existence of the following solid phases were reported: $NaHg_4$, $NaHg_2$, $Na_{12}Hg_{13}$, $NaHg$, Na_3Hg_2, Na_5Hg_2, Na_3Hg.

AUXILIARY INFORMATION

METHOD/APPARATUS/PROCEDURE:	SOURCE AND PURITY OF MATERIALS:
The amalgams were prepared by addition of mercury to melted sodium under vaseline. The freezing points were determined with thermometers and thermocouples.	Pure sodium from Merck. Mercury from Merck was double-distilled.
	ESTIMATED ERROR: Soly: precision ± 0.3 %. Temp: precision ± 0.5 K.
	REFERENCES:

COMPONENTS:	ORIGINAL MEASUREMENTS:
(1) Sodium; Na; [7440-23-5] (2) Mercury; Hg; [7439-97-6]	Vanstone, E. *Trans. Faraday Soc.* 1911, 7, 42-64.

VARIABLES:	PREPARED BY:
Temperature: (-47)-354°C	C. Guminski; Z. Galus

EXPERIMENTAL VALUES:

The freezing points for sodium amalgams over the complete composition range were reported; two sets of data, I and II, were presented for amalgams from different methods of preparation.

I.

t/°C	Soly/at %	t/°C	Soly/at %	t/°C	Soly/at %
-40.8	1.17	288.5	24.03	120.9	63.2
-42.3	1.42	315.2	26.11	117.1	63.9
-42.8	1.77	333.4	28.10	115.1	64.3
-46.8	2.38	340.5	29.02	102.0	66.9
-46.8	2.76	345.8	30.26	86.5	69.0
-46.8	3.30	347.2	30.6	69.8	70.6
-5.5	4.12	350.4	31.8	69.6	71.6
18.8	5.04	352.4	32.4	65.4	73.91
22.3	5.18	353.6	33.4	64.1	74.23
27.6	5.45	351.6	34.8	63.9	75.0
33.5	6.03	347.5	35.88	61.4	76.1
54.6	7.67	335.4	38.5	60.0	76.74
62.9	7.71	328.5	39.5	55.2	79.24
75.9	8.83	331.7	39.7	54.2	79.67
83.5	9.03	323.3	40.1	43.0	81.8
91.0	9.90	323.8	40.14	32.6	83.4
105.7	10.97	305.5	41.3	27.2	84.8
111.1	11.49	291.6	42.8	21.35	86.7
122.2	12.49	251.0	47.4	51.4	90.47
139.1	14.2	220.8	48.4	58.5	91.68
148.2	15.4	220.6	48.7	67.8	93.6
154.9	16.57	219.6	49.1	74.5	95.25
155.4	17.6	220.	49.16	83.0	97.2
156.2	18.0	217.7	49.77	86.1	97.7
156.2	18.34	218.4	50.71	93.1	99.06
182.4	18.84	208.9	53.1	93.8	99.25
200.4	19.09	202.4	55.0	95.1	99.52
234.2	20.7	189.6	56.8	96.8	99.86
267.2	22.50	169.4	59.2		
274.0	23.05	142.1	61.7		

(continued next page)

AUXILIARY INFORMATION

METHOD/APPARATUS/PROCEDURE:	SOURCE AND PURITY OF MATERIALS:
(I) Na was melted in a current of dry CO_2 and made to flow via a glass tube into a preweighed tube filled with CO_2. Known weight of Hg was added to the known quantity of molten Na and stirred to form homogeneous liquid; the glass tubes containing Na and amalgams were always flushed with CO_2. Freezing points were determined by heating and cooling amalgam tubes in various types of baths and with use of gaseous and liquid thermometers, depending upon temperature range. (continued next page)	Nothing specified.

	ESTIMATED ERROR:
	Soly: nothing specified. Temp: precision ± 0.2 K.

	REFERENCES:

COMPONENTS:	ORIGINAL MEASUREMENTS:
(1) Sodium; Na; [7440-23-5] (2) Mercury; Hg; [7439-97-6]	Vanstone, E. *Trans. Faraday Soc.* 1911, *7*, 42-64.
VARIABLES:	PREPARED BY:
Temperature: (-47)-354°C	C. Guminski; Z. Galus

EXPERIMENTAL VALUES: (continued)

II.

$t/°C$	Soly/at %	$t/°C$	Soly/at %	$t/°C$	Soly/at %
234.6	47.26	152.2	60.05	67.0	75.55
222.1	47.88	134.2	61.70	67.2	77.49
221.4	48.38	119.4	63.04	53.9	80.20
219.2	49.27	113.6	66.48	34.4	82.18
217.1	49.89	113.2	66.73	37.7	83.26
214.7	50.52	91.8	70.57	31.8	84.13
210.4	52.58	77.6	72.51	35.6	87.62
203.8	54.26	75.6	72.65	49.2	89.99
188.8	56.54	65.1	74.88	75.2	95.06

The following phases were reported: $NaHg_4$, $NaHg_2$, Na_7Hg_8, $NaHg$, Na_3Hg_2, Na_3Hg.

AUXILIARY INFORMATION

METHOD/APPARATUS/PROCEDURE: (continued)	SOURCE AND PURITY OF MATERIALS:
(II) Na was freed of oxide by pipetting molten Na at 403 K, then discharging the liquid by dipping the glass-wool covered tip of pipette under liquid vaseline contained in the experimental tube; Na had not been in contact with air or moisture at any time. Hg was added to molten Na as in method (I), and freezing points determined similarly.	
	ESTIMATED ERROR:
	REFERENCES:

COMPONENTS:	ORIGINAL MEASUREMENTS:
(1) Sodium; Na; [7440-23-5] (2) Mercury; Hg; [7439-97-6]	Bent, H.E.; Swift, E. J. Am. Chem. Soc. 1936, 58, 2216-20.
VARIABLES:	PREPARED BY:
Temperature: 5-25°C	C. Guminski; Z. Galus; M. Salomon

EXPERIMENTAL VALUES:

The experimental EMF (E_1 and E_2 for the cells given in eqs [1] and [2] below) were added algebraically to give all potentials in terms of E_2. These data were fitted by least squares to the following smoothing equation:

$$\log(a_1/x_1) = a + bx_1 + cx_1^2 \qquad [3]$$

where a_1 and x_1 are, respectively, the activity and mole fraction of Na in the amalgam. Eq [3] was used to compute the soly of Na (see below), and the results are summarized in the following table.

$t/°C$	x_1(sat)	f_1^*	$-a^{**}$	b^{**}	c^{**}
5.00	0.043955	5.282	13.86807	16.1820	5.970
15.00	0.04870	6.164	13.32030	15.87260	7.110
25.00	0.05380	7.274	12.81441	15.6130	7.530

*Rational activity coefficient of Na in satd slns calcd by compilers from eq [3].
**Constants of eq [3].

AUXILIARY INFORMATION

METHOD/APPARATUS/PROCEDURE:	SOURCE AND PURITY OF MATERIALS:
EMF's were measured for eight amalgams and solid Na using the following cells: $$Na(Hg)_{a_1} | NaI, DMA | Na(Hg)_{a_2} \qquad [1]$$ and $$Na(s) | NaI, DMA | Na(Hg)_{a_2} \qquad [2]$$ where DMA is dimethylamine. The concn of NaI was not specified, but the EMF's of these cells are independent of NaI concn. Amalgams prepared by distilling Hg into Na. Details on manipulation of amalgams and filling of the cells not given, but probably as in (1). Amalgams analyzed by titrn with stnd H_2SO_4 using brom thymol blue indicator. Titrns were performed in quartz flasks under a stream of CO_2-free air, and said to be reproducible to 0.02%. The authors state that the EMF's of cell [2], E_2, using the two phase amalgams were used in eq [3] to compute the soly of Na. Details on this calcn were not given by the authors.	Nothing specified for Hg and Na, but probably as in (1) and (2); i.e., Hg washed with HNO_3 and filtered, and Na melted and filtered. Dimethylamine distilled onto CaO and then onto sodium and benzophenone. NaI prepared by fusion under vac as in (2).
	ESTIMATED ERROR: Soly: av dev 0.1% (authors); 0.5% (compilers). EMF's: reproducibility 0.01 to 0.03 mV. Temp: ± 0.01°C.
	REFERENCES: 1. Bent, H.E.; Gilfillan, E.S. J. Am. Chem. Soc. 1933, 55, 3989. 2. Bent, H.E.; Forziati, A.F. J. Am. Chem. Soc. 1936, 58, 2220. 3. Dietrick, H.; Yeager, E.; Hovorka, F. Tech. Rpt No. 3. O.N.R. Contract No. 581(00). Western Reserve Univ. 1953.

COMPONENTS:	ORIGINAL MEASUREMENTS:
(1) Sodium; Na; [7440-23-5] (2) Mercury; Hg; [7439-97-6]	Inoue, Y.; Osugi, A. *J. Electrochem. Soc. Japan* 1952, 20, 502-4.
VARIABLES: Temperature: (-48)-190°C	PREPARED BY: C. Guminski; Z. Galus

EXPERIMENTAL VALUES:

Solubility of sodium in mercury:

$t/°C$	Soly mass %	at %[a]	$t/°C$	Soly mass %	at %[a]
-48	0.22	1.89	70	0.94	7.65
-40	0.10	0.88	80	1.00	8.10
-40	0.26	2.22	90	1.05	8.47
-30	0.32	2.72	100	1.11	8.92
-20	0.37	3.14	110	1.20	9.58
-10	0.42	3.55	120	1.24	9.87
0	0.48	4.04	130	1.36	10.7
10	0.52	4.36	140	1.61	12.5
20	0.56	4.68	150	1.64	12.7
30	0.61	5.08	160	1.85	14.1
40	0.69	5.71	170	2.16	16.1
50	0.74	6.11	180	2.31	17.1
60	0.82	6.73	190	2.36	17.4

[a] by compilers

The phase diagram proposed by the authors is not smooth and contains a number of inflections that are not in agreement with other works. The authors attribute the inflections to the solid phases: $NaHg_5$, $NaHg_6$, $NaHg_7$, $NaHg_8$, $NaHg_9$, $NaHg_{10}$, $NaHg_{12}$ and $NaHg_{14}$. The disagreement with other published works is attributed to experimental inaccuracy (by the compilers).

AUXILIARY INFORMATION

METHOD/APPARATUS/PROCEDURE:	SOURCE AND PURITY OF MATERIALS:
The amalgams were obtained by mixing the two metals or by electrolysis of saturated NaCl solutions with a mercury cathode. No further details were given.	Nothing specified.
	ESTIMATED ERROR: Soly: nothing specified. Temp: nothing specified.
	REFERENCES:

COMPONENTS:	ORIGINAL MEASUREMENTS:
(1) Sodium; Na; [7440-23-5] (2) Mercury; Hg; [7439-97-6]	Filippova, L.M.; Gayfullin, A.Sh.; Zebreva, A.I. *Prikl. Teoret. Khim.*, Alma-Ata 1974, No. 5, 76-82.
VARIABLES:	PREPARED BY:
One temperature: 25°C	C. Guminski; Z. Galus

EXPERIMENTAL VALUES:

Solubility of sodium in mercury was reported to be 5.15 ± 0.03 at % at 25°C.

The same result was also obtained in (1) and a slightly higher value of 3.66 mol dm^{-3}, corresponding to 5.42 at % (calculated by compilers), in (2).

AUXILIARY INFORMATION

METHOD/APPARATUS/PROCEDURE:	SOURCE AND PURITY OF MATERIALS:
The heterogeneous amalgam containing 8.52 at % Na was obtained by electrolysis. Content of Na in the amalgam was estimated by chemical analysis by acid decomposition. All operations were carried out in an argon atmosphere. Enthalpy of dilution (Q) of the amalgams of various composition was measured. A break in the curve relating Q to the Na concentration in the amalgam corresponded to the composition of the saturated amalgam.	Nothing specified.
	ESTIMATED ERROR: Soly: precision ± 0.6%. Temp: nothing specified.
	REFERENCES: 1. Filippova, L.M.; Zebreva, A.I.; Espenbetov, A.A. *Izv. Vyssh. Ucheb. Zaved., Khim. Khim. Tekhnol.* 1977, *20*, 1468. 2. Filippova, L.M.; Zebreva, A.I.; Zhumakanov, V.Z. *Ibid.* 1982, *25*, 827.

COMPONENTS:	ORIGINAL MEASUREMENTS:
(1) Sodium; Na; [7440-23-5] (2) Mercury; Hg; [7439-97-6]	1. Lange, A.A.; Bukhman, S.P.; Makarova, I.A. *Izv. Akad. Nauk Kaz. SSR, Ser. Khim.* 1977, *27*, No. 6, 61-3. 2. Same authors. *Elektrokhimiya* 1979, *15*, 618-23.
VARIABLES:	PREPARED BY:
Temperature: 20-80°C	C. Guminski; Z. Galus

EXPERIMENTAL VALUES:

Solubility of mercury:

$t/°C$	Soly mass %	Soly at %[a]	Reference
20	0.58 ± 0.02	4.84	1
20	0.58	4.84	2
40	0.75 ± 0.03	6.18	1
40	0.76	6.26	2
60	0.86 ± 0.07	7.04	1
60	1.03	8.32	2
80	1.00 ± 0.07	8.10	1

[a] by compilers

AUXILIARY INFORMATION

METHOD/APPARATUS/PROCEDURE:	SOURCE AND PURITY OF MATERIALS:
Amalgams in both works prepared by electrolysis of 2 mol dm^{-3} NaOH with Hg cathode. (1) The amalgams were kept for 2.5-18 hrs under cathodic polarization in $(CH_3)_4NI$ at 20 and 40°C, or 1:1 water-ethanol at 60 and 80°C, in a burette-type vessel. Fractions of amalgams were separated through the stopcock. Samples were analyzed by addition of excess stnd. acid and back-titration with stnd. base. (2) Solubility measurements made by polarization measurements: polarization current vs. Na-concentration curves were drawn, and a break in the curves corresponded to the concentration of the saturated amalgam. It was observed that the concentration of Na drops only 1% when the amalgam was aged for less than 2 hrs.	NaOH was analytical grade. Pure $(CH_3)_4NI$ was twice recrystallized. Hg purity not specified.
	ESTIMATED ERROR:
	Soly: precision ±3-7% in (1); nothing specified in (2), but precision better than few percent (compilers). Temp: nothing specified in (1); precision ± 0.5 K in (2).
	REFERENCES:

COMPONENTS:	ORIGINAL MEASUREMENTS:
(1) Sodium; Na; [7440-23-5] (2) Mercury; Hg; [7439-97-6]	Balej, J. *Chemicke Zvesti* 1979, *33*, 585-93.
VARIABLES: Temperature: 225-421 K	PREPARED BY: C. Guminski; Z. Galus

EXPERIMENTAL VALUES:

Solubility of sodium in mercury:

T/K	Soly/at %	T/K	Soly/at %
232.0	0.8435[a]	320.95	6.534
229.82	1.483[a]	338.15	7.65
225.4	2.5524[a]	363.15	9.42
227.4	2.829[a]	368.15	9.766
288.15	4.870	382.35	11.002
298.15	5.40	393.15	12.05
306.65	5.763	397.35	12.473
313.15	6.25	421.65	17.134

[a] Results also presented in ref. (1).

AUXILIARY INFORMATION

METHOD/APPARATUS/PROCEDURE:	SOURCE AND PURITY OF MATERIALS:
The first four results in table obtained by direct thermal analysis with a differential scanning calorimeter. The other results were obtained from potentiometric measurements of concentration cells (2,3). For measurements at 15 and 25°C, electrolyte of extra dry (<0.1 mg/10^3 g H_2O) $NaClO_4$ in propylene carbonate was used (2). 2% MgO-doped β-alumina was used at higher temperatures (3). All measurements conducted in atmosphere of purified, dry nitrogen. Sodium amalgams prepared by dissolving filtered, molten Na into Hg under vacuum.	Sodium was reagent grade from Lachema, Brno. Mercury was redistilled.
	ESTIMATED ERROR: Soly: not specified; precision better than few tenths of a percent (compilers). Temp: Precision ± 0.1 K at T/K < 363; ± 0.2 K at T/K > 363.
	REFERENCES: 1. Balej, J.; Biros, J. *Coll. Czech. Chem. Commun.* 1978, *43*, 2834. 2. Balej, J.; Dousek, F.P.; Jansta, J. *Coll. Czech. Chem. Commun.* 1977, *42*, 2737. 3. Balej, J.; Dousek, F.P.; Jansta, J. *Coll. Czech. Chem. Commun.* 1978, *43*, 3123.

COMPONENTS:	EVALUATOR:
(1) Potassium; K; [7440-09-7] (2) Mercury; Hg; [7439-97-6]	C. Guminski; Z. Galus Department of Chemistry University of Warsaw Warsaw, Poland July, 1985

CRITICAL EVALUATION:

Tammann (1) observed that addition of up to 0.693 at % of potassium to mercury progressively lowered the melting point of the mercury by 1.24 K. Kerp (2) reported potassium solubilities of 2.27 and 1.27 at % at room temperature and 273 K, respectively. Kerp and coworkers (4) made further determinations between 261 and 373 K, and observed that the solubility of potassium increased from 1.07 to 9.83 at % in this temperature range. These results agree only partly with those of subsequent workers. Kurnakov (5) applied thermal analysis and determined the phase diagram of this system over the concentration range of 3.11 to 86.73 at % potassium. Guntz and Fèrèe (6) used a filtration method and determined a solubility of 1.99 at % at room temperature, but this value is slightly too low. Smith and Bennett (7) obtained a solubility of 2.37 at % at 293 K; this solubility agrees with that of Kerp and coworkers.

Very precise potentiometric measurements of the solubility of potassium in mercury at 273.2 to 300.0 K were reported by Bent and Gilfillan (8). Armbruster and Crenshaw (9) also made potentiometric measurements of the K-Hg system, and their results on the potassium solubilities at 273.2 to 308.2 K are in good agreement with those of (8). Empirical equations relating the potassium solubility to temperature in the measured composition ranges were derived in the latter two papers. Roeder and Morawietz (10) found that the eutectic in the K-rich region was situated at 94.1 at % potassium and 320.70 K. Schuhmann and Kaltwasser (12) investigated the K-Hg phase diagram between 22 and 30 at % potassium, and these authors confirmed the earlier results of Kurnakov (5). Filippova and coworkers (14-18) employed calorimetric titration and reported potassium solubilities of 3.0 ± 0.1 and 4.0 ± 0.1 at % at 298 and 313 K, respectively; these values are slightly higher than those of (8,9).

There have been other determinations of potassium solubility in mercury, but these are rejected in the evaluation because of errors in the determinations (3,11,13), or because of insufficient definition of the experimental procedure (19,20). Kozin's (21) estimated solubility of 94.2 at % at 298 K is clearly too high.

As shown in Fig. 1, the saturated potassium amalgams are in equilibrium with various compounds in this system.

Recommended (r) and tentative solubility of potassium in mercury:

T/K	Soly/at %	Reference
273.2	1.27 (r)	[8,9]
293.2	2.25 (r)	[7-9]
298.2	2.53 (r)	[8,9]
323	4.5	[5]
373	11[a]	[5]
473	24 (r)	[5,12]
543	33.3	[5]

[a] Interpolated value from data of (5).

(continued next page)

COMPONENTS:	EVALUATOR:
(1) Potassium; K; [7440-09-7] (2) Mercury; Hg; [7439-97-6]	C. Guminski; Z. Galus Department of Chemistry University of Warsaw Warsaw, Poland July, 1985

CRITICAL EVALUATION:

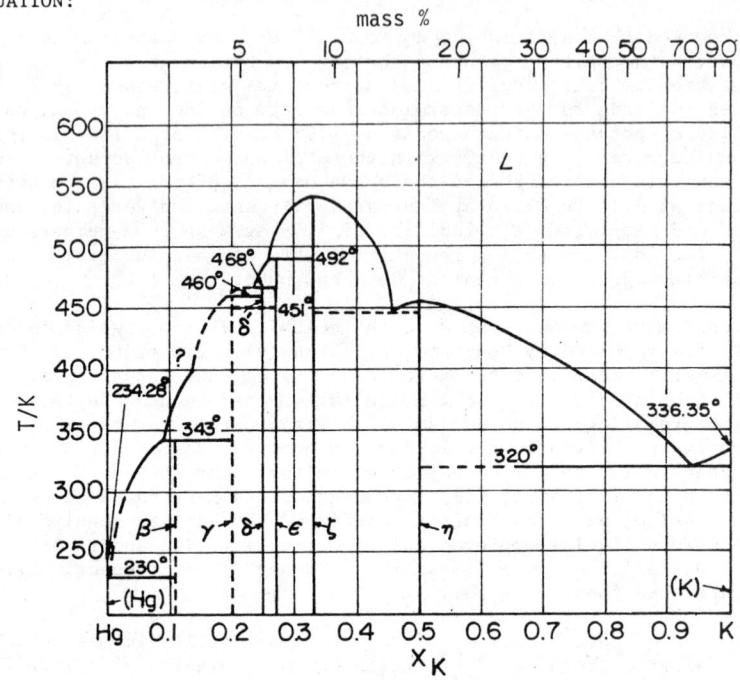

Fig. 1. The K-Hg phase diagram (22).

REFERENCES:

1. Tammann, G. *Z. Phys. Chem.* 1889, *3*, 441.
2. Kerp, W. *Z. Anorg. Chem.* 1898, *17*, 284.
3. Maey, E. *Z. Phys. Chem.* 1899, *29*, 119.
4. Kerp, W.; Böttger, W.; Winter, H. *Z. Anorg. Chem.* 1900, *25*, 1.
5. Kurnakov, N.S. *Z. Anorg. Chem.* 1900, *23*, 439; *Zh. Russ. Fiz. Khim. Obshch., Ser. Khim.* 1889, *31*, 927.
6. Guntz, A.; Férée, J. *C.R. Acad. Sci., Ser. 2* 1900, *131*, 182.
7. Smith, G.McP.; Bennett, H.C. *J. Am. Chem. Soc.* 1909, *31* 799; 1910, *32*, 622.
8. Bent, H.E.; Gilfillan, E.S. *J. Am. Chem. Soc.* 1933, *55*, 3989.
9. Armbruster, M.H.; Crenshaw, J.L. *J. Am. Chem. Soc.* 1934, *56*, 2525.
10. Roeder, A.; Morawietz, W. *Z. Metallk.* 1956, *47*, 734.
11. Strachan, J.F.; Harris, N.L. *J. Inst. Metals* 1956-57, *85*, 17.
12. Schuhmann, H.; Kaltwasser, K. *Z. Phys. Chem.* 1962, *219*, 168.
13. Smith, G.McP., Ball, T.R. *J. Am. Chem. Soc.* 1917, *32*, 179.
14. Filippova, L.M.; Zebreva, A.I.; Omarova, N.D.; Korobkina, N.P. *Izv. Vyssh. Ucheb. Zaved., Khim. Khim. Tekhnol.* 1978, *21*, 316.
15. Filippova, L.M.; Zebreva, A.I.; Espenbetov, A.A. *ibid* 1977, *20*, 1468.
16. Filippova, L.M.; Zhumakanov, V.Z.; Zebreva, A.I. *ibid* 1978, *21*, 1450.
17. Filippova, L.M.; Zebreva, A.I.; Korobkina, N.P. *Ukr. Khim. Zh.* 1978, *44*, 791.
18. Filippova, L.M.; Zebreva, A.I.; Zhumakanov, V.Z. *Izv. Vyssh. Ucheb. Zaved., Khim. Khim. Tekhnol.* 1982, *25*, 827.
19. Korshunov, V.N.; Kuznetsova, N.K.; Gradkikh, I.P.; Volkov, A.G. *Elektrokhimiya* 1971, *7*, 1501.
20. Ruban, L.M.; Gladyshev, V.P.; Zebreva, A.I.; cited by Kozlovskii, M.T.; Zebreva, A.I.; Gladyshev, V.P. *Amalgamy i Ikh Primenenie*, Nauka, Alma-Ata, 1971, p. 19.
21. Kozin, L.F. *Fiziko-Khimicheskie Osnovy Amalgamnoi Metallurgii*, Nauka, Alma-Ata, 1964.
22. Hultgren, R.; Desai, P.D.; Hawkins, D.T.; Gleiser, M.; Kelley, K.K. *Selected Values of the Thermodynamic Properties of Binary Alloys*, Am. Soc. Metals, Metals Park, OH, 1973, pp. 949-51.

COMPONENTS:	ORIGINAL MEASUREMENTS:
(1) Potassium; K; [7440-09-7] (2) Mercury; Hg; [7439-97-6]	Tammann, G. *Z. Phys. Chem.* 1889, *3*, 441-9.

VARIABLES:	PREPARED BY:
Temperature	C. Guminski; Z. Galus

EXPERIMENTAL VALUES:

Depression of the freezing point of mercury, $\Delta T/K$, by small additions of potassium:

	Potassium Content	
$\Delta T/K$	mass %	at %[a]
0.27	0.018	0.092
0.42	0.030	0.15
0.73	0.091	0.46
1.04	0.111	0.567
1.24	0.136	0.693

[a] by compilers.

AUXILIARY INFORMATION

METHOD/APPARATUS/PROCEDURE:	SOURCE AND PURITY OF MATERIALS:
Freezing points were determined thermometrically. Details of experiment were not given.	Nothing specified.
	ESTIMATED ERROR: Soly: nothing specified. Temp: precision $\pm$ 0.05 K.
	REFERENCES:

COMPONENTS:	ORIGINAL MEASUREMENTS:
(1) Potassium; K; [7440-09-7] (2) Mercury; Hg; [7439-97-6]	Kurnakov, N.S. Z. Anorg. Chem. 1900, 23, 439-62; Zh. Russ. Fiz. Khim. Obshch., Ser. Khim. 1899, 31, 921-48.
VARIABLES:	PREPARED BY:
Temperature: 33-269°C	C. Guminski; Z. Galus

EXPERIMENTAL VALUES:

Freezing points of the amalgams were reported; the solubilities corresponding to the liquidus concentrations are as follows:

t/°C	Soly/at %	t/°C	Soly/at %	t/°C	Soly/at %	t/°C	Soly/at %
33.0	3.11	76.5	9.52	195.0	23.35	249.5	39.45
45.0	3.91	80.5	9.77	198.7	23.53	215.0	43.39
52.0	4.90	89.5	10.42	203.5	24.24	175.0	45.24
56.7	5.32	106.0	11.35	216.5	25.73	151.0	61.74
63.5	6.39	112.5	11.70	239.5	27.64	148.7	62.48
66.0	6.76	~121.0	12.53	254.0	29.73	145.9	63.44
67.3	7.31	129.0	13.61	268.0	32.11	142.7	64.28
68.3	7.53	151.0	14.27	269.7	33.34	141.9	65.18
69.4	7.71	165.0	15.41	269.2	34.19	135.4	67.70
69.9	8.15	174.0	16.53	269.5	34.45	115.4	76.09
70.3	8.65	189.5	20.57	263.0	37.11	88.4	85.09
73.5	9.03	194.5	22.38	251.5	39.04	82.4	86.73

Composition of the crystalline phases was also discussed.

AUXILIARY INFORMATION

METHOD/APPARATUS/PROCEDURE:	SOURCE AND PURITY OF MATERIALS:
Amalgams were prepared by dissolution of potassium in mercury in hydrogen atmosphere. Amalgam was covered with paraffin and heated. Freezing points determined from cooling curves with the use of mercury thermometers.	Nothing specified.
	ESTIMATED ERROR: Soly: nothing specified. Temp: precision ± 0.2 K below 473 K; precision ± 1 K above 474 K.
	REFERENCES:

COMPONENTS:	ORIGINAL MEASUREMENTS:
(1) Potassium; K; [7440-09-7] (2) Mercury; Hg; [7439-97-6]	Guntz, A.; Fèrèe, J. *C.R. Acad. Sci.*, Ser. 2 <u>1900</u>, *131*, 182-4.
VARIABLES:	PREPARED BY:
One temperature: room temperature	C. Guminski; Z. Galus

EXPERIMENTAL VALUES:

Room temperature solubility of potassium in mercury was reported to be 0.395 mass %. The corresponding atomic % solubility calculated by the compilers is 1.99 at %.

Solid phase analysis suggested the existence of the compounds, KHg_{10}, KHg_{12} and KHg_{18}.

AUXILIARY INFORMATION

METHOD/APPARATUS/PROCEDURE:	SOURCE AND PURITY OF MATERIALS:
Amalgams were prepared by dissolution of potassium in mercury. The solid phase was separated by filtration through a chamois leather after equilibration.	Nothing specified.
	ESTIMATED ERROR: Soly: nothing specified; precision no better than few percent (compilers). Temp: nothing specified.
	REFERENCES:

COMPONENTS:	ORIGINAL MEASUREMENTS:
(1) Potassium; K; [7440-09-7] (2) Mercury; Hg; [7439-97-6]	Kerp, W.; Böttger, W.; Winter, H. Z. Anorg. Chem. 1900, 25, 2-71.
VARIABLES:	PREPARED BY:
Temperature: (-12)-200°C	C. Guminski; Z. Galus

EXPERIMENTAL VALUES:

Solubility of potassium in mercury:

$t/°C$	Soly/mass %	Soly/at %[a]
-12	0.21	1.07
0	0.31 ± 0.01	1.57
20	0.47 ± 0.01	2.37
25	0.53 ± 0.01	2.66
30	0.56 ± 0.01	2.81
45.8	0.81 ± 0.06	4.02
56.1	0.88 ± 0.06	4.35
60	1.02 ± 0.01	5.02
65	1.23 ± 0.03	6.00
71	1.41 ± 0.02	6.84
73.5	1.64 ± 0.03	7.88
74	1.71 ± 0.03	8.19
75	1.85 ± 0.02	8.82
81	1.89 ± 0.02	9.00
90	2.01 ± 0.03	9.52
99.8	2.08 ± 0.02	9.83

[a] by compilers.

Analysis of solid phases suggested the existence of the compounds KHg_{12} and KHg_{10}.

Comments:
The data in the temperature ranges of 20-46 and 75-100°C are in good agreement with other workers; the data in the other temperature ranges are in poor agreement with other workers (compilers).

AUXILIARY INFORMATION

METHOD/APPARATUS/PROCEDURE:	SOURCE AND PURITY OF MATERIALS:
Amalgams were obtained by electrolysis of saturated KCl solution with circulating Hg as cathode. The amalgams were filtered through chamois skin in a Gooch crucible. The experimental operations were performed in dry hydrogen atmosphere. The filtrate and the crystals were analyzed by addition of excess standard HCl and back-titration with standard baryta water.	Nothing specified.
	ESTIMATED ERROR: Soly: precision no better than ± 7%. Temp: nothing specified.
	REFERENCES:

COMPONENTS:	ORIGINAL MEASUREMENTS:
(1) Potassium; K; [7440-09-7] (2) Mercury; Hg; [7439-97-6]	Smith, G.McP.; Bennett, H.C. *J. Am. Chem. Soc.* 1909, *31*, 799-806. ibid. 1910, *32*, 622-26.
VARIABLES: One temperature: 20°C	PREPARED BY: C. Guminski; Z. Galus

EXPERIMENTAL VALUES:

The solubility of potassium in mercury at 20°C was reported to be 0.46 mass %. The corresponding atomic % solubility calculated by the compilers is 2.32 at %.

Analysis of the crystals showed 8.14 at % K; this corresponded to the formula KHg_{11} or KHg_{12}.

AUXILIARY INFORMATION

METHOD/APPARATUS/PROCEDURE:	SOURCE AND PURITY OF MATERIALS:
The amalgam was prepared by electrolysis of saturated KCl solution with Hg as the cathode. The amalgam was washed with water and dried between filter paper, then filtered through chamois skin. The filtrate and solid were analyzed alkacimetrically by decomposition with 0.1 mol dm^{-3} HCl, with subsequent addition of excess 0.1 mol dm^{-3} NaOH and back-titration with 0.1 mol dm^{-3} HCl.	Very pure salts were obtained from Kahlbaum. Mercury purity not specified.
	ESTIMATED ERROR: Soly: nothing specified; precision no better than few percent (compilers). Temp: nothing specified.
	REFERENCES:

COMPONENTS:	ORIGINAL MEASUREMENTS:
(1) Potassium; K; [7440-09-7] (2) Mercury; Hg; [7439-97-6]	Bent, H.E.; Gilfillan, E.S. J. Am. Chem. Soc. 1933, 55, 3989-4001.
VARIABLES:	PREPARED BY:
Temperature: 0-27°C	C. Guminski; Z. Galus

EXPERIMENTAL VALUES:

Solubility of potassium in mercury:

$t/°C$	Soly/at %	Soly/Mass %
0	1.271	0.250
6.15	1.544	0.305
14.35	1.951	0.386
15.00	1.986	0.393
18.82	2.191	0.435
23.09	2.428	0.483
25.00	2.536	0.505
26.79	2.638	0.525

AUXILIARY INFORMATION

METHOD/APPARATUS/PROCEDURE:	SOURCE AND PURITY OF MATERIALS:
EMF were measured for nine amalgams of different concentrations with the cell, $K(Hg)_y$ | KI in ethylamine | $K(Hg)_x$. Concentration of KI was not specified. The amalgam was prepared by distilling mercury onto distilled potassium, and the electrolyte was prepared from purified materials in the glass cell system without exposure to the ambient atmosphere. Amalgams were analyzed by titration with standard H_2SO_4, with bromothymol blue indicator, in an atmosphere of CO_2-free air.	Purity of materials not specified, but probably of high purity as in (1).
	ESTIMATED ERROR: Soly: ave. dev. $\pm$ 0.4%. Temp: precision $\pm$ 0.01 K.
	REFERENCES: 1. Bent, H.E.; Forziati, A.F. J. Am. Chem. Soc. 1936, 58, 2200.

COMPONENTS:	ORIGINAL MEASUREMENTS:
(1) Potassium; K; [7440-09-7] (2) Mercury; Hg; [7439-97-6]	Armbruster, M.H.; Crenshaw, J.L. J. Am. Chem. Soc. 1934, 56, 2525-34.
VARIABLES: Temperature: 0-35°C	PREPARED BY: C. Guminski; Z. Galus

EXPERIMENTAL VALUES:

Solubility of potassium in mercury:

$t/°C$	g K/100 g Hg	Soly/at %[a]
0.00	0.2508	1.270
5.00	0.2945	1.488
10.00	0.3427	1.728
15.00	0.3945	1.984
20.00	0.4490	2.253
25.00	0.5054	2.527
30.00	0.5654	2.819
35.00	0.6248	3.106

[a] by compilers.

AUXILIARY INFORMATION

METHOD/APPARATUS/PROCEDURE:	SOURCE AND PURITY OF MATERIALS:
Amalgams were prepared by the electrolysis of a saturated aqueous solution of K_2CO_3 with a pool of Hg as the cathode; the amalgam was then filtered into an evacuated glass bulb for storage until used. The solubility of potassium in the saturated amalgam was determined by drawing off a sample of the liquid after equilibration at each temperature. The potassium concentration was determined by adding an excess of standard HCl, then back-titrating with standard $Ba(OH)_2$ in a CO_2-free atmosphere, using rosolic acid indicator. The residual Hg was determined gravimetrically. EMF were determined as a function of temperature with the concentration cell, $K(Hg)_{sat} | KCl\ soln | K(Hg)_x$.	Mercury was chemically purified and distilled twice. Other chemicals of original high purity were further purified by recrystallization.
	ESTIMATED ERROR: Soly: precision ± 0.05 %. Temp: precision ± 0.01 K.
	REFERENCES:

COMPONENTS:	ORIGINAL MEASUREMENTS:
(1) Potassium; K; [7440-09-7] (2) Mercury; Hg; [7439-97-6]	Schuhmann, H.; Kaltwasser, K. Z. Phys. Chem. 1962, 219, 168-70.
VARIABLES: Temperature: 189-265°C	PREPARED BY: C. Guminski; Z. Galus

EXPERIMENTAL VALUES:

A partial phase diagram was presented by the authors; the liquidus data points were read from the curve by the compilers.

t/°C	Soly/at %	t/°C	Soly/at %	t/°C	Soly/at %
265	30.2	218	25.7	203	24.0
258	28.7	214	25.4	204	23.8
246	28.0	215	25.3	201	23.6
237	27.3	210	25.1	199	23.5
235	27.2	213	25.1	197	23.3
240	27.0	211	24.7	195	23.0
236	26.7	206	24.6	194	22.9
230	26.7	209	24.4	193	22.7
231	26.3	206	24.2	192	22.5
225	26.0			189	22.3

The existence of the compounds, KHg_3 and KHg_2, was confirmed by thermal analysis.

AUXILIARY INFORMATION

METHOD/APPARATUS/PROCEDURE:	SOURCE AND PURITY OF MATERIALS:
Amalgams were prepared by dissolution of potassium in mercury in nitrogen atmosphere. The potassium content was determined by decomposing the amalgam with 0.05 mol dm^{-3} sulfuric acid, and the excess acid was back-titrated. The residual Hg was weighed to determine its concentration. The liquidus temperatures of the amalgams were determined with copper-constantan thermoelement. The thermal analyses were made in an evacuated glass vessel heated by an electric oven.	Potassium purity higher than 99.5%. Mercury was purified chemically, then twice distilled.
	ESTIMATED ERROR: Soly: nothing specified. Temp: precision $\pm$ 1 K.
	REFERENCES:

COMPONENTS:	ORIGINAL MEASUREMENTS:
(1) Potassium; K; [7440-09-7] (2) Mercury; Hg; [7439-97-6]	1. Filippova, L.M.; Zebreva, A.I.; Omarova, N.D.; Korobkina, N.P. *Izv. Vyssh. Ucheb. Zaved., Khim. Khim. Tekhnol.* 1978, *21*, 316-9.
VARIABLES: Temperature: 25-40°C	PREPARED BY: C. Guminski; Z. Galus

EXPERIMENTAL VALUES:

Solubility of potassium in mercury:

$t/°C$	Soly/mol dm^{-3}	Soly/at %	Reference
25	2.1	3.0 ± 0.1	1,2,3,4
25	---	3.23	5
40	2.86	4.0 ± 0.1	1

AUXILIARY INFORMATION

METHOD/APPARATUS/PROCEDURE:	SOURCE AND PURITY OF MATERIALS:
Potassium amalgams were prepared by electrolysis. Potassium content in the amalgam was determined by chemical analysis. All operations were carried out in an argon atmosphere (2). Enthalpy of dilution, Q, of the various heterogeneous and homogeneous amalgams was determined by calorimetric titration. A breakpoint on the curve relating Q to the potassium concentration in the amalgam denoted the saturation point.	Nothing specified.
	ESTIMATED ERROR: Soly: precision no better than ± 3%. Temp: nothing specified.

REFERENCES:

2. Filippova, L.M.; Zebreva, A.I.; Espenbetov, A.A. *Izv. Vyssh. Ucheb. Zaved., Khim. Khim. Tekhnol.* 1977, *20*, 1468-71.
3. Filippova, L.M.; Zhumakanov, V.Z.; Zebreva, A.I. *Izv. Vyssh. Ucheb. Zaved., Khim. Khim. Tekhnol.* 1978, *21*, 1450-3.
4. Filippova, L.M.; Zebreva, A.I.; Korobkina, N.P. *Ukr. Khim. Zh.* 1978, *44*, 791-3.
5. Filippova, L.M.; Zebreva, A.I.; Zhumakanov, V.Z. *Izv. Vyssh. Ucheb. Zaved., Khim. Khim. Tekhnol.* 1982, *25*, 827-9.

COMPONENTS:	EVALUATOR:
(1) Rubidium; Rb; [7440-17-7] (2) Mercury; Hg; [7439-97-6]	C. Guminski; Z. Galus Department of Chemistry University of Warsaw Warsaw, Poland July, 1985

CRITICAL EVALUATION:

Kerp and coworkers (1) reported the first determination of the solubility of rubidium in mercury by a method of filtration and alkacimetric analysis of the rubidium. At 273 and 298 K the solubilities were 2.13 and 3.16 at %, respectively. These results in the range of low rubidium concentration are in good agreement with later measurements of Kurnakov and Zukovsky (2) and of Smith and Bennett (3); the rubidium concentration in both of these works were determined alkacimetrically. The thermoanalytical data of Kurnakov and Zukovsky determined the partial phase diagram of this system up to approximately 15 at % rubidium. Biltz and coworkers (4) investigated the equilibria over the complete concentration range by thermoanalysis and alkacimetric determination of the rubidium content. However, the latter authors' liquidus temperatures in the mercury-rich region, below 393 K, were significantly lower than those obtained by the previous three workers (1,2,3). The discrepancy is attributed to the lower precision of the thermal analysis of Biltz and coworkers at the lower temperatures.

Other determinations of the solubility of rubidium have been reported, but these values are rejected because of erroneous values (6) or because of incomplete experimental description (7). Kozin (8) estimated a 298 K solubility of 96.8 at %; this value is very near to the experimental value in the Rb-rich amalgams.

Figure 1 (5) shows the phase diagram based on the data of refs. (2) and (4).

Recommended (r) and tentative values of the solubility of Rb in Hg:

T/K	Soly/at %	Reference
227	0.7[a]	[4]
273	2[b]	[4,1]
293	3.0[c]	[3,7]
298	3.2 (r)	[1,2]
323	4.8[d]	[2]
373	9.6	[2]
459	20[a]	[4]
470	22	[4]
530	33.3[b]	[4]

[a] Eutectic point.
[b] Extrapolated value from data of cited references.
[c] Mean value from data of cited references.
[d] Interpolated value from data of cited reference.

(continued next page)

COMPONENTS:	EVALUATOR:
(1) Rubidium; Rb; [7440-17-7] (2) Mercury; Hg; [7439-97-6]	C. Guminski; Z. Galus Department of Chemistry University of Warsaw Warsaw, Poland July, 1985

CRITICAL EVALUATION: (continued)

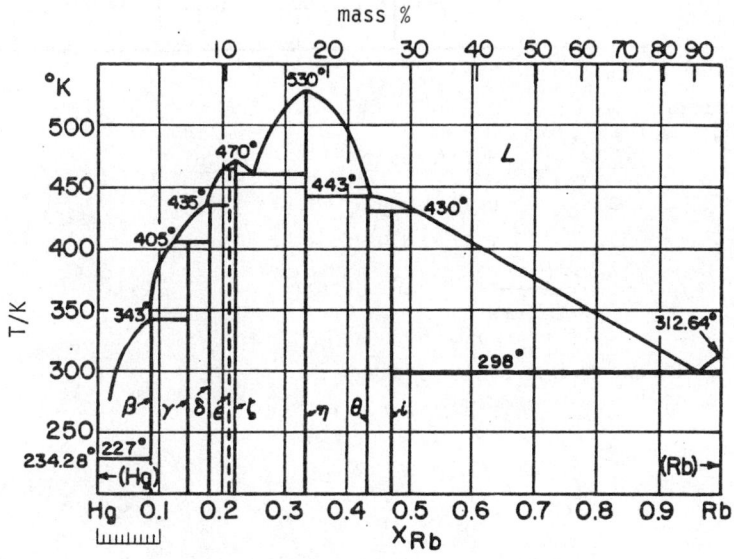

Fig. 1. The Rb-Hg phase diagram (5).

References

1. Kerp, W.; Böttger, W.; Winter, H. *Z. Anorg. Chem.* 1900, *25*, 1.
2. Kurnakòw, N.S.; Zukowsky, G.J. *Z. Anorg. Chem.* 1907, *52*, 416.
3. Smith, G.McP.; Bennett, H.C. *J. Am. Chem. Soc.* 1910, *32*, 622, 1909, *31*, 799.
4. Biltz, W.; Weibke, F.; Eggers, H. *Z. Anorg. Chem.* 1934, *219*, 119.
5. Hultgren, R.; Desai, P.D.; Hawkins, D.T.; Gleiser, M.; Kelley, K.K. *Selected Values of the Thermodynamic Properties of Binary Alloys*, Am. Soc. Metals, Metals Park, OH, 1973, p. 977.
6. Strachan, J.F.; Harris, N.L. *J. Inst. Metals* 1956-57, *85*, 17.
7. Korshunov, V.N.; Kuznetsova, N.K.; Gradkih, I.P.; Volkov, A.G. *Elektrokhimiya*, 1971, *7*, 1501.
8. Kozin, L.F. *Fiziko-Khimicheskie Osnovy Amalgamnoi Metallurgii*, Nauka, Alma-Ata, 1964.

COMPONENTS:	ORIGINAL MEASUREMENTS:
(1) Rubidium; Rb; [7440-17-7] (2) Mercury; Hg; [7439-97-6]	Kerp, W.; Böttger, W.; Winter, H. *Z. Anorg. Chem.* 1900, *25*, 1-71.
VARIABLES: Temperature: 0-25°C	PREPARED BY: C. Guminski; Z. Galus

EXPERIMENTAL VALUES:

Solubility of rubidium in mercury:

t/°C	Soly/mass %	Soly/at %[a]
0	0.92 ± 0.02	2.13
25	1.37 ± 0.02	3.16

[a] by compilers

Solid phase analysis showed the presence of $RbHg_{11}$.

AUXILIARY INFORMATION

METHOD/APPARATUS/PROCEDURE:	SOURCE AND PURITY OF MATERIALS:
The amalgams were obtained by electrolysis of RbCl solution with circulating mercury as the cathode. The amalgam was then filtered through chamois skin placed inside of a Gooch crucible. An excess of acid was added to the separated phases and the solution was back-titrated with standard baryta water.	Nothing specified.
	ESTIMATED ERROR: Soly: precision ± 2%, but appears to be less precise to compilers. Temp: nothing specified.
	REFERENCES:

COMPONENTS:	ORIGINAL MEASUREMENTS:
(1) Rubidium; Rb; [7440-17-7] (2) Mercury; Hg; [7439-97-6]	Kurnakov, N.S.; Zukovsky, G.J. Z. Anorg. Chem. 1907, 52, 416-28.
VARIABLES: Temperature: 26-148°C	PREPARED BY: C. Guminski; Z. Galus

EXPERIMENTAL VALUES:

Solubility of rubidium which corresponds to the concentration at the crystallization temperatures of the amalgams:

$t/°C$	Soly/at %	$t/°C$	Soly/at %
147.7	14.64	69.1	7.87
138.8 (136.5)	13.37	70.2	7.80
132.3	12.59	69.4	7.55
127.5	11.95	68.5	7.32
117.2	10.71	66.3	6.88
104.6	9.85	62.7	6.30
91.4	9.04	56.9	5.64
78.3	8.31	48.5	4.77
74.5	8.10	36.6	3.97
70.5	7.95	26.4	3.31

AUXILIARY INFORMATION

METHOD/APPARATUS/PROCEDURE:	SOURCE AND PURITY OF MATERIALS:
The method of amalgam preparation was not specified. The samples were analyzed alkacimetrically to determine the rubidium content. Solidification temperatures were determined as the samples were cooled.	Nothing specified.
	ESTIMATED ERROR: Soly: nothing specified. Temp: precision $\pm$ 0.1 K.
	REFERENCES:

COMPONENTS:	ORIGINAL MEASUREMENTS:
(1) Rubidium; Rb; [7440-17-7] (2) Mercury; Hg; [7439-97-6]	Smith, G.McP.; Bennett, H.C. *J. Am. Chem. Soc.* 1909, *31*, 799-806; *Ibid.* 1910, *32*, 622-26.
VARIABLES:	PREPARED BY:
One temperature: 19.5°C	C. Guminski; Z. Galus

EXPERIMENTAL VALUES:

The solubility of rubidium in mercury at 19.5°C was reported to be 1.21 ± 0.01 mass %. The corresponding atomic % solubility calculated by the compilers is 2.79 at %.

Solid phase analyses suggest the compounds $RbHg_{11}$ or $RbHg_{12}$.

AUXILIARY INFORMATION

METHOD/APPARATUS/PROCEDURE:	SOURCE AND PURITY OF MATERIALS:
The amalgams were obtained by electrolysis of concentrated RbCl solution. After 24 hours of equilibration the amalgam was filtered through chamois skin with a suction pump. The analysis of the amalgam was carried out alkacimetrically: an excess of 0.1 mol dm^{-3} HCl was added to the sample, then back-titrated with 0.1 mol dm^{-3} NaOH.	"Very pure salts" from Kahlbaum were used. Mercury purity not specified.
	ESTIMATED ERROR: Soly: precision ± 1%. Temp: nothing specified.
	REFERENCES:

COMPONENTS:	ORIGINAL MEASUREMENTS:
(1) Rubidium; Rb; [7440-17-7] (2) Mercury; Hg; [7439-97-6]	Biltz, W.; Weibke, F.; Eggers, H. *Z. Anorg. Chem.* <u>1934</u>, *219*, 119-28.
VARIABLES:	PREPARED BY:
Temperature: (-6)-255°C	C. Guminski; Z. Galus

EXPERIMENTAL VALUES:

Solubility of rubidium in mercury which corresponds to the concentration at the crystallization temperatures of the amalgams:

t/°C	Soly/at %	t/°C	Soly/at %	t/°C	Soly/at %	t/°C	Soly/at %
-6	2.1	161.5	17.5	250	37.0	124	63.0
11	3.2	174	18.4	228	39.8	110	68.0
26	4.3	183	19.3	206	41.2	97	72.2
39	5.9	188	20.1	200	41.6	84.5	76.5
61	9.0	193	21.0	169	44.0	70.5	80.7
88	12.0	196	22.0	168	45.2	60	83.9
103	11.0	194	23.4	167	46.2	46.5	89.0
113.5	12.4	197	26.0	164	47.7	34.5	93.0
123.5	13.0	221	27.6	162	48.9	26	96.2
131	13.6	236	29.3	157	51.3	29	97.1
143.5	15.0	246.5	30.9	150	54.3	33	98.0
150	15.5	252.5	32.0	145	55.8	35	98.7
158.5	16.9	255	34.7	138	58.4		

AUXILIARY INFORMATION

METHOD/APPARATUS/PROCEDURE:	SOURCE AND PURITY OF MATERIALS:
The amalgams were prepared by dropping mercury into fused rubidium in an argon atmosphere. The amalgam was heated and cooling curves were recorded with a thermoelement. The amalgam composition was determined alkacimetrically.	95% purity Rb_2SO_4 was first purified to a product with 0.01% impurities in other alkali metals. The salt was then used to prepare metallic Rb. Mercury was purified by vacuum distillation. Test for calcium in the amalgams was negative.
	ESTIMATED ERROR: Soly: precision better than ± 0.5%. Temp: precision ± 0.5%.
	REFERENCES:

COMPONENTS:	EVALUATOR:
(1) Cesium; Cs; [7440-46-2] (2) Mercury; Hg; [7439-97-6]	C. Guminski; Z. Galus Department of Chemistry University of Warsaw Warsaw, Poland July, 1985

CRITICAL EVALUATION:

The first and most comprehensive study of the cesium-mercury equilibria was reported by Kurnakov and Zukovsky (1). The authors determined the liquidus curve by thermal analysis over the complete composition range. However, it appears that the results in the Hg-rich region, between -226 and 346 K, are 10-20% too high by comparison with other determinations of the solubilities. Smith and Bennett (2,3) determined the liquid equilibrium amalgam composition at 273 to 299 K by acid-base titration of the equilibrated liquid amalgam which was separated from the solid by filtration and by centrifugation. The results of the latter authors were only in rough agreement with those of Kurnakov and Zukovsky. Although the analytical method used by Smith and Bennett is capable of yielding accurate analysis of the amalgam, there is some doubt in regard to the solubility at the temperatures reported by these authors because of the method of separation of the liquid amalgam from the equilibrium solid phase. Kozin (4) reported a calculated solubility of cesium in mercury of 99.7 at % at 298 K; this value is very near that found by Kurnakov and Zukovsky in the Cs-rich region. However, the calculation of Kozin will tend to be too high because of the neglect of the strong interaction between these metals. Korshunov and coworkers (5) reported a concentration of 4.5 at % cesium in mercury at about 293 K, a value in agreement with that of Smith and Bennett (2), but no experimental details were presented by these authors.

Hultgren et al. (6) reported the phase diagram for this system, Fig. 1; these authors based their phase diagram on the data of Kurnakov and Zukovsky (1). A critical evaluation of the enthalpy of solution also is presented by (6).

Tentative values of the solubility of cesium in mercury:

T/K	Soly/at %	Reference
227	2[a]	[1]
273	3.0	[3]
293	4.1	[2]
298	4.4	[3]
323	5.5[b]	[1,3]
373	7.4	[1]
413	10	[1]
473	31	[1]
481	33.3	[1]

[a] Eutectic point.
[b] Interpolated from data of cited references.

(continued next page)

COMPONENTS:	EVALUATOR:
(1) Cesium; Cs; [7440-46-2] (2) Mercury; Hg; [7439-97-6]	C. Guminski; Z. Galus Department of Chemistry University of Warsaw Warsaw, Poland July, 1985

CRITICAL EVALUATION: (continued)

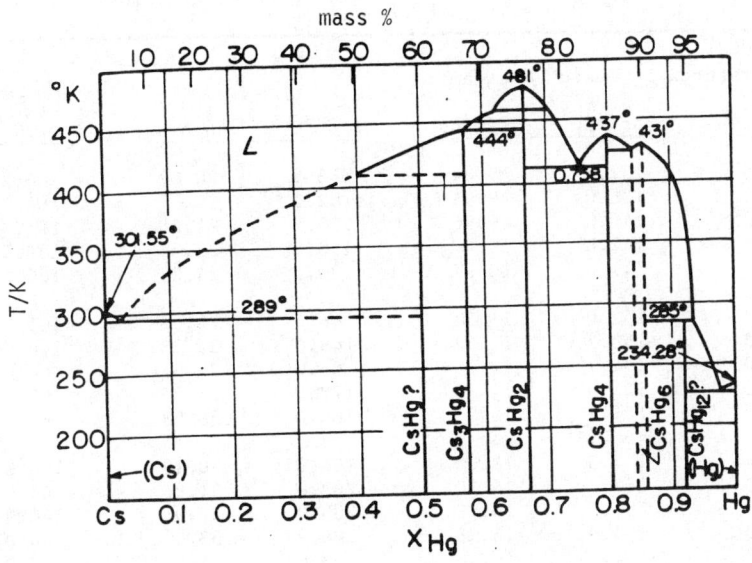

Fig. 1. The Cs-Hg phase diagram (6).

References

1. Kurnakov, N.S.: Zukovsky, G.J. *Z. Anorg. Chem.* 1907, *52*, 416.
2. Smith, G.McP.; Bennett, H.C. *J. Am. Chem. Soc.* 1909, *31*, 799.
3. Smith, G.McP.; Bennett, H.C. *J. Am. Chem. Soc.* 1910, *32*, 622.
4. Kozin, L.F. *Fiziko-Khimicheskie Osnovy Amalgamnoi Metallurgii*, Nauka, Alma-Ata, 1964.
5. Korshunov, V.N.; Kuznetsova, N.K.; Gradkih, J.P.; Volkov, A.G. *Elektrokhimiya* 1971, *7*, 1501.
6. Hultgren, R.; Desai, P.D.; Hawkins, D.T.; Gleiser, M.; Kelley, K.K. *Selected Values of the Thermodynamic Properties of Binary Alloys*, Am. Soc. Metals, Metals Park, OH, 1973, p. 727.

COMPONENTS:	ORIGINAL MEASUREMENTS:
(1) Cesium; Cs; [7440-46-2] (2) Mercury; Hg; [7439-97-6]	Kurnakov, N.S.; Zukovsky, G.J. Z. Anorg. Chem. 1907, 52, 416-29.
VARIABLES:	PREPARED BY:
Temperature: (-47)-208°C	C. Guminski; Z. Galus

EXPERIMENTAL VALUES:

Liquidus temperatures of cesium amalgams:

t/°C	at % Cs	t/°C	at % Cs	t/°C	at % Cs	t/°C	at % Cs
-41.7	0.45	132.8	9.84	163.5	20.07	186.2	38.24
-43.5	0.99	136.0	9.91	162.0	20.57	184.0	38.73
-44.8	1.28	140.0	9.96	159.5	21.60	171.0	40.42
-46.6	2.25	142.0	10.47	150.5	23.78	169.5	41.83
-26.5	3.71	147.0	10.77	139.5	24.25	166.0	43.15
-9.2	4.12	152.0	11.34	146.6	24.62	165.2	45.56
7.1	4.90	153.7	11.87	149.0	25.23	165.0	45.65
6.7	5.90	154.7	12.28	165.0	27.54	164.0	46.12
73.3	6.36	156.0	12.53	172.6	27.75	163.0	46.88
77.9	6.60	156.7	12.97	185.0	28.88	161.0	48.67
86.1	6.64	157.3	13.66	194.0	30.12	160.0	51.60
93.6	7.13	157.7	14.20	202.8	31.88	146.6	56.68
101.2	7.43	156.8	14.78	205.8	32.99	140.3	60.93
97.3	7.50	155.0	16.28	208.2	33.60	128.0	61.83
108.2	7.90	154.1	16.46	207.6	34.21	112.0	67.89
112.6	8.21	153.3	16.94	206.0	34.81	103.0	73.14
118.1	8.70	152.0	17.40	204.5	35.48	26.3	97.57
125.1	8.78	156.7	18.40	199.0	36.51	19.3	98.43
132.0	9.24	161.0	19.09	192.0	37.58	19.3	99.37

AUXILIARY INFORMATION

METHOD/APPARATUS/PROCEDURE:	SOURCE AND PURITY OF MATERIALS:
Metallic cesium was obtained by reduction of Cs_2CO_3 with magnesium in very pure hydrogen atmosphere, then mercury was added to form the amalgam. The amalgams were analyzed by alkacimetry. The liquidus temperatures were determined from cooling curves.	Cs_2CO_3 was supplied by Kahlbaum. The magnesium was free of cesium. Mercury purity not specified.
	ESTIMATED ERROR: Soly: nothing specified. Temp: precision ± 0.1 K.
	REFERENCES:

COMPONENTS:	ORIGINAL MEASUREMENTS:
(1) Cesium; Cs; [7440-46-2] (2) Mercury; Hg; [7439-97-6]	Smith, G.McP.; Bennett, H.C. *J. Am. Chem. Soc.* 1909, *31*, 799-807.
VARIABLES: One temperature: 17°C	PREPARED BY: C. Guminski; Z. Galus

EXPERIMENTAL VALUES:

Solubility of cesium in mercury at 17°C was reported to be 2.75 ± 0.01 mass %. The corresponding atomic % solubility calculated by the compilers is 4.09 at %. Analysis of the solid phase led to the formula $CsHg_{13}$.

AUXILIARY INFORMATION

METHOD/APPARATUS/PROCEDURE:	SOURCE AND PURITY OF MATERIALS:
The amalgam was obtained by electrolysis of concentrated aqueous CsCl solution with a pool of Hg as the cathode. The amalgam was then washed with water and dried with a filter paper. After standing in a glass-stoppered bottle for several days, the amalgam was rapidly suction-filtered through chamois skin on a Gooch crucible. The analyses of the filtrate and solid residue were made by acid-base titration: an excess of 0.1 mol dm^{-3} HCl was added to the sample, then made alkaline with an excess of 0.1 mol dm^{-3} NaOH. The excess NaOH was back-titrated with 0.1 mol dm^{-3} HCl.	"Very pure salts" from Kahlbaum were used. Mercury purity not specified.
	ESTIMATED ERROR: Soly: nothing specified; precision no better than few percent (compilers). Temp: nothing specified.
	REFERENCES:

MM-C*

COMPONENTS:	ORIGINAL MEASUREMENTS:
(1) Cesium; Cs; [7440-46-2] (2) Mercury; Hg; [7439-97-6]	Smith, G.McP.; Bennett, H.C. *J. Am. Chem. Soc.* 1910, *32*, 622-26.
VARIABLES: Temperature: 0-26°C	PREPARED BY: C. Guminski; Z. Galus

EXPERIMENTAL VALUES:

Solubility of cesium in mercury:

t/°C	Soly/mass %	Soly/at %
0	1.96	2.93
18	2.61	3.89
26	2.98	4.43

Analyses of the equilibrated solid phases separated by filtration and by centrifugation suggested the formulae $CsHg_{13}$ or $CsHg_{12}$.

AUXILIARY INFORMATION

METHOD/APPARATUS/PROCEDURE:	SOURCE AND PURITY OF MATERIALS:
The amalgam was prepared as in the previous study (1), but the solid residue, after filtration of the equilibrated amalgam through chamois skin, was sealed into glass tubes after remelting and removal of air. The latter samples were subsequently equilibrated at room temperature and at 0°C, then the contents of the tube were rapidly centrifuged and the solids were analyzed by alkacimetry: an excess of 0.1 mol dm^{-3} HCl was added to the sample, then an excess of 0.1 mol dm^{-3} NaOH was added to the acidified solution, and the excess base was finally back-titrated with 0.1 mol dm^{-3} HCl.	"Very pure salts" from Kahlbaum were used. Mercury purity not specified. ESTIMATED ERROR: Soly: precision no better than a few percent at best (compilers). Temp: nothing specified. REFERENCES: 1. Smith, G.McP.; Bennett, H.C. *J. Am. Chem. Soc.* 1909, *31*, 799.

COMPONENTS:	EVALUATOR:
(1) Beryllium; Be; [7440-41-7] (2) Mercury; Hg; [7439-97-6]	C. Guminski; Z. Galus Department of Chemistry University of Warsaw Warsaw, Poland July, 1985

CRITICAL EVALUATION:

Nerad (1) reported that the solubility of beryllium in mercury increases uniformly from 2×10^{-5} to 8×10^{-4} at % at 373 and 1073 K, respectively; however, no experimental details were given by the author. Wang (2) determined the solubility of beryllium at 644 K and reported a value of 1.3×10^{-4} at %. This value appears to be in agreement with the estimates by Nerad (1). Strachan and Harris (3) could not detect the dissolution of beryllium in mercury at room temperature, and these authors estimated that the solubility was below 2×10^{-2} at %.

Zucker (4) heated a mixture of mercury and beryllium powder at 923 K for one hour, and this author also reduced Be(II) on a mercury cathode from various solvents; the content of beryllium in the amalgams from these studies was never higher than 5×10^{-2} at %. Zucker stated that the latter concentration is the upper limit of the solubility at room temperature, but in the opinion of the evaluators this value is much too high. Kozin calculated that the solubility of beryllium at 298 K is 8.7×10^{-3} (4) and 1.5×10^{-2} at % (5); these estimated values appear too high, as were the predicted solubilities of a number of other amalgam systems. The formation of $BeHg_2$ has been reported for the Be-Hg system (6).

Tentative value of the solubility of Be in Hg at 644 K is 1×10^{-4} at % (2).

References

1. Nerad, A.J.; as cited by Kelman, L.R.; Wilkinson, W.D.; Yaggee, F.L. U.S. At. Ener. Comm. Rep., ANL-4417, 1950.
2. Wang, J.Y.N. *Nucl. Sci. Eng.* 1964, *18*, 18.
3. Strachan, J.F.; Harris, N.L. *J. Inst. Metals* 1956-57, *85*, 17.
4. Zucker, D. U.S. At. Ener. Comm. Rep., ORNL-3488, 1963, p. 28.
5. Kozin, L.F. *Tr. Inst. Khim. Nauk Akad. Kaz. SSR* 1962, *9*, 101.
6. Kozin, L.F. *Fiziko-Khimicheskie Osnovy Amalgamnoĭ Metallurgii*, Nauka, Alma-Ata, 1964.
7. Kells, M.C.; Holden, R.B.; Whitman, C.I. *J. Am. Chem. Soc.* 1957, *79*, 3925.

COMPONENTS:	ORIGINAL MEASUREMENTS:
(1) Beryllium; Be; [7440-41-7] (2) Mercury; Hg; [7439-97-6]	Wang, J.Y.N. *Nucl. Sci. Eng.* 1964, *18*, 18-30.
VARIABLES: Temperature: 644 K	PREPARED BY: C. Guminski; Z. Galus

EXPERIMENTAL VALUES:

The solubility of beryllium in mercury at 644 K was reported to be 0.06 mg/Kg.

The corresponding solubility in atom % calculated by the compilers is 1.3×10^{-4} at %.

AUXILIARY INFORMATION

METHOD/APPARATUS/PROCEDURE:	SOURCE AND PURITY OF MATERIALS:
Sheet of Be, which was cleaned, degreased, and vacuum-dried, was presumably equilibrated with Hg in a quartz capsule; the capsule was contained in a stainless steel autoclave. The Be content in the liquid was determined by an unspecified acid extraction analysis.	Fresh, triple-distilled mercury and beryllium of "high purity" were used.
	ESTIMATED ERROR: Soly: nothing specified; precision about 10% (compilers). Temp: precision $\pm$ 5 K.
	REFERENCES:

COMPONENTS:	EVALUATOR:
(1) Magnesium; Mg; [7439-95-4] (2) Mercury; Hg; [7439-97-6]	C. Guminski; Z. Galus Department of Chemistry University of Warsaw Warsaw, Poland July, 1985

CRITICAL EVALUATION:

Kerp and coworkers (1) reported the first determination of magnesium solubility in mercury; they found solubilities of 2.52 at % at room temperature and approximately 8 at % at 573 K. The room temperature solubility agrees with subsequent measurements by other workers, but the 573 K value is much too low, probably because of oxidation of the magnesium. Cambi and Speroni (2) determined a partial phase diagram in the Hg-rich region and they showed that the solubility of magnesium increases monotonically from 2.5 to 29 at % in the temperature range of 290 to 643 K. Smits and Beck (3) and Beck (4) determined the phase diagram for the composition range above 15 at % Mg by thermo-analytical and potentiometric measurements. Loomis (5) equilibrated the saturated amalgam at 295.6 K and precisely determined the magnesium content in the liquid phase to be 2.60 at %. At 277 K, Williams (6) reported a solubility of 2.15 at %. Danilchenko (7) redetermined the complete phase diagram and obtained solubilities which were slightly higher at low temperatures, and the solubilities were slightly different between 17 and 33 at %, as compared to the data of (2) and (4). Dergacheva and Kozin (8) determined a solubility of 2.82 at % at 298 K.

Other solubility determinations of magnesium, which were reported, gave only solubility limits: less than 2.5 at % (9) and less than 8×10^{-3} at % (10) at room temperature. Also Kozin's (11) predicted value of 0.86 at % at 298 K is too low.

The saturated magnesium amalgams are in equilibrium with various intermediate solid phases, as shown by the phase diagram (12) in Fig. 1.

Recommended (r) and tentative solubilities of magnesium in the Hg-rich region.

T/K	Soly/at %	Reference
293	2.50 (r)	1,2,5
298	2.7	5,8
323	4.5[a]	2
373	9.3	2,7
473	20 (r)	2,4,7
573	26	4,7
673	31[a]	4
773	37[a]	4
873	45	4
900	50.0	4,7

[a] Interpolated from data of cited reference

COMPONENTS:	EVALUATOR:
(1) Magnesium; Mg; [7439-95-4] (2) Mercury; Hg; [7439-97-6]	C. Guminski; Z. Galus Department of Chemistry University of Warsaw Warsaw, Poland July, 1985

CRITICAL EVALUATION:

References

1. Kerp, W.; Böttger, W.; Iggena, H. *Z. Anorg. Chem.* 1900, *25*, 1.
2. Cambi, L.; Spereni, G. *Atti Reale Accad. Lincei, Ser. 5* 1915, *24*, 734.
3. Smits, A.; Beck, R.P. *Proc. Kong. Akad. Wetensch, Amsterdam* 1921, *23*, 975.
4. Beck, R.P. *Rec. Trav. Chim.* 1922, *41*, 353.
5. Loomis, A.G. *J. Am. Chem. Soc.* 1922, *44*, 8.
6. Williams, E.J. *Phil. Mag., Ser. 6* 1925, *50*, 589.
7. Danilchenko, P.T. *Zh. Russ. Fiz. Khim. Obshch., Ser. Khim.* 1930, *62* (1), 975.
8. Dergacheva, M.B.; Kozin, L.F. *Vestn. Akad. Nauk Kaz. SSR* 1974, No. 6, 56.
9. Kremann, R.; Müller, R. *Z. Metallk.* 1920, *12*, 303.
10. Strachan, J.F.; Harris, N.L. *J. Inst. Metals* 1956-57, *85*, 17.
11. Kozin, L.F. *Fiziko-Khimicheskie Osnovy Amalgamnoi Metallurgii*, Nauka, Alma-Ata, 1964.
12. Hansen, M.; Anderko, K. *Constitution of Binary Alloys*, McGraw-Hill, New York, 1958, p. 823.

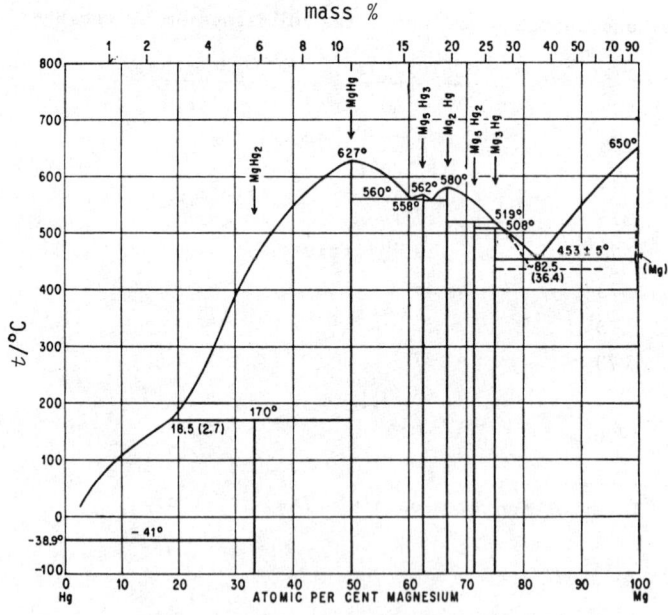

Fig. 1. The Mg-Hg system (12).

COMPONENTS:	ORIGINAL MEASUREMENTS:
(1) Magnesium; Mg; [7439-95-4] (2) Mercury; Hg; [7439-97-6]	Kerp, W.; Böttger, W.; Iggena, H. Z. Anorg. Chem. 1900, 25, 1-71.
VARIABLES: Room temperature	PREPARED BY: C. Guminski; Z. Galus

EXPERIMENTAL VALUES:

The solubility of magnesium in mercury was reported to be 0.313 mass %. The solubility in atomic % calculated by the compilers is 2.52 at %. At about 300°C the solubility was estimated to be around 1 mass %. This value is much too low (compilers). The compound, $MgHg_6$, was found in the equilibrium solid phase at room temperature. However, this compound has not been confirmed by later workers.

AUXILIARY INFORMATION

METHOD/APPARATUS/PROCEDURE:	SOURCE AND PURITY OF MATERIALS:
Bands of Mg were cleaned in alcohol and ether, then equilibrated with Hg in a glass container. The amalgam was filtered and the Mg content in the saturated filtrate was determined as magnesium phosphate.	Nothing specified.
	ESTIMATED ERROR: Soly: nothing specified; precision no better than ± 10% (compilers). Temp: nothing specified.
	REFERENCES:

COMPONENTS:	ORIGINAL MEASUREMENTS:
(1) Magnesium; Mg; [7439-95-4] (2) Mercury; Hg; [7439-97-6]	Cambi, L.; Speroni, G. *Atii Reale Accad. Lincei, Ser. 5*, 1915, 24, 734-38.
VARIABLES: Temperature: 17-370°C	PREPARED BY: G. Cuminski; Z. Galus

EXPERIMENTAL VALUES:

Freezing points in the Mg-Hg system were reported for concentrations up to 29 at % Mg.

t/°C	Mg/at %
17	2.5
55	5
89	8
106	10
119	12
145	14
168	18
207	21
230	23
277	25
335	27
370	29

At the higher magnesium concentrations it was impossible to record the liquidus curves because of the boiling of the amalgams at about 412°C. The solid phase in equilibrium with the saturated amalgams was determined to be $MgHg_2$.

AUXILIARY INFORMATION

METHOD/APPARATUS/PROCEDURE:	SOURCE AND PURITY OF MATERIALS:
Appropriate amounts of magnesium were dissolved in boiling mercury in an atmosphere of pure nitrogen for a period of up to 2 days. Cooling curves were then recorded on the amalgams. The samples of the amalgams were analyzed alkacimetrically.	Pure mercury was redistilled. 99% pure magnesium contained 0.36% of Fe and Al.
	ESTIMATED ERROR: Soly: precision ± 3%. Temp: precision ± 1 K.
	REFERENCES:

COMPONENTS:	ORIGINAL MEASUREMENTS:
(1) Magnesium; Mg; [7439-95-4] (2) Mercury; Hg; [7439-97-6]	Loomis, A.G. *J. Am. Chem. Soc.* 1922, *44*, 8-19.
VARIABLES: Temperature: 22°C	PREPARED BY: C. Guminski; Z. Galus

EXPERIMENTAL VALUES:

Solubility of magnesium in mercury at 22.4°C was determined to be 0.323 ± 0.001 mass %. The corresponding solubility in atomic % calculated by the compilers is 2.60 at %.

AUXILIARY INFORMATION

METHOD/APPARATUS/PROCEDURE:	SOURCE AND PURITY OF MATERIALS:
Amalgams were prepared in vacuo by warming Hg with an excess of Mg. The amalgams were allowed to stand for several days with frequent shaking, then they were filtered through a plug of glass wool under a pressure of hydrogen. The magnesium content in the filtrate was determined as magnesium phosphate.	Mercury was purified chemically and distilled in vacuo. Magnesium of high quality was carefully freed from all oxides.
	ESTIMATED ERROR: Soly: accuracy ± 0.3%. Temp: not specified.
	REFERENCES:

COMPONENTS:	ORIGINAL MEASUREMENTS:
(1) Magnesium; Mg; [7439-95-4] (2) Mercury; Hg; [7439-97-6]	Beck, R.P. *Rec. Trav. Chim.* 1922, *41*, 353-61.
VARIABLES:	PREPARED BY:
Temperature: 151-637°C	C. Guminski; Z. Galus

EXPERIMENTAL VALUES:

Crystallization temperatures of magnesium amalgams were reported.

t/°C	at % Mg	t/°C	at % Mg
637 ± 1	99.0	569 ± 3	70.0
623 ± 1	97.0	578 ± 1	68.0
609	95.0	579	67.5
529	90.0	576	66.67
488	85.0	566	65.0
435 ± 1	82.5	562 ± 1	62.0
462	82.0	562	60.0
482 ± 1	80.0	587	57.5
489 ± 1	79.0	607	55.0
485 ± 1	77.41	624 ± 1	50.0
505	77.0	603	45.0
508 ± 1	76.0	549 ± 1	40.0
518 ± 1	75.5	477	35.0
517 ± 3	75.0	388	30.0
529	74.0	290	25.0
550 ± 3	72.5	200 ± 1	20.0
		151	16.0

These results were previously reported only in graphical form (1).

AUXILIARY INFORMATION

METHOD/APPARATUS/PROCEDURE:	SOURCE AND PURITY OF MATERIALS:
The amalgams were prepared by dissolution, in vacuo, of weighed amounts of magnesium in mercury. The crystallization temperatures were determined from heating and cooling curves. Temperatures were determined with a thermocouple.	Magnesium from Kahlbaum was free of alkali metals. Mercury was purified with the "Ostwald pipette."
	ESTIMATED ERROR: Soly: Nothing specified. Temp: precision ± 2 K.
	REFERENCES: 1. Smits, A.; Beck, R.P. *Proc. Kong. Akad. Wetensch.*, Amsterdam, 1921, *23*, 975.

COMPONENTS:	ORIGINAL MEASUREMENTS:
(1) Magnesium; Mg; [7439-95-4] (2) Mercury; Hg; [7439-97-6]	Williams, E.J. *Phil. Mag. Ser. 6*, 1925, *50*, 589-99.
VARIABLES: Temperature: 4°C	PREPARED BY: C. Guminski; Z. Galus

EXPERIMENTAL VALUES:

Solubility of magnesium in mercury at 4°C was determined to be 0.2654 mass %. The atomic % solubility calculated by the compilers is 2.149 at %.

It is possible that the amalgams were slightly supersaturated, so that the solubility value is several percent too high.

AUXILIARY INFORMATION

METHOD/APPARATUS/PROCEDURE:	SOURCE AND PURITY OF MATERIALS:
The preparation of the amalgam and the measurements were made in an evacuated cell. The electrical resistance was measured at decreasing temperatures on an amalgam which contained 0.2654 mass % Mg. The resistance decreased suddenly as the temperature was lowered to about 4°C, thus indicating the point of saturation.	Nothing specified.
	ESTIMATED ERROR: Soly: nothing specified. Temp: $\pm$ 0.2 K.
	REFERENCES:

COMPONENTS:	ORIGINAL MEASUREMENTS:
(1) Magnesium; Mg; [7439-95-4] (2) Mercury; Hg; [7439-97-6]	Danilchenko, P.T. *Zh. Russ. Fiz. Khim. Obshch., Ser Khim.* <u>1930</u>, *62*, 975-88.
VARIABLES: Temperature: 15-620°C	PREPARED BY: C. Guminski; Z. Galus

EXPERIMENTAL VALUES:

The freezing points of magnesium amalgams were reported.

$t/°C$	at % Mg	$t/°C$	at % Mg
15	2.74	558	64.42
66	6.62	570	66.67
107	10.16	569	67.52
112	10.91	558	70.84
136	13.88	552	71.84
155	16.55	544	72.58
171	19.34	534	73.50
203	21.10	518	74.09
219	22.81	508	75.24
241	23.00	502	76.25
289	24.82	487	78.44
308	25.95	472	79.95
305	26.72	461	81.09
346	28.60	454	81.47
366	29.86	448	81.91
630	50.46	487	85.14
608	54.85	497	86.04
601	55.34	537	89.64
590	56.67	560	91.77
567	59.05	590	93.25
553	61.70	620	96.53

AUXILIARY INFORMATION

METHOD/APPARATUS/PROCEDURE:	SOURCE AND PURITY OF MATERIALS:
The amalgams with 0-8 mass % of Mg were prepared by dissolution of magnesium chips in mercury in a glass tube at temperature of 350 to 420°C. Further heating under vacuum or in hydrogen atmosphere yielded the alloy with 12.5 mass % of Mg. Such alloys were melted with mercury or magnesium under layer of carnalyte. Samples of the liquid amalgams were analyzed: Mg as MgO or $Mg_2P_2O_7$ and Hg probably gravimetrically. Cooling and heating curves were recorded with the help of a calibrated Nichrome-constantan thermocouple.	Magnesium purity was 99.81%. Mercury was double-distilled.
	ESTIMATED ERROR: Soly: nothing specified. Temp: nothing specified; no better than $\pm$ 1 K (compilers).
	REFERENCES:

COMPONENTS:	ORIGINAL MEASUREMENTS:
(1) Magnesium; Mg; [7439-95-4] (2) Mercury; Hg; [7439-97-6]	Dergacheva, M.B.; Kozin, L.F. *Vestn. Akad. Nauk Kaz. SSR* 1974, No. 6, 56-60.
VARIABLES: Temperature: 25°C	PREPARED BY: C. Guminski; Z. Galus

EXPERIMENTAL VALUES:

Solubility of magnesium in mercury at 25°C was reported to be 1.955 mol dm^{-3}. The solubility in mass % and atomic % calculated by the compilers are 0.35 mass % and 2.82 at %, respectively.

AUXILIARY INFORMATION

METHOD/APPARATUS/PROCEDURE:	SOURCE AND PURITY OF MATERIALS:
The amalgams were prepared electrolytically and used to determine the potentials of the cell: Mg(Hg) | Mg^{2+} | Mg(Hg)$_x$ The electrolyte was an ether solution of MgBrC$_2$H$_5$. The solubility of the magnesium was determined from the breakpoint in the plot of EMF against the logarithm of magnesium concentration.	Nothing specified.
	ESTIMATED ERROR: Soly: nothing specified; precision probably several percent (compilers). Temp: nothing specified.
	REFERENCES:

COMPONENTS:	EVALUATOR:
(1) Calcium; Ca; [7440-70-2] (2) Mercury; Hg; [7439-97-6]	C. Guminski; Z. Galus Department of Chemistry University of Warsaw Warsaw, Poland July, 1985

CRITICAL EVALUATION:

The most reliable solubilities for the Ca-Hg system are the most recent determinations by Bruzzone and Merlo (1) who reported the complete phase diagram for this system. Unfortunately, these authors reported their results as the phase diagram only, and no numerical data were presented. Data points on the liquidus were determined in the range of 10 to 100% Ca in (1), and the authors combined their data with those of an early report by Eilert (2) for the liquidus in the range of 4.5 to 13.4 at % Ca to draw the complete phase diagram. The data in the overlapping region in (1) and (2) were in satisfactory agreement. The eutectic at 759 K in the Ca-rich region was confirmed by Hilpert (3). There were other early efforts to determine the solubility of calcium in mercury at lower temperatures (4-7), but only Cambi and Speroni (5) presented solubility data which are acceptable. The latter authors found that the solubility increased from 2.86 to 13.81 at % in the temperature range of 382 to 573 K. Also, Cambi (6) showed from potentiometric measurements that the solubility of calcium at 298 K is slightly higher than 1 at %. Kozin's (8) predicted solubility of 0.62 at % at 298 K appears to be of the correct magnitude.

The saturated calcium amalgams are in equilibrium with various intermediate phases, as shown in Figure 1 (1). Only the compounds $CaHg$, $CaHg_2$ and $CaHg_3$ have been established with certainty in the Hg-rich region (1); other reported compounds (1,2,5, 9) are still questionable. The system needs further investigation in this region.

Recommended (r) and tentative values of the solubility of calcium in mercury in the Hg-rich region. See phase diagram, Figure 1, for complete solubility range.

T/K	Soly/at %	Source
373	4	2
473	7.7 (r)	2,5
573	11 (r)	2,5,1
673	14.5[a] (r)	1,2
773	19[a]	1
873	25	1
973	29	1
985	30	1

[a] Interpolated from data of cited references.

References

1. Bruzzone, C.; Merlo, F. *J. Less-Common Met.* 1973, *32*, 237.
2. Eilert, A. *Z. Anorg. Chem.* 1926, *151*, 96.
3. Hilpert, K. *Ber. Kernforschungsanlage, Jülich* 1981, JUEL-1744, pp. 121, 132.
4. Smith, G.McP.; Bennett, H.C. *J. Am. Chem. Soc.* 1909, *31*, 799; 1910, *32*, 622.
5. Cambi, L.; Speroni, G. *Atti Reale Accad. Lincei, Ser. 5* 1914, *23*, (2), 599.
6. Cambi, L. *Atti Reale Accad. Lincei, Ser. 5* 1915, 24, (1), 817.
7. Strachan, J.F.; Harris, N.L. *J. Inst. Metals* 1956-57, *85*, 17.
8. Kozin, L.F. *Fiziko-Khimicheskie Osnovy Amalgamnoi Metallurgii*, Nauka, Alma-Ata, 1964.
9. Jangg, G.; Weihs, G. *Monatsh. Chem.* 1975, *106*, 1149.

COMPONENTS:	EVALUATOR:
(1) Calcium; Ca; [7440-70-2] (2) Mercury; Hg; [7439-97-6]	C. Guminski; Z. Galus Department of Chemistry University of Warsaw Warsaw, Poland July, 1985

CRITICAL EVALUATION: (Continued)

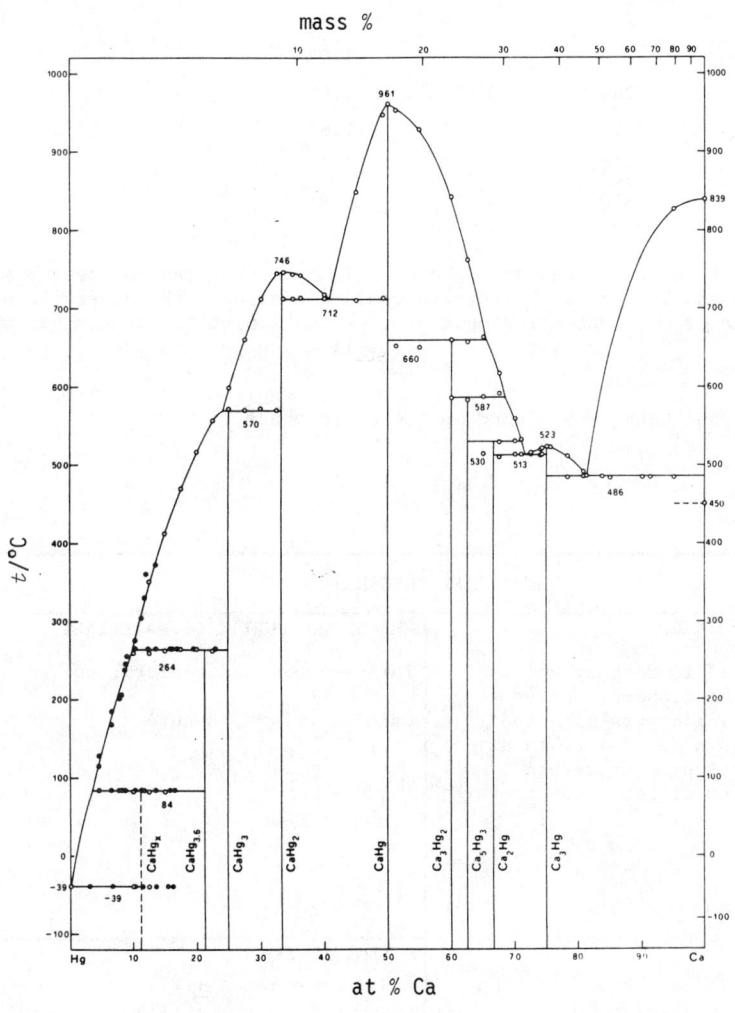

Fig. 1. The Ca-Hg system (1).

COMPONENTS:	ORIGINAL MEASUREMENTS:
(1) Calcium; Ca; [7440-70-2] (2) Mercury; Hg; [7439-97-6]	Cambi, L.; Speroni, G. *Atti Reale Accad. Lincei*, Ser. 5 <u>1914</u>, 23 (2), 599-611.
VARIABLES: Temperature: 109-300°C	PREPARED BY: C. Guminski; Z. Galus

EXPERIMENTAL VALUES:

Crystallization temperatures of calcium amalgams were reported.

$t/°C$	at % Ca
109	2.86
184	6.55
195	7.75
225	9.07
244	9.65
252	11.87
264	13.0
300	13.81

Additional experiments at lower and at higher calcium content than in the above were performed. At 0.48 and 1.82 at % the authors could not observe the crystallization temperature. In the higher concentration range, up to 32.8 at %, the amalgam was observed to boil at 377°C. The last three crystallization temperatures in the table are too low.

The compounds $CaHg_4$ and $CaHg_2$ were found in the solid phase.

AUXILIARY INFORMATION

METHOD/APPARATUS/PROCEDURE:	SOURCE AND PURITY OF MATERIALS:
Calcium was dissolved in mercury and the amalgams were kept at temperatures up to 300°C for 3 days. The crystallization temperatures were then determined in an atmosphere of dry nitrogen or carbon dioxide. The samples of the amalgams were analyzed alkacimetrically.	Pure mercury was redistilled. Calcium was 99.8% pure.
	ESTIMATED ERROR: Soly: precision ± 2%. Temp: nothing specified.
	REFERENCES:

COMPONENTS:	ORIGINAL MEASUREMENTS:
(1) Calcium; Ca; [7440-70-2] (2) Mercury; Hg; [7439-97-6]	Eilert, A. Z. Anorg. Chem. 1926, 151, 96-104.

VARIABLES:	PREPARED BY:
Temperature: 115-372°C	C. Guminski; Z. Galus

EXPERIMENTAL VALUES:

Crystallization temperatures of calcium amalgams were determined.

t/°C	mass % Ca	at % Ca[a]
372	3.00	13.4
359.5	2.67	12.1
330	2.58	11.7
304	2.46	11.2
274.5	2.23	10.2
256	1.93	8.97
246	1.90	8.84
238	1.84	8.57
206	1.73	8.09
201	1.64	7.70
185	1.41	6.68
168	1.37	6.50
128	0.94	4.53
115	0.93	4.49

[a] by compilers

Analyses of the solid phases showed the presence of the compounds $CaHg_{10}$, $CaHg_5$, and $CaHg_3$.

AUXILIARY INFORMATION

METHOD/APPARATUS/PROCEDURE:	SOURCE AND PURITY OF MATERIALS:
The amalgams were obtained by dissolution of calcium in mercury at temperatures up to 340°C. The cooling curves of the samples were determined with a copper-constantan thermocouple. The experiments were performed in an atmosphere of pure, dry carbon dioxide. The samples were analyzed alkacimetrically: an excess of standard HCl was added and back-titrated with standard NaOH.	Calcium purity was 99.2%; the metal contained 0.8% CaO. Mercury was purified with Hg_2SO_4-H_2SO_4 solution and was distilled under vacuum.
	ESTIMATED ERROR: Soly: nothing specified. Temp: precision ± 0.25 K.
	REFERENCES:

COMPONENTS:	ORIGINAL MEASUREMENTS:
(1) Calcium; Ca; [7440-70-2] (2) Mercury; Hg; [7439-97-6]	Bruzzone, G.; Merlo, F. J. Less-Common Met. 1973, 32, 237-41.
VARIABLES: Temperature: 533-1234 K	PREPARED BY: C. Guminski; Z. Galus

EXPERIMENTAL VALUES:

The data were presented as points on the phase diagram. The points from the liquidus were read from the curve by the compilers.

T/K	Soly/at %	T/K	Soly/at %
533	10	1226	51.3
623	12.5	1199	55
686	15	1114	60
743	17.5	1036	62.5
789	20	935	65
830	22.5	891	67.5
872	25	842	70
934	27.5	804	71.2
986	30	789	72.5
1018	23.5	793	74.1
1019	33.3	796	75
1017	35	794	75.7
1015	36.1	789	78.2
991	40	764	80.8
1121	45	759	81.4
1221	49.2	1100	95
1234	50		

AUXILIARY INFORMATION

METHOD/APPARATUS/PROCEDURE:	SOURCE AND PURITY OF MATERIALS:
Appropriate amounts of both metals, to yield approximately 25 grams of amalgam, were placed in iron crucibles and the iron lids were sealed onto the crucibles. The latter were heated to melt the amalgams, then continuously shaken while they were cooled in air. Thermal analyses were made from heating and cooling curves with Chromel-Alumel thermocouples. X-ray analyses and metallographic examination were made on the solid phases. Sample handling of the amalgams was done in argon atmosphere.	Calcium from Fluka was further purified by method in (1). Mercury was 99.99% pure.
	ESTIMATED ERROR: Soly: nothing specified. Temp: precision ± 2 K (compilers).
	REFERENCES: 1. Peterson, D.T.; Fattore, V.G. J. Phys. Chem. 1961, 65, 2052.

COMPONENTS:	EVALUATOR:
(1) Strontium; Sr; [7440-24-6] (2) Mercury; Hg; [7439-97-6]	C. Guminski; Z. Galus Department of Chemistry University of Warsaw Warsaw, Poland July, 1985

CRITICAL EVALUATION:

Kerp (1) reported the first investigation on the solubility of strontium in mercury, and he determined solubilities of 3.4 and 3.6 at % at 338 and 354 K, respectively. Subsequently, Kerp and coworkers (2) used the same method of filtration and chemical analysis of the amalgams which had been equilibrated at temperatures ranging from 273 to 354 K; the solubilities of Sr at 338 and 354 K in the second work were higher than in (1). Smith and Bennett (3) employed a similar method at 296 K and reported a solubility of 2.53 at %. Kozin's (4) predicted solubility of 0.49 at % at 298 K is too low because the strong interaction between the metals were neglected.

Most recently, Bruzzone and Merlo (5) determined the complete phase diagram of the Sr-Hg system. However, in the region of low strontium content the results were in only qualitative agreement with earlier determinations (2). As shown in the phase diagram (5), the saturated amalgams are in equilibrium with various intermediate solid phases.

Tentative solubility of strontium in the Hg-rich region. See the phase diagram, Figure 1, for complete solubility range.

T/K	Soly/at %	Reference
273	1.6	2
293	2.3	2,3
298	2.5[a]	2,3
323	3.2[b]	2
373	4[b]	2,5
473	6[b]	2,5
573	9	5
673	13	5
773	22[b]	5
873	24[b]	5
969	27	5

[a] Extrapolated from data of cited references.
[b] Interpolated from data of cited references.

References

1. Kerp, W. *Z. Anorg. Chem.* 1898, *17*, 284.
2. Kerp, W.; Böttger, W.; Iggena, H. *Z. Anorg. Chem.* 1900, *25*, 1,
3. Smith, G.McP.; Bennett, H.C. *J. Am. Chem. Soc.* 1910, *32*, 622; 1909, *31*, 799.
4. Kozin, L.F. *Fiziko-Khimicheskie Osnovy Amalgamnoi Metallurgii*, Nauka, Alma-Ata, 1964.
5. Bruzzone, G.; Merlo, F. *J. Less-Common Metals* 1974, *35*, 153.

COMPONENTS:	EVALUATOR:
(1) Strontium; Sr; [7440-24-6] (2) Mercury; Hg; [7439-97-6]	C. Guminski; Z. Galus Department of Chemistry University of Warsaw Warsaw, Poland July, 1985

CRITICAL EVALUATION:

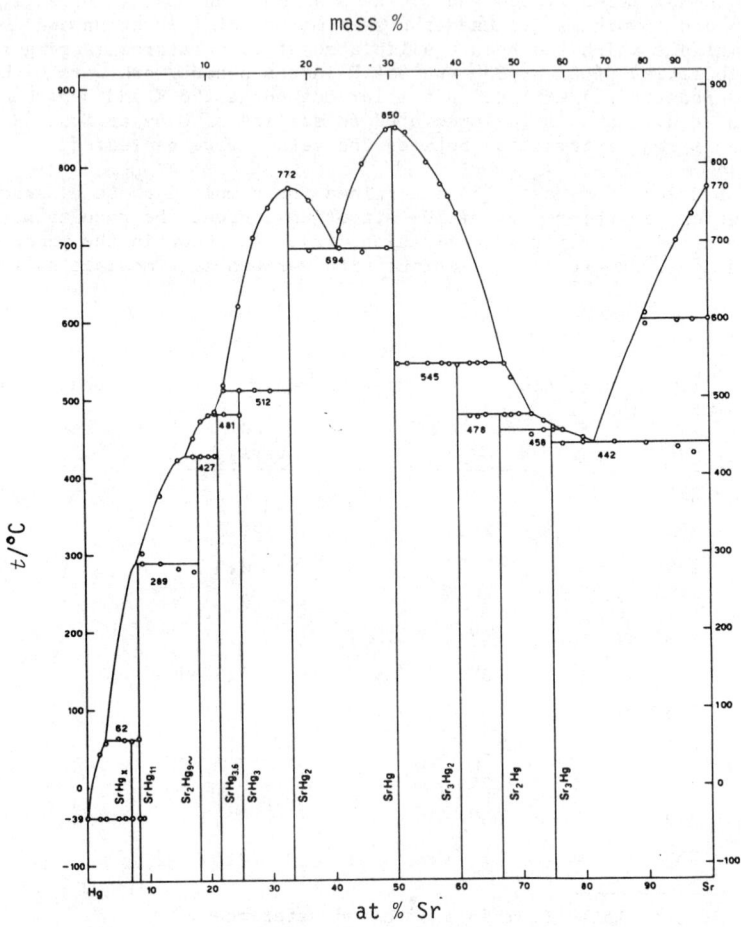

Fig. 1. The Sr-Hg system (5).

COMPONENTS:	ORIGINAL MEASUREMENTS:
(1) Strontium; Sr; [7440-24-6] (2) Mercury; Hg; [7439-97-6]	Kerp, W.; Böttger, W.; Iggena, H. Z. Anorg. Chem. 1900, 25, 1-71.
VARIABLES: Temperature: 0-81°C	PREPARED BY: C. Guminski; Z. Galus

EXPERIMENTAL VALUES:

The solubility of strontium in mercury was determined at various temperatures.

$t/°C$	Soly/mass %	Soly/at %[a]
0	0.73 ± 0.02	1.65
20	1.02 ± 0.05	2.30
30	1.25 ± 0.05	2.82
46	1.33 ± 0.02	2.99
56	1.52 ± 0.06	3.41
64.5	1.76 ± 0.12	3.94
81	1.77 ± 0.19	3.96

[a] by compilers

It is possible that the amalgams were not saturated above 30°C and supersaturated at 0°C. The solid compound in equilibrium with the amalgam was determined to be $SrHg_{12}$.

AUXILIARY INFORMATION

METHOD/APPARATUS/PROCEDURE:	SOURCE AND PURITY OF MATERIALS:
The amalgams were prepared by electrolysis of a saturated solution of $SrCl_2$ with circulating amalgam as the cathode. The electrolyte was renewed several times during the electrolysis. The equilibrated amalgams were filtered through a Gooch crucible at the equilibration temperature. The strontium contents were determined alkacimetrically. All experiments were performed in an atmosphere of dry hydrogen.	Mercury was treated with HNO_3, then washed, dried and filtered. $SrCl_2$ purity not specified.
	ESTIMATED ERROR: Soly: precision better than ± 10%. Temp: nothing specified.
	REFERENCES:

COMPONENTS:	ORIGINAL MEASUREMENTS:
(1) Strontium; Sr; [7440-24-6] (2) Mercury; Hg; [7439-97-6]	Bruzzone, G.; Merlo, F. J. Less-Common Met. 1974, 35, 153-7.
VARIABLES: Temperature: 316-1122 K	PREPARED BY: C. Guminski; Z. Galus

EXPERIMENTAL VALUES:

The data were presented as points on the phase diagram. The liquidus points were read from the curve by the compilers.

T/K	Soly/at %	T/K	Soly/at %
316	2.0	1119	49
330	3.0	1122	50.3
575	10	1113	52
650	12.5	1076	55.3
696	15	1050	57.5
723	16.5	1034	58.7
745	18.7	1012	60
753	20	817	67.5
756	21	1076	68.3
791	22.5	751	71.2
894	25	743	73.7
981	27.5	733	75.2
1021	30	731	77
1045	33.3	721	80
1030	36.5	881	90
989	41	975	95
1076	45	1008	97.5

AUXILIARY INFORMATION

METHOD/APPARATUS/PROCEDURE:	SOURCE AND PURITY OF MATERIALS:
Appropriate amounts of both metals, to yield approximately 25 grams of amalgam, were placed in iron crucibles and the iron lids were sealed onto the crucibles. The crucibles were heated to melt the amalgams, then continuously shaken while they were cooled in air. Thermal analyses were made from heating and cooling curves with Chromel-Alumel thermocouples. X-ray analyses and metallographic examinations were made on the solid phases. Sample handling of the amalgams was done in argon atmosphere.	Strontium from Fluka was 99.8% pure. Mercury was 99.99% pure.
	ESTIMATED ERROR: Soly: nothing specified. Temp: $\pm$ 2 K (compilers).
	REFERENCES:

COMPONENTS:	ORIGINAL MEASUREMENTS:
(1) Strontium; Sr; [7440-24-6] (2) Mercury; Hg; [7439-97-6]	1. Smith, G. McP.; Bennett, H.C. *J. Am. Chem. Soc.* <u>1909</u>, *31*, 799-806. 2. Same authors, ibid. <u>1910</u>, *32*, 622-26.
VARIABLES:	PREPARED BY:
Temperature: 23°C	C. Guminski; Z. Galus

EXPERIMENTAL VALUES:

At 23°C the solubility of strontium in mercury was reported to be 1.12 mass %. The atomic % solubility calculated by the compilers is 2.53 at %.

Chemical analysis of the solid phase suggested the compound $SrHg_{12-13}$.

AUXILIARY INFORMATION

METHOD/APPARATUS/PROCEDURE:	SOURCE AND PURITY OF MATERIALS:
The amalgam was obtained by electrolysis of a saturated solution of $SrCl_2$. The resulting amalgam was washed and dried, then kept for 3 days in a glass-stoppered bottle, then again washed, dried, and filtered. Both the solid and filtrate were analyzed alkacimetrically: an excess of 0.1 mol-dm^{-3} HCl was added to the filtrate then back-titrated with 0.1 mol-dm^{-3} NaOH.	"Very pure salts" from Kahlbaum. Mercury purity not specified.
	ESTIMATED ERROR: Soly: nothing specified; no better than few percent (compilers). Temp: nothing specified.
	REFERENCES:

COMPONENTS:	EVALUATOR:
(1) Barium; Ba; [7440-39-3] (2) Mercury; Hg; [7439-97-6]	C. Guminski; Z. Galus Department of Chemistry University of Warsaw Warsaw, Poland July, 1985

CRITICAL EVALUATION:

The first determinations of the barium content in its saturated amalgams were reported by Kerp (1) for the temperature range of 273 to 354 K; it was found that the barium solubility increased from 0.25 to 1.18 at % in this temperature range. Subsequently, Kerp and coworkers (2) determined the solubility up to 322 K to verify the earlier results; these authors observed that the solubility did not increase smoothly over their temperature range, but that there was a break at 303 K. Smith and Bennett (3) reported a barium solubility of 0.47 at % at 297 K, a value which was in good agreement with a solubility of 0.50 at % at 298 K which was reported by Kerp et al. In all of these early works the solubilities were determined by filtration and chemical analysis of the equilibrated amalgams.

More recently, the complete phase diagram for the Ba-Hg system was determined by thermal analysis and X-ray crystallography by Bruzzone and Merlo (4). These authors reported their data as a phase diagram only, but their solubilities for barium in the Hg-rich region were higher than those reported by Kerp et al. (2), and the liquidus was a smooth curve near 303 K, contrary to that observed by (2). Makarova and coworkers (5) also observed a smooth curve at 293 to 333 K where the solubility increased from 0.63 to 1.09 at % over this range. However, the solubilities reported by (5) at 293 and 313 K appear to be too high. Filipova et al. (6,7) reported a solubility of 0.63 at % at 298 K; this value lies between those of (2) and (5).

Rejected values for the solubility of barium at room temperature were reported by Strachan and Harris (8) and by Kozin (9); the latter predicted a solubility of 1.9 at % at 298 K.

As shown in the phase diagram in Figure 1 (4), the saturated liquid is in equilibrium with various intermediate solid phases.

Tentative solubility of barium in the Hg-rich region. See Figure 1 for complete solubility range.

T/K	Soly/at %	Reference
273	0.23	1,2
293	0.46	1,2
298	0.49	2,3
323	0.9	2
373	2	2
473	6	4
573	9	4
673	11	4
763	16	4

Barium

COMPONENTS:

(1) Barium; Ba; [7440-39-3]
(2) Mercury; Hg; [7439-97-6]

EVALUATOR:

C. Guminski; Z. Galus
Department of Chemistry
University of Warsaw
Warsaw, Poland

July, 1985

CRITICAL EVALUATION:

References

1. Kerp, W. *Z. Anorg. Chem.* 1898, *17*, 284.
2. Kerp, W.; Böttger, W.; Iggena, H. *Z. Anorg. Chem.* 1900, *25*, 1.
3. Smith, G.McP.; Bennett, H.C. *J. Am. Chem. Soc.* 1910, *32*, 622; 1909, *31*, 799.
4. Bruzzone, G.; Merlo, F. *J. Less-Common Metals* 1975, *39*, 27.
5. Makarova, I.A.; Lange, A.A.; Bukhman, S.P. *Izv. Akad. Nauk Kaz. SSR, Ser. Khim.* 1980, No. 6, 37.
6. Filippova, L.M.; Zhumakanov, V.Z.; Klyukas, Yu.E.; Zebreva, A.I. *Izv. Vyssh. Ucheb. Zaved., Khim. Khim. Tekhnol.* 1984, *27*, 1241.
7. Filippova, L.M.; Zhumakanov, V.Z.; Zebreva, A.I.; Smurigina, T.V. *Fiz. Khim. Issled. v Rastvorakh*, Alma-Ata, 1982, 40.
8. Strachan, J.F.; Harris, N.L. *J. Inst. Metals* 1956-57, *85*, 17.
9. Kozin, L.F. *Fiziko-Khimicheskie Osnovy Amalgamnoi Metallurgii*, Nauka, Alma-Ata, 1964.

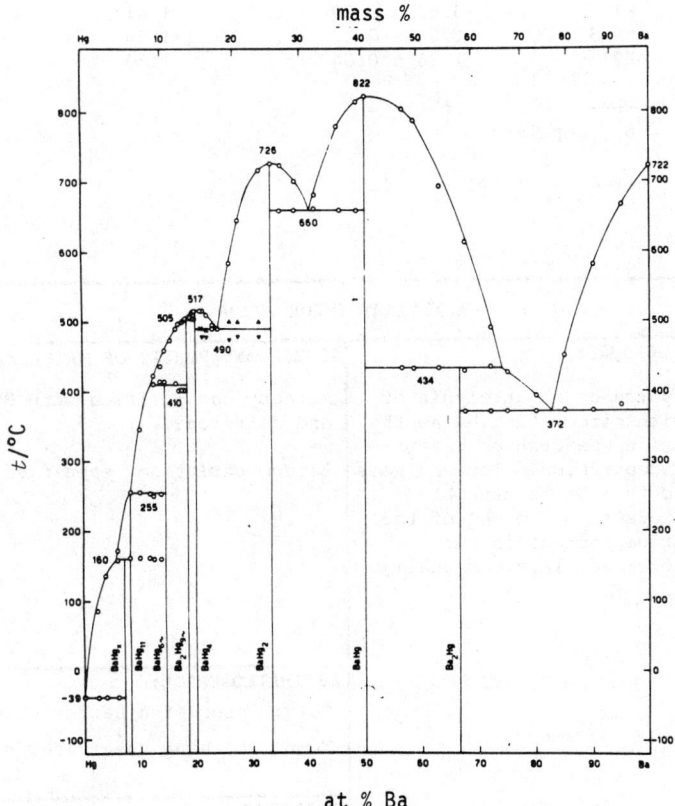

Fig. 1. The Ba-Hg system (4).

COMPONENTS:	ORIGINAL MEASUREMENTS:
(1) Barium; Ba; [7440-39-3] (2) Mercury; Hg; [7439-97-6]	Kerp, W.; Böttger, W.; Iggena, H. *Z. Anorg. Chem.* <u>1900</u>, *25*, 1-71.
VARIABLES: Temperature: 0-99°C	PREPARED BY: C. Guminski; Z. Galus

EXPERIMENTAL VALUES:

Solubility of barium in mercury.

t/°C	Soly/Mass %	Soly at %[a]
0	0.15 ± 0.01	0.22
0[b]	0.17 ± 0.01	0.25
20	0.32 ± 0.02	0.47
21[b]	0.32 ± 0.02	0.47
25	0.34	0.50
27.6	0.35	0.51
28.1	0.36	0.52
29.2	0.38 ± 0.01	0.55
30	0.43 ± 0.02	0.63
35	0.46	0.67
46	0.52 ± 0.01	0.76
56	0.68 ± 0.02	0.99
64.7[b]	0.81 ± 0.02	1.18
65	0.83 ± 0.02	1.21
81	0.97 ± 0.03	1.41
89.5	1.06 ± 0.03	1.54
99	1.26 ± 0.04	1.83

[a] by compilers

[b] from ref. (1)

AUXILIARY INFORMATION

METHOD/APPARATUS/PROCEDURE:	SOURCE AND PURITY OF MATERIALS:
Amalgams were prepared by electrolysis of saturated $BaCl_2$ with circulating Hg as the cathode; the solution was renewed several times during the preparation. The amalgams were filtered through a Gooch crucible after various periods from the end of the electrolysis. Barium content in the filtrates was determined alkacimetrically.	Mercury was purified with HNO_3, then dried and filtered. Barium purity not specified.
	ESTIMATED ERROR: Soly: precision better than ± 5%. Temp: nothing specified.
	REFERENCES: 1. Kerp, W. *Z. Anorg. Chem.* <u>1898</u>, *17*, 284.

COMPONENTS:	ORIGINAL MEASUREMENTS:
(1) Barium; Ba; [7440-39-3] (2) Mercury; Hg; [7439-97-6]	1. Smith, G.McP.; Bennett, H.C. J. Am. Chem. Soc. 1910, 32, 622-26. 2. Same authors, ibid. 1909, 31, 799-806.
VARIABLES: Temperature: 24°C	PREPARED BY: C. Guminski; Z. Galus

EXPERIMENTAL VALUES:

Solubility of barium in mercury at 24°C was reported to be 0.32 mass %. The solubility in atomic % calculated by the compilers is 0.47 at %. Solid phase chemical analysis suggested the compound $BaHg_{12}$.

AUXILIARY INFORMATION

METHOD/APPARATUS/PROCEDURE:	SOURCE AND PURITY OF MATERIALS:
The amalgam was obtained by electrolysis of saturated solution of $BaCl_2$ at 6-7 V, then the resulting amalgam was washed and dried, and the solid phase was separated by suction filtration through Chamois skin. The filtrate and the crystals were analyzed alkacimetrically by adding an excess of 0.1 mol dm^{-3} HCl to a weighed portion of the amalgam then back-titrating with 0.1 mol dm^{-3} NaOH to determine the Ba content.	"Very pure salts" from Kahlbaum. Mercury purity not specified.
	ESTIMATED ERROR: Soly: nothing specified; probably no better than few percent (compilers). Temp: nothing specified.
	REFERENCES:

COMPONENTS:	ORIGINAL MEASUREMENTS:
(1) Barium; Ba; [7440-39-3] (2) Mercury; Hg; [7439-97-6]	Bruzzone, G.; Merlo, F. J. Less-Common Metals 1975, 39, 271-6.
VARIABLES: Temperature: 360-1095 K	PREPARED BY: C. Guminski; Z. Galus

EXPERIMENTAL VALUES:

The data were reported graphically as points on the phase diagram. The points from the liquidus line were read from the curve by the compilers.

T/K	Soly/at %	T/K	Soly/at %
360	2.3	918	27.5
410	4.0	990	31.5
446	6.2	999	33.3
528	8.5	996	35
619	10	974	37.5
696	12.5	957	41.2
710	13.7	1052	45
731	14.5	1087	48.5
745	15.5	1095	50
762	16.5	1079	56.5
770	17.0	1060	58.5
773	17.4	967	62.5
777	17.8	888	67.5
781	18.2	764	72
789	19.5	701	75
790	21	667	80
782	22	725	85
769	23	855	90
763	24	939	95
857	26		

AUXILIARY INFORMATION

METHOD/APPARATUS/PROCEDURE:	SOURCE AND PURITY OF MATERIALS:
Appropriate amounts of both metals, to yield approximately 25 grams of amalgam, were placed in iron crucibles and the iron lids were sealed onto the crucibles. The latter were heated to melt the amalgams, then continuously shaken while they were cooled in air. Thermal analyses were made from heating and cooling curves, with Chromel-Alumel thermocouples. X-ray analyses and metallographic examinations were made on the solid phases. Sample handling of the amalgams was done in argon atmosphere.	Barium from Fluka was 99.6% pure. Mercury was 99.99% pure.
	ESTIMATED ERROR: Soly: nothing specified. Temp: precision ± 2 K (compilers).
	REFERENCES:

COMPONENTS:	ORIGINAL MEASUREMENTS:
(1) Barium; Ba; [7440-39-3] (2) Mercury; Hg; [7439-97-6]	Makarova, I.A.; Lange, A.A.; Bukhman, S.P. *Izv. Akad. Nauk Kaz. SSR, Ser. Khim.* 1980, No. 6, 37-41.
VARIABLES:	PREPARED BY:
Temperature: 293-333 K	C. Guminski; Z. Galus

EXPERIMENTAL VALUES:

Solubility of barium in mercury.

T/K	Soly/mass %	Soly/at %[a]
293	0.43	0.63
313	0.64	0.93
333	0.75	1.09

[a] by compilers

AUXILIARY INFORMATION

METHOD/APPARATUS/PROCEDURE:	SOURCE AND PURITY OF MATERIALS:
The amalgam was obtained by electro-reduction of Ba(II) on Hg from a solution of 0.05 mol-dm^{-3} BaCl$_2$ in 0.5 mol-dm^{-3} LiCl. Barium content in the amalgam was determined by decomposition of the amalgam with 0.1 mol-dm^{-3} HCl and gravimetric analysis as BaSO$_4$. Voltammetric oxidation of the stirred amalgam was performed; a bend on the curve relating limiting current to concentration corresponded to the solubility of barium in mercury.	BaCl$_2$ and LiCl were chemically pure. Hg purity not specified.
	ESTIMATED ERROR: Soly: nothing specified. Temp: precision ± 0.5 K.
	REFERENCES:

COMPONENTS:	ORIGINAL MEASUREMENTS:
(1) Barium; Ba; [7440-39-3] (2) Mercury; Hg; [7439-97-6]	Filippova, L.M.; Zhumakanov, V.Z.; Klyukas, Yu.E.; Zebreva, A.I. *Izv. Vyssh. Ucheb. Zaved., Khim. Khim. Tekhnol.* 1984, *27*, 1241-2.
VARIABLES:	PREPARED BY:
Temperature: 25°C	C. Guminski; Z. Galus

EXPERIMENTAL VALUES:

The solubility of barium in mercury was reported to be 0.42 mol-dm^{-3}. The atomic % solubility calculated by the compilers is 0.63 at %.

These results also were reported in ref. (1).

AUXILIARY INFORMATION

METHOD/APPARATUS/PROCEDURE:	SOURCE AND PURITY OF MATERIALS:
The heterogenous barium amalgam was obtained by an electrolytic method, but the details were not specified. Barium content (N_{Ba}) was determined by an unspecified analysis. The amalgams were titrated with mercury and the heat of dilution (Q) was determined. A breakpoint in the curve of Q vs. N_{Ba} corresponds to the composition of the saturated amalgam. All experiments were carried out in an argon atmosphere.	Nothing specified.
	ESTIMATED ERROR: Soly: precision probably ± 10% (compilers). Temp: stability of ± 0.005 K.
	REFERENCES: 1. Filippova, L.M.; Zhumakanov, V.Z.; Zebreva, A.I.; Smurigina, T.V. *Fiz.-Khim. Issled v Rastvorakh*, Alma-Ata, 1982, 40.

COMPONENTS:	EVALUATOR:
(1) Boron; B; [7440-42-8] (2) Mercury; Hg; [7439-97-6]	C. Guminski; Z. Galus Department of Chemistry University of Warsaw Warsaw, Poland July, 1985

CRITICAL EVALUATION:

There are no experimental data on the solubility of boron in mercury, but the solubility is expected to be very low. From his semiempirical equations Kozin first estimated (1) a 298 K solubility of 3.1×10^{-12} at %, and he subsequently estimated (2) a solubility of 4.75×10^{-9} at % at the same temperature. Neither of the estimated solubilities can be recommended by the evaluators.

Based on the experimental observations of Wald and Stormont (3), Moffatt (4) constructed a schematic phase diagram of the B-Hg system. No stable compounds or solid solutions of boron and mercury are formed in this system.

References

1. Kozin, L.F. *Tr. Inst. Khim. Nauk Akad. Nauk Kaz. SSR* 1962, *9*, 101.
2. Kozin, L.F. *Fiziko-Khimicheskie Osnovy Amalgamnoi Metallurgii*, Nauka, Alma-Ata, 1964.
3. Wald, F.; Stormont, R.W. *J. Less-Common Metals* 1965, *9*, 423.
4. Moffatt, W.G. *The Handbook of Binary Phase Diagrams*, Vol. I, Genium Publishing Corp., Schenectady, NY 1978.

COMPONENTS:	EVALUATOR:
(1) Aluminum; Al; [7429-90-5] (2) Mercury; Hg; [7439-97-6]	C. Guminski; Z. Galus Department of Chemistry University of Warsaw Warsaw, Poland July, 1985

CRITICAL EVALUATION:

The solubility of aluminum in mercury near room temperature is low and some early reports (1-3) indicated only that the solubility limit is below 10^{-2} at %. The first precise determination of the solubility was reported by Fogh (4) who found 1.4×10^{-2} and 2.79 at % aluminum in the saturated amalgam at room temperature and at the boiling point, respectively; more recent measurements confirm these estimates. Shalaevskaya and coworkers (5-7) reported that the solubility increased from 8.9×10^{-3} to 1.63×10^{-2} at % in the temperature range of 293 to 323 K. These values are of the proper magnitude but their dependence on temperature is too low. The potentiometric measurements of Ziegel and coworkers (8) resulted in a solubility of 1.3×10^{-2} at % at 303 K; this value lies between the results of (4) and (5-7). If aluminum interacts with the amalgamated silver (5-7) and platinum (8) of the working electrodes in the potentiometric measurements, then the results of (5-8) may be slightly understated. Kozin's (9) predicted solubility of 0.22 at % at 298 K is much too high. Smits and De Gruyter (10,11) conducted thermoanalytical measurements at higher temperatures and reported the phase diagram for this system; the numerical data for the liquidus were reported by De Gruyter (12). Klemm and Weiss (13) determined the solubility between 695 and 868 K by equilibration of the metals and chemical analysis of the saturated liquids; these authors found that the solubility increased from 7.5 to 82.7 at % in this temperature range. The latter solubilities were in good agreement with those reported by De Gruyter. In a lower temperature range of 333 to 573 K, Schmidt (14) reported that the solubilities increased from 4.5×10^{-2} to 1.25 at %, respectively. The latter results are in good agreement with those determined by Liebhafsky (15) at 349 to 585 K. Jangg and Palman (16), without presenting their data, stated that the solubility of aluminum from their measurements agreed to within ±5% with those of (12), (13) and (15).

The saturated aluminum amalgams are in equilibrium with solid aluminum, and no Al-Hg phases are known to exist (12). The phase diagram for this system is shown in Fig. 1 (17).

Tentative and recommended (r) values of aluminum solubility in mercury:

T/K	Soly/at %	Reference
293	0.014	[4]
298	0.016[a]	[4,14]
373	0.10[b]	[14,15]
473	0.51	[14]
573	1.3[b]	[14,15]
673	5.6	[15]
773	17	[13]
873	84 (r)	[12,13]

[a] Interpolated value from data of cited references.
[b] Mean value from cited references.

(Continued next page)

COMPONENTS:	EVALUATOR:
(1) Aluminum; Al; [7429-90-5] (2) Mercury; Hg; [7439-97-6]	C. Guminski; Z. Galus Department of Chemistry University of Warsaw Warsaw, Poland July, 1985

CRITICAL EVALUATION: (continued)

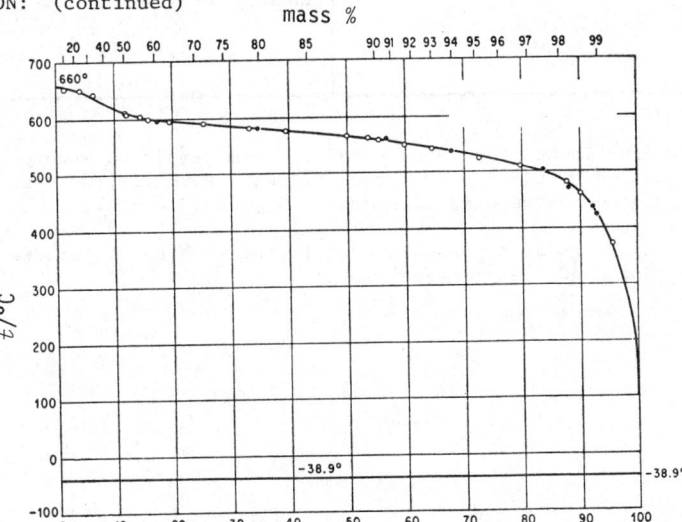

Fig. 1. The Al-Hg system (17).

References

1. Mylius, F.; Rose, F. *Z. Instrumentenk.* 1893, *13*, 81.
2. Kremann, R.; Müler, R. *Z. Metallk.* 1920, *12*, 311.
3. Strachan, J.F.; Harris, N.L. *J. Inst. Metals* 1956-57, *85*, 17.
4. Fogh, I. *Kgl. Dansk. Vidensk. Selsk. Mat. Fys. Medd.* 1921, *III*, No. 15.
5. Shalaevskaya, V.N.; Igolinskii, V.A.; Kataev, G.A. *Dep. VINITI*, 588-75, 1975;
   Abstracted in *Zh. Fiz. Khim.* 1975, *49*, 1587; *Uspekhi Polarogr. s Nakopl.*, Tomsk, 1973, p. 115.
6. Shalaevskaya, V.N.; Igolinskii, V.A. *Zh. Prikl. Khim.* 1975, *48*, 1152.
7. Igolinskii, V.A.; Shalaevskaya, V.N.; Guryanova, O.N.; Igolinskaya, I.M.;
   Kotova, N.A. *Sovr. Probl. Polarografii s Nakopleniem*, Tomsk, 1975, p. 150.
8. Ziegel, G.; Peled, E.; Gileadi, E. *Electrochim. Acta* 1978, *23*, 363.
9. Kozin, L.F. *Fiziko-Khimicheskie Osnovy Amalgamnoi Metallurgii*, 1964.
10. Smits, A.; De Gruyter, C.J. *Proc. Acad. Sci. Amsterdam* 1921, *23*, 966;
    Versl. Akad. Wetensch. 1921, *29*, 747.
11. Smits, A. *Z. Elektrochem.* 1924, *30*, 424.
12. De Gruyter, C.J. *Rec. Trav. Chim.* 1925, *44*, 937.
13. Klemm, W.; Weiss, P. *Z. Anorg. Chem.* 1940, *245*, 285.
14. Schmidt, W. *Metall.* 1949, *3*, 10.
15. Liebhafsky, H.A. *J. Am. Chem. Soc.* 1949, *71*, 1468.
16. Jangg, G.; Palman, H. *Z. Metallk.* 1963, *54*, 364.
17. Hansen, M.; Anderko, K. *Constitution of Binary Alloys*, McGraw-Hill, N.Y., 1958, p. 99.

COMPONENTS:	ORIGINAL MEASUREMENTS:
(1) Aluminum; Al; [7429-90-5] (2) Mercury; Hg; [7439-97-6]	Fogh, I. *Kgl. Dansk. Vidensk. Selsk. Mat. Fys. Medd.* 1921, III, No. 15.
VARIABLES:	PREPARED BY:
Temperature	C. Guminski; Z. Galus

EXPERIMENTAL VALUES:

Solubility of aluminum in boiling mercury and at room temperature were determined to be 0.385 ± 0.002 and 0.0019 ± 0.0001 mass %, respectively. The respective atomic % solubilities calculated by the compilers are 2.79 and 0.014 at %.

The author reported Al_2Hg_3 as a phase in equilibrium with the saturated amalgams. However, this was not confirmed in later works.

AUXILIARY INFORMATION

METHOD/APPARATUS/PROCEDURE:	SOURCE AND PURITY OF MATERIALS:
A piece of aluminum was heated in hydrogen atmosphere in Jena-glass tube. Then this piece was placed under the surface of mercury and the system was boiled for 2-3 hours. The amalgams were filtered with the use of glass-wool. The samples were weighed, then treated with HCl. Aluminum was determined as Al_2O_3.	Nothing specified.
	ESTIMATED ERROR: Soly: precision better than ± 5%. Temp: nothing specified.
	REFERENCES:

COMPONENTS:	ORIGINAL MEASUREMENTS:
(1) Aluminum; Al; [7429-90-5] (2) Mercury; Hg; [7439-97-6]	De Gruyter, C.J. Rec. Trav. Chim. 1925, 44, 937-48.
VARIABLES: Temperature: 369-652°C	PREPARED BY: C. Guminski; Z. Galus

EXPERIMENTAL VALUES:

Crystallization temperatures were reported as a function of aluminum concentration:

t/°C	at % Al	t/°C	at % Al	t/°C	at % Al
652	98.6	595	80.46	550	40.17
650	95.87	590	74.56	542	35.46
643	93.5	582	66.7	524	27.38
613	88.16	576	60.55	510	20.36
610	87.99	566	50.0	479	12.42
604	85.36	561	46.54	460	10.0
600	84.17	558	44.73	369	4.55

AUXILIARY INFORMATION

METHOD/APPARATUS/PROCEDURE:	SOURCE AND PURITY OF MATERIALS:
Aluminum and mercury were mixed in glass tubes, then the tubes were sealed and heated and cooling curves were recorded.	Aluminum supplied by Kahlbaum; purity not specified. Mercury purity not specified.
	ESTIMATED ERROR: Soly: nothing specified. Temp: nothing specified; precision no better than few degrees (compilers).
	REFERENCES:

COMPONENTS:	ORIGINAL MEASUREMENTS:
(1) Aluminum; Al; [7429-90-5] (2) Mercury; Hg; [7439-97-6]	Klemm, W.; Weiss, P. Z. Anorg. Chem. 1940, 245, 285-7.
VARIABLES: Temperature: 422-595°C	PREPARED BY: C. Guminski; Z. Galus

EXPERIMENTAL VALUES:

Solubility of aluminum in mercury:

t/°C	Soly/at %	Soly/mass %
422	7.5	1.01
435	7.9	1.14
470	12.2	1.83
502	16.5	2.20
537	32.2	5.99
560	43.4	9.33
581	65.4	20.61
595	82.7	39.15

AUXILIARY INFORMATION

METHOD/APPARATUS/PROCEDURE:	SOURCE AND PURITY OF MATERIALS:
The metals were sealed in evacuated quartz tubes then heated for 24 hours at the desired temperatures. After equilibration, each tube was turned up and the amalgam was filtered through a narrow constriction in the tube. The filtrate was treated with dilute HCl, and the mercury was dried and weighed. Aluminum in the solution was determined as Al_2O_3 after precipitation with ammonium hydroxide.	Nothing specified.
	ESTIMATED ERROR: Soly: precision ± 1%. Temp: precision ± 2 K.
	REFERENCES:

COMPONENTS:	ORIGINAL MEASUREMENTS:
(1) Aluminum; Al; [7429-90-5] (2) Mercury; Hg; [7439-97-6]	Liebhafsky, H.A. J. Am. Chem. Soc. 1949, 71, 1468-70.
VARIABLES: Temperature: 76-400°C	PREPARED BY: C. Guminski; Z. Galus

EXPERIMENTAL VALUES:

Solubility of aluminum in mercury:

$t/°C$	Soly/mass %	Soly/at %[a]
76	9.0×10^{-3}	0.067
101	1.5×10^{-2}	0.11
103	1.7×10^{-2}	0.13
125	2.4×10^{-2}	0.18
160	3.5×10^{-2}	0.26
260	0.11	0.81
307[b]	--	1.8
312	0.18	1.32
400[b]	--	5.6

[a] by compilers.
[b] Unpublished data of Norton and Harrington (1).

AUXILIARY INFORMATION

METHOD/APPARATUS/PROCEDURE:	SOURCE AND PURITY OF MATERIALS:
The saturated amalgam was obtained by rotating an Al rod, which was used as the stirrer, in the amalgam which was always flushed with hydrogen to prevent oxidation of the amalgam. Samples of the amalgam were extracted with a glass sampling tube at the equilibration temperatures. The amalgam was then treated with 2 mol dm^{-3} HCl and the evolved H_2 was measured with a gas burette to determine the Al content. The Hg was determined volumetrically. Norton and Harrington used the procedure of Klemm and Weiss (2).	Aluminum purity was 99+%. Mercury purity not specified.
	ESTIMATED ERROR: Soly: precision no better than several percent (compilers). Temp: nothing specified.
	REFERENCES: 1. Norton, F.H.; Harrington, R.H. Unpublished work. 2. Klemm, W.; Weiss, P. Z. Anorg. Chem. 1940, 245, 285.

COMPONENTS:	ORIGINAL MEASUREMENTS:
(1) Aluminum; Al; [7429-90-5] (2) Mercury; Hg; [7439-97-6]	Schmidt, W. *Metall.* 1949, *3*, 10-13.
VARIABLES: Temperature: 60-300°C	PREPARED BY: C. Guminski; Z. Galus

EXPERIMENTAL VALUES:

Solubility of aluminum in mercury:

t/°C	Soly/mass %	Soly/at %[a]
60	6×10^{-3}	0.045
100	1.2×10^{-2}	0.089
150	3.4×10^{-2}	0.25
200	6.9×10^{-2}	0.51
300	0.17	1.25

[a] by compilers.

AUXILIARY INFORMATION

METHOD/APPARATUS/PROCEDURE:	SOURCE AND PURITY OF MATERIALS:
No experimental details were given, but the results compare favorably with other published measurements. The determinations were performed in the laboratory of Firma W. Schmidt, Leichtmetallhütte, in München, W. Germany.	Nothing specified.
	ESTIMATED ERROR: Nothing specified.
	REFERENCES:

COMPONENTS:	ORIGINAL MEASUREMENTS:
(1) Aluminum; Al; [7429-90-5] (2) Mercury; Hg; [7439-97-6]	Shalaevskaya, V.N.; Igolinskii, V.A. *Zh. Prikl. Khim.* 1975, *48*, 1152-4.
VARIABLES: Temperature: 20-50°C	PREPARED BY: C. Guminski; Z. Galus

EXPERIMENTAL VALUES:
Solubility of aluminum in mercury:

$t/°C$	Soly/10^3 mass %	Soly/10^3 at %
20	1.18	8.85
30	1.28	9.74
40	1.40	10.77
50	3.10[a]	16.33

[a] This value should be 2.22×10^{-3}; the compilers attribute this error to a misprint in the paper.

The above data were reported in (1) and (2).

AUXILIARY INFORMATION

METHOD/APPARATUS/PROCEDURE:	SOURCE AND PURITY OF MATERIALS:
Small aluminum cylinder pressed exactly into a silver tube was polished in 0.5% solution of $Hg_2(NO_3)_2$ for subsequent amalgamation of the surface with a drop of mercury. The thickness of the mercury film on the aluminum was measured. The tube was then placed in an electrolyte (0.5 mol dm^{-3} $KAlO_2$, 1 mol dm^{-3} KOH, 1.5 mol dm^{-3} KCl) and was polarized anodically. The stationary oxidation current was recorded and the solubility was calculated from the slope of the curve relating the current to the thickness of the mercury film. The measurements were performed in an argon atmosphere.	Aluminum was of high purity. Mercury purity not specified. ESTIMATED ERROR: Soly: nothing specified. Temp: precision $\pm$ 0.1 K. REFERENCES: 1. Igolinskii, V.A.; Shalaevskaya, V.N.; Guyanova, O.N.; Igolinskaya, I.M.; Kotova, N.A. *Sovr. Probl. Polarografii s Nakopleniem*, Tomsk, 1975, p. 150. 2. Shalaevskaya, V.N.; Igolinskii, V.A.; Kataev, G.A. *Dep. VINITI*, 588-75, 1975; abstracted in *Zh. Fiz. Khim.* 1975, *49*, 1587.

COMPONENTS:	ORIGINAL MEASUREMENTS:
(1) Aluminum; Al; [7429-90-5] (2) Mercury; Hg; [7439-97-6]	Ziegel, S.; Peled, E.; Gileadi, E. *Electrochim. Acta* 1978, *23*, 363-8.
VARIABLES: One temperature: 303 K	PREPARED BY: C. Guminski; Z. Galus

EXPERIMENTAL VALUES:

The solubility of aluminum in mercury at 303 K was reported to be $(17-18) \times 10^{-4}$ mass %. The atom % solubility calculated by the compilers is 1.3×10^{-2} at %.

This result may be understated (see below under Method).

AUXILIARY INFORMATION

METHOD/APPARATUS/PROCEDURE:	SOURCE AND PURITY OF MATERIALS:
Aluminum amalgam was prepared into mercury drop electrodes suspended on the tip of a platinum wire. The electroreduction was carried out in a solution of 1.3 mol dm^{-3} AlBr$_3$ + 0.52 mol dm^{-3} KBr in toluene at constant current. Then the open circuit potentials were measured at times longer than 300 s. The inflection on the curve relating reversible potential vs. logarithm of the charge passed corresponds to the saturation point of aluminum in mercury. All experiments were performed in a glove-box under an atmosphere of purified nitrogen or argon. It is possible part of Al reacted with Pt surface, so that concentration of Al in the Hg drop was decreased.	Toluene was dried by refluxing on Na, followed by 2 steps of vacuum distillation and drying by molecular sieves. AlBr$_3$ was purified by double vacuum sublimation. KBr was dried by heating overnight at 523 K in vac. Final purification of solution achieved by placing Al wire in Hg pool in cell and stirring several hours. Hg (Frutarom AR) was cleaned first by washing with conc. H$_2$SO$_4$, then rinse with 10% HNO$_3$ and triple distilled H$_2$O, and vac. distilled.
	ESTIMATED ERROR: Soly: precision about $\pm$ 10% (compilers). Temp: nothing specified.
	REFERENCES:

COMPONENTS:	EVALUATOR:
(1) Gallium; Ga; [7440-55-3] (2) Mercury; Hg; [7439-97-6]	C. Guminski; Z. Galus Department of Chemistry University of Warsaw Warsaw, Poland July, 1985

CRITICAL EVALUATION:

Gallium melts at 303 K, but the solubility of the liquid metal in mercury is only a few atom percent between 273 and 373 K. Gilfillan and Bent (1), from freezing point depression measurements, found that the solubility at 233.5 K is 0.37 at %.

Spicer and Bartholomay (2) equilibrated the saturated amalgams at 308 and 373 K, and they determined the solubility of gallium by chemical analysis of the liquid phase. At 308 K the solubility in the mercury-rich and the gallium-rich liquid was 3.6 and 97.6 at %, respectively; at 373 K the corresponding solubilities were 3.9 and 96.8 at %. Although the solubilities at 308 K are satisfactory the values at 373 K are erroneous. This study suggested that a critical miscibility temperature is non-existent at normal pressures.

Predel (3) determined the phase diagram of the Ga-Hg system by thermal analysis and found the critical miscibility point at 477 K at 50 at % Ga and the monotectic point at 300.88 K at 98.49 at % Ga. Nizhnik and Zvagolskaya (4), from potentiometric and analytical measurements, determined a solubility of 3.81 at % at 303 K; this value is in good agreement with Predel's solubility curve. Yatsenko and Druzhinina (5) determined the solubility of gallium at 283 to 368 K by equilibration and chemical analysis of the liquid phases. At 308 K the latter authors were in agreement with ref. (2), but the solubilities at higher temperatures were lower than those reported by Predel (3). It should be noted here that the critical temperature first reported by Predel was confirmed by Shürmann and Parks (6) and by D'Abramo et al. (7) who determined the temperatures at 476.48 and 475.58 K, respectively. Schürmann and Parks employed high-precision electrical resistivity measurements, while D'Abramo et al. utilized neutron radiography. A comparison of the data of ref. (3) with those of (6) and (7) shows that the liquidus curve of Predel should be slightly modified toward lower temperatures to give a better fit to the solubility data of Yatsenko and Druzhinina. More recently, Gaune-Escard and Bros (8) employed calorimetric measurements to redetermine the liquidus line of the Ga-Hg system; these authors also incorporated some unpublished data of Amarell (9) and confirmed the earlier liquidus reported by Predel (3).

Grosse (10,11) determined the solubility of liquid Ga at 293 K and of solid Ga at 273 and 254 K; because the melting point of Ga is 303 K, the liquid Ga system was metastable. In spite of the apparently high precision of Grosse's measurements, Lindauer (12) expressed skepticism because the difference between the solubilities of solid and liquid Ga, i.e., 3.1 ± 0.05 and 3.28 ± 0.05 at %, respectively, is not significantly higher than the precision of the method used for the measurements.

Kozin (13) predicted a solubility of 98.6 at % at 298 K, but this solubility is clearly too high for the Hg-rich region. However, the calculated solubility is nearly the same as that of the supercooled amalgam in the Ga-rich region; in this region no liquid phase is stable below 300.9 K. The solubility of 3.7×10^{-2} at % reported by Stepanova and Zakharov (14) at 298 K is too low and is rejected.

At temperatures below 301.0 K the saturated amalgams are in equilibrium with solid gallium which is saturated with a small amount of mercury. Between 301.0 and 475.6 K two immiscible phases are in equilibrium, as shown in Fig. 1 (3).

(Continued next page)

COMPONENTS:

(1) Gallium; Ga; [7440-55-3]
(2) Mercury; Hg; [7439-97-6]

EVALUATOR:

C. Guminski; Z. Galus
Department of Chemistry
University of Warsaw
Warsaw, Poland

July, 1985

CRITICAL EVALUATION: (Continued)

Recommended (r) and tentative solubility of gallium in mercury; see Fig. 1 for complete solubility.

Hg-rich region

T/K	Soly/at %		Reference
254	1.1		11
273	1.9		10
293	3.1	Ga (solid)	5
	3.3	Ga (liquid)	10
298	3.4[a]		5
301	3.8		4
323	4.8[a]		3,8,9
373	8.2		9
473	42		3
476	50.0 (r)		3,6,7

[a] Interpolated value from cited references.

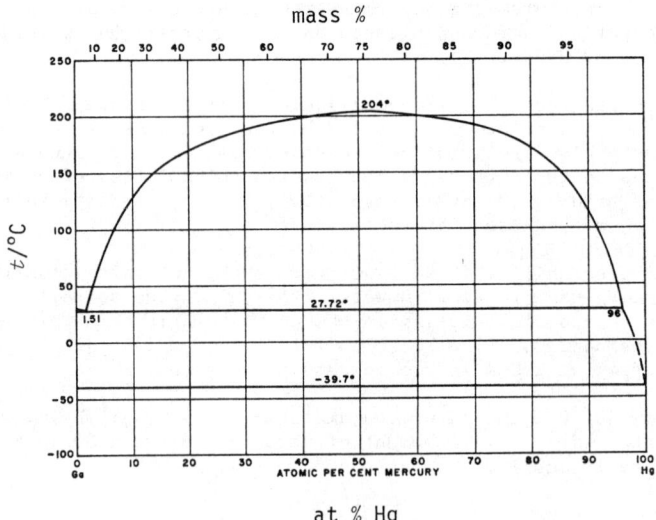

Fig. 1. The Ga-Hg system (3).

COMPONENTS:	EVALUATOR:
(1) Gallium; Ga; [7440-55-3] (2) Mercury; Hg; [7439-97-6]	C. Guminski; Z. Galus Department of Chemistry University of Warsaw Warsaw, Poland July, 1985

CRITICAL EVALUATION: (continued)

References

1. Gilfillan, E.S.; Bent, H.E. *J. Am. Chem. Soc.* 1934, *56*, 1661.
2. Spicer, W.M.; Bartholomay, H.W. *J. Am. Chem. Soc.* 1951, *73*, 868.
3. Predel, B. *Z. Phys. Chem., N.F.* 1960, *24*, 206.
4. Nizhnik, A.T.; Zvagolskaya, E.V. *Zh. Neorg. Khim.* 1961, *6*, 1006.
5. Yatsenko, S.P.; Druzhinina, E.P. *Zh. Neorg. Khim.* 1961, *6*, 1902.
6. Schürmann, H.K.; Parks, R.D. *Phys. Rev. Letters* 1971, *26*, 367, 835.
7. D'Abramo, G.; Ricci, F.P.; Menzinger, F. *Phys. Rev. Letters* 1972, *28*, 22.
8. Gaune-Escard, M.; Bros, J.P. *Thermochim. Acta* 1979, *31*, 323.
9. Amarell, G. *Ph.D. Thesis*, Karlsruhe, 1958; as cited in ref. (8).
10. Grosse, A.V. *U.S. At. Ener. Comm. Rep.*, *NYO-2082-4*, 1966.
11. Grosse, A.V. *U.S. At. Ener. Comm. Rep.*, *NYO-2082-12*, 1967.
12. Lindauer, G.C. *U.S. At. Ener. Comm. Rep.*, *BNL-50048*, 1967.
13. Kozin, L.F. *Fiziko-Khimicheskie Osnovy Amalgamnoi Metallurgii*, Nauka, Alma-Ata, 1964.
14. Stepanova, O.S.; Zakharov, M.S. *Izv. Tomsk. Politekhn. Inst.* 1966, *151*, 21.

COMPONENTS:	ORIGINAL MEASUREMENTS:
(1) Gallium; Ga; [7440-55-3] (2) Mercury; Hg; [7439-97-6]	Nizhnik, A.T.; Zvagolskaya, E.V. *Zh. Neorg. Khim.* 1961, *6*, 1006-8.
VARIABLES: Temperature: 30°C	PREPARED BY: C. Guminski; Z. Galus

EXPERIMENTAL VALUES:

The solubility of gallium in mercury at 30°C was determined to be 1.3-1.5 mass % by a potentiometric method; by equilibration and chemical analysis the solubility was determined to be 1.36 mass %. The atomic % solubility calculated by the compilers is 3.81 at %.

AUXILIARY INFORMATION

METHOD/APPARATUS/PROCEDURE:	SOURCE AND PURITY OF MATERIALS:
The amalgams were prepared by mixing accurately weighed specimens of the metals in hot water which was slightly acidified with HCl. The amalgams were kept in a closed vessel under a solution of acidified $GaCl_3$. The potentiometric solubility measurements were presumably made on concentration cells under a protective atmosphere of nitrogen. The solubility was determined from the break-point in the plot of EMF against Ga concentration. The solubility from chemical analysis was determined by equilibrating amalgams with 0.1-10 mass % Ga in weighed, glass tubes. Samples were taken from the equilibrated amalgam, then treated with HCl to dissolve the Ga. At low concentrations of Ga this metal was determined colorimetrically, while at high concentrations it was determined gravimetrically.	Gallium was 99.99% pure. Mercury was polarographic grade.
	ESTIMATED ERROR: Soly: no better than few percent (by compilers). Temp: nothing specified.
	REFERENCES:

COMPONENTS:	ORIGINAL MEASUREMENTS:
(1) Gallium; Ga; [7440-55-3] (2) Mercury; Hg; [7439-97-6]	Spicer, W.M.; Bartholomay, H.W. J. Am. Chem. Soc. 1951, 73, 868-9.
VARIABLES: Temperature: 35-100°C	PREPARED BY: C. Guminski; Z. Galus

EXPERIMENTAL VALUES:

The solubility of gallium in mercury-rich and in gallium-rich regions was determined at 35 and 100°C.

	Mercury-rich		Gallium-rich	
t/°C	Soly/mass %	Soly/at %[a]	Soly/mass %	Soly/at %[a]
35	1.3 ± 0.1	3.6	93.3 ± 0.4	97.6
100	1.4	3.9	91.4 ± 0.2	96.8

[a] by compilers

The result at 100°C for the Hg-rich region is too low, probably because part of the gallium was oxidized (compilers).

AUXILIARY INFORMATION

METHOD/APPARATUS/PROCEDURE:	SOURCE AND PURITY OF MATERIALS:
Weighed portions of both metals were placed in a glass tube and covered with a solution of $GaCl_3$ in dilute HCl, then the samples were equilibrated with frequent shaking at constant temperature. After equilibration, several small samples were taken from each layer and weighed, then the gallium was extracted with HCl and the mercury was washed, dried and reweighed.	Gallium from Aluminum Company of America was 99.95% pure. Mercury was purified by washing with nitric acid and water, then dried and distilled.
	ESTIMATED ERROR: Soly: precision better than ± 7%. Temp: nothing specified.
	REFERENCES:

COMPONENTS:	ORIGINAL MEASUREMENTS:
(1) Gallium; Ga; [7440-55-3] (2) Mercury; Hg; [7439-97-6]	Predel, B. Z. Phys. Chem. N.F. 1960, 24, 206-16.
VARIABLES: Temperature: 337-477 K	PREPARED BY: C. Guminski; Z. Galus

EXPERIMENTAL VALUES:

The data were reported graphically as a phase diagram for the Ga-Hg system. The liquidus data points were read from the curve by the compilers.

T/K	at % Hg	T/K	at % Hg
337	4.1	469	38.4
370	6.6	473	43.8
385	8.0	477	50.6
391	8.7	475	55.4
407	11.0	471	61.0
404	11.2	468	66.2
419	13.3	463	72.0
428	15.0	456	75.8
447	20.3	443	81.0
449	21.2	429	84.5
458	25.7	397	89.3
457	26.4	356	93.0
464	31.1	337	94.2

The eutectic point was determined at 1.51 at % Hg and 27.72°C.

AUXILIARY INFORMATION

METHOD/APPARATUS/PROCEDURE:	SOURCE AND PURITY OF MATERIALS:
The freezing points were determined from cooling curves on amalgam samples which were protected from oxidation by an atmosphere of nitrogen. The temperatures were determined with a NiCr-Ni thermocouple. Heating curves also were obtained to ascertain the cooling curve data. Although not reported, the amalgams were presumably prepared by mixing desired amounts of the metals.	Gallium purity was 99.999%. Mercury was purified by vacuum distillation.
	ESTIMATED ERROR: Soly: nothing specified. Temp: precision $\pm$ 0.01 K in measurements, but $\pm$ 1 K in read-out values by compilers.
	REFERENCES:

COMPONENTS:	ORIGINAL MEASUREMENTS:
(1) Gallium; Ga; [7440-55-3] (2) Mercury, Hg; [7439-97-6]	Yatsenko, S.P.; Druzhinina, E.P. *Zh. Neorg. Khim.* 1961, *6*, 1902-4.
VARIABLES: Temperature: 10-95°C	PREPARED BY: C. Guminski; Z. Galus

EXPERIMENTAL VALUES:

The solubility of gallium in the Ga- and Hg-rich regions were determined:

	Soly in Hg-rich region		Soly in Ga-rich region	
t/°C	Mass %	at %[a]	Mass %	at %[a]
10	0.86	2.44	(96.25)[b]	(98.68)
22	1.13	3.19	(95.0)[b]	(98.17)
30.5	1.20	3.38	94.6	98.06
35	1.30	3.65	93.9	97.73
50	1.49	4.17	93.1	97.43
65	1.72	4.82	92.0	97.07
80	1.90	5.29	91.0	96.67
95	2.22	6.14	89.7	96.16

[a] by compilers.

[b] The values in parentheses are for the metastable region of liquid Ga layer.

AUXILIARY INFORMATION

METHOD/APPARATUS/PROCEDURE:	SOURCE AND PURITY OF MATERIALS:
Gallium amalgams were obtained by electrolysis of Ga from $Ga_2(SO_4)_3$ onto a mercury cathode. The mixture was agitated and equilibrated in a thermostat under a $Ga_2(SO_4)_3$ solution. A steel ball on top of the Hg layer indicated the phase boundary. After equilibration, samples of amalgams were taken from both layers and the analysis made by: 1) dissolving the weighed sample in HCl and determination of Ga by titration with EDTA; 2) anodic oxidation of the amalgam where the end point of the dissolution was controlled potentiometrically.	Mercury was specified as "pure". Gallium purity not specified.
	ESTIMATED ERROR: Soly: precision ± 1.5%. Temp: nothing specified.
	REFERENCES:

COMPONENTS:	ORIGINAL MEASUREMENTS:
(1) Gallium; Ga; [7440-55-3] (2) Mercury; Hg; [7439-97-6]	Grosse, A.V. *U.S. At. Ener. Comm. Rep.*, NYO-2082-4, 1966.
VARIABLES:	PREPARED BY:
Temperature: (-18.8)-20°C	C. Guminski; Z. Galus

EXPERIMENTAL VALUES:

The solubility of liquid gallium in mercury by equilibration at 20°C was determined to be 1.16 mass %. The solubility determined by first heating the amalgam, followed by equilibration at 20°C, was found to be 1.12 mass %. The average value was 1.14 $\pm$ 0.02 mass %, or 3.28 $\pm$ 0.05 at % for this metastable equilibrium.

The solubility of solid gallium in mercury at -18.8 and 0.0°C were determined to be 1.15 and 1.90 at %, respectively.

The determination at 0.0 and -18.8°C were probably made by the same method; the result at 0.0°C was published in (1).

AUXILIARY INFORMATION

METHOD/APPARATUS/PROCEDURE:	SOURCE AND PURITY OF MATERIALS:
In the first determination the unsaturated Ga amalgam was contacted with supercooled liquid Ga for about 30 h, and the area of the liquid blister on the amalgam remained constant for 2 weeks. The amount of Ga dissolved was found from the mass balance with the help of the blister-area vs. volume relationship established in separate experiments. In the second determination a blister of Ga was immersed in unsaturated Ga amalgam and was warmed to about 40°C with stirring in order to quickly dissolve the Ga. The mixture was cooled and allowed to stand for many hours at 20°C. The undissolved Ga blister was weighed to determine the solubility. The heterogeneous Ga amalgam was cooled to -18.8°C. It was filtered and Ga content was determined in the filtrate by addition of HCl at room temperature. Concentration of Ga was calculated from amount of H_2 evolved by the reaction.	Gallium from Aluminum Company of America was 99.99% pure. Mercury from Bethlehem Apparatus Co. was triply vacuum distilled material; impurity content was less than 2×10^{-5}%.
	ESTIMATED ERROR: Soly: precision $\pm$ 2%. Temp: precision $\pm$ 0.5 K.
	REFERENCES: 1. Grosse, A.V. *U.S. At. Ener. Comm. Rep.*, NYO-2082-12, 1967.

COMPONENTS:	ORIGINAL MEASUREMENTS:
(1) Gallium; Ga; [7440-55-3] (2) Mercury; Hg; [7439-97-6]	1. Schürmann, H.K.; Parks, R.D. *Phys. Rev. Letters* 1971, *26*, 367-70. 2. D'Abramo, G.; Ricci, F.P.; Menzinger, F. *Phys. Rev. Letters* 1972, *28*, 22-4.
VARIABLES: Temperature: 203°C	PREPARED BY: C. Guminski; Z. Galus

EXPERIMENTAL VALUES:

The critical temperature for miscibility at 50 at % Ga was determined to be 203.32 ± 0.50°C in (1). The critical temperature reported by (2) was 475.48 ± 0.01 K (202.33°C by compilers).

AUXILIARY INFORMATION

METHOD/APPARATUS/PROCEDURE:	SOURCE AND PURITY OF MATERIALS:
(1) Weighed amounts of gallium and mercury were mixed in a Pyrex tube provided with ten tungsten electrodes at 15 mm intervals. The tube was placed in a thermostat and stirred at 250°C, then the temperature was slowly lowered. The resistance between different layers of the amalgam was measured as a function of temperature and R/Rc was plotted against temperature, where R is the measured resistance and Rc is the critical resistance at the critical temperature, $\partial(R/Rc)/\partial T=0$. (2) The amalgams were prepared by mixing the metals in near to equimolar ratios in a stainless steel cell. Neutron transmission measurements were made at decreasing temperatures starting at a temperature about 10°C above the critical temperature. There was a sharp change in transmission at the critical temperature for complete miscibility.	(1) Gallium from Eagle-Picher Co. was 99.99% pure. Mercury from Beckman Instruments was 99.99999% pure. (2) Gallium from Fluka AG was 99.99% pure. Mercury from Rudipoint was 99.9% pure.
	ESTIMATED ERROR: Temp: precision ± 0.50 K in (1); ± 0.01 K in (2).
	REFERENCES:

COMPONENTS:	ORIGINAL MEASUREMENTS:
(1) Gallium; Ga; [7440-55-3] (2) Mercury; Hg; [7439-97-6]	Gaune-Escard, M.; Bros, J.P. *Thermochim. Acta* 1979, *31*, 323-39.
VARIABLES: Temperature: 313-466 K	PREPARED BY: C. Guminski; Z. Galus

EXPERIMENTAL VALUES:

Liquidus points were determined from microcalorimetric measurements:

	This work		Unpublished work (1)	
T/K	x(Na)		x(Na)	
313	0.042	0.976	0.041	0.980
353	0.063	0.954	0.066	0.950
373	0.935[a]	0.938	0.082	0.935
423	0.150	0.870	0.150	0.860
466	0.360	0.720	0.300	0.672

[a] obvious misprint in publication.

AUXILIARY INFORMATION

METHOD/APPARATUS/PROCEDURE:	SOURCE AND PURITY OF MATERIALS:
The liquidus points were determined microcalorimetrically: The metals mixed by breaking ampule, presumably containing Ga, in the Hg at equilibrated temperature and enthalpy of mixing determined from heat effect. Enthalpy determined as function of composition, x_{Ga}, at each temperature, and breakpoint in plot of ΔH_M vs. x_{Ga} is liquidus at that temperature. Calorimeter calibrated by Joule effect. Measurements made under pressurized argon; both metals were protected from oxidation with layer of oil.	Ga and Hg purity not specified. Ar was grade "U" from Air Liquide Co.
	ESTIMATED ERROR: ΔH_M: precision 2-6%. Temp: not specified.
	REFERENCES: 1. Amarell, G. *Dissert. Dokt. Naturwiss.*, Karlsruhe, 1958.

COMPONENTS:	EVALUATOR:
(1) Indium; In; [7440-74-6] (2) Mercury; Hg; [7439-97-6]	C. Guminski; Z. Galus Department of Chemistry University of Warsaw Warsaw, Poland July, 1985

CRITICAL EVALUATION:

The phase equilibria of the In-Hg system have been studied extensively, but most of the data have been reported graphically as phase diagrams only. In some of the reports the phase diagram appeared in relatively small figures and it was not possible to precisely read the numerical values of the liquidus from these phase diagrams.

Parks and Moran (1) reported the first study of the solubility of In in Hg, but these authors reported the indium solubilities of only 2.15 to 2.27 at % at 273 to 323 K; these very low values are rejected. Ito and coworkers (2) reported the In-Hg phase diagram and showed that indium has an appreciable solubility at room temperatures. Although the shape of their phase diagram was similar to those reported by subsequent authors, the liquidus temperatures of Ito et al. were too low by a few degrees, probably because of impurities in the indium which was used. Spicer and Banick (3), from thermoanalytical measurements, reported more accurate liquidus temperatures in the region of 68.05 to 100 at % In; the liquidus temperature increased monotonically from 283.5 to 429.2 K, respectively, in this range, and the authors fitted an equation for the solubility as a function of the temperature. Kozin and coworkers determined the phase diagram from thermoanalytical (4,5) and from potentiometric (6) measurements. Several other determinations of the phase diagram were reported during the years 1962-1964 (7-13), but Chiaranzelli and Brown (8) were the only authors to report numerical liquidus data. Robert and Thibault (14) also reported a phase diagram for the In-Hg system, but the liquidus between 7 and 25 at % In by these authors is not in agreement with those of the other accepted measurements; five different In-Hg compounds were proposed in this range by these authors. More recently, Franck (15,16) determined the liquidus in the In-rich region from vapor pressure measurements, while Hilpert (17) applied thermal analysis to confirm the liquidus temperature of 352 K at 80 at % In.

Kozin's (18) calculated solubility of 67.95 at % at 298 K is in good agreement with the accepted experimental solubility of 70 at %. Liebl (19) also has reported an indium solubility of 68 at % at room temperature, determined by coulometry, but no other details were reported for this measurement.

From potentiometric measurements at 293 K, Sundén (20) reported a solubility of 68 at %. The solubility measurement of Strachan and Harris (21) at room temperature is inconsequential.

Table 1 summarizes the congruently melting and the eutectic points which were derived from the phase diagrams reported by the various authors. The variation of the composition is approximately ±1 at %, while that for the temperature is ±1 K. In spite of the high precision in each data set reported in the literature, these variations arise because of the difficulty in exactly reading the data from the graphical presentations of the phase diagrams.

The saturated indium amalgams are in equilibrium with In-Hg intermetallic solid phases. The compounds, $In_{11}Hg$, $InHg$, $InHg_4$ and $InHg_5$ or $InHg_6$, have been identified with some certainty, but others, such as $InHg_{11}$, $InHg_9$, $InHg_3$, In_5Hg_7 and In_7Hg, are of doubtful existence. Fig. 1 shows the phase diagram reported by (5).

(Continued next page)

COMPONENTS:	EVALUATOR:
(1) Indium; In; [7440-74-6] (2) Mercury; Hg; [7439-97-6]	C. Guminski; Z. Galus Department of Chemistry University of Warsaw Warsaw, Poland July, 1985

CRITICAL EVALUATION: (continued)

Tentative and recommended (r) values of In solubility in Hg:

	Hg-rich Region		In-rich Region	
T/K	Soly/at %	Ref.	Soly/at %	Ref.
235.6	34.4 (r)[a]			
242.6			62.3 (r)[a]	
254.7	7.5	[4]	64 (r)[b]	[7,13]
258.7	14.3 (r)[a]			
273.2			66.5 (r)[b]	[7,8]
293.2			68.0 (r)	[6,20]
298.2			70.0	[4]
323.2			75.3	[6]
373			85 (r)[b]	[4,5,10,15]

[a] Average of all reported data.
[b] Interpolated from data in cited references.

TABLE 1

Summary of Melting Points of Congruently Melting Compounds and Eutectic Points

Melting Points, T/K			Eutectic Points				
$InHg_6$	$InHg_5$	$InHg$	T/K	at % In	T/K	at % In	Ref.
	256	250	235.5	34.3	240.5	63.6	[2]
258.8		254.6	236.5	32.8	243.1	63.0	[4]
	258.6±0.3	254.0±0.3	235.8±0.3	34.0	242.4±0.3	61.7	[7]
260.8±0.4		256.6±0.1	240.1±0.±	34.3±0.5	244.6±0.1	63.6±0.5	[8]
	260.0	254.7	236.4	33.3	243.2	62.7	[9]
259.0±0.2		253.9±0.2	236.0±0.2	34.7±0.2	242.1±0.2	61.5±0.3	[10]
260±1		255±1	237±1	35	242.5±1	60	[11]
258.2		254.0	236.4	34.1	242.6	61.2	[12]
257.5		254.5	236	34.7	243	63.0	[13]
259.2±0.5		255.0±0.5	236.4±0.5	34.9	243.5±0.5	62.5	[14]
258.2±0.5		254.6±0.5	235.7±0.5	35.0	241.7±0.5	63.0	[5]

(Continued next page)

COMPONENTS:	EVALUATOR:
(1) Indium; In; [7440-74-6] (2) Mercury; Hg; [7439-97-6]	C. Guminski; Z. Galus Department of Chemistry University of Warsaw Warsaw, Poland July, 1985

CRITICAL EVALUATION: (continued)

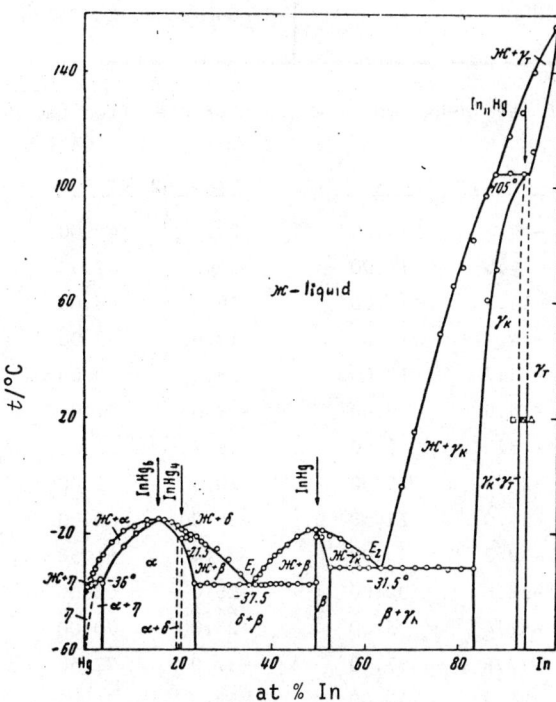

Fig. 1. The In-Hg system (5)

References

1. Parks, W.G.; Moran, W.G. *J. Phys. Chem.* 1937, *41*, 343.
2. Ito, H.; Ogawa, E.; Yanagaze, T. *Nippon Kinzoku Gakkaishi* 1951, *15B*, 382.
3. Spicer, W.M.; Banick, C.J. *J. Am. Chem. Soc.* 1953, *75*, 2268.
4. Kozin, L.F.; Tannanaeva, N.N. *Zh. Neorg. Khim.* 1961, *6*, 909.
5. Kozin, L.F.; Sudakov, V.A. *Izv. Akad. Nauk SSSR, Metally* 1970, No. 5, 197; *Izv. Akad. Nauk Kaz. SSR, Ser. Khim.* 1970, No. 1, 50.
6. Kozin, L.F. *Tr. Inst. Khim. Nauk Akad. Nauk Kaz. SSR* 1962, *9*, 71, 81.
7. Eggert, G.L. *Trans. A.S.M.* 1962, *55*, 891.
8. Chiaranzelli, R.V.; Brown, O.L. *J. Chem. Eng. Data* 1962, *7*, 477.
9. Jangg, G. *Z. Metallk.* 1962, *53*, 612.
10. Coles, B.R.; Merriam, M.F.; Fisk, Z. *J. Less-Common Metals* 1963, *5*, 41.
11. Knütter, R. *Tech.-Wiss. Abh. Osram Gesel.* 1963, *8*, 112.
12. Morawietz, W. *Chem. Ing. Tech.* 1964, *36*, 638.
13. Cusack, N.E.; Kendall, P.; Fielder, M. *Phil. Mag., Ser. 8*, 1964, *10*, 871.
14. Robert, J.; Thibault, M. *J. Chim. Phys.* 1966, *63*, 345.
15. Franck, G. *Tech.-Wiss. Abh. Osram Gesel.* 1973, *11*, 101.
16. Franck, G. *Z. Naturforsch., A* 1971, *26*, 150.
17. Hilpert, K. *Ber. Kernforschungsanlage, Jülich* 1981, JUEL-1744, p. 95.
18. Kozin, L.F. *Fiziko-Khimicheskie Osnovy Amalgamnoi Metallurgii*, Nauka, Alma-Ata, 1964.
19. Liebl, G. quoted by H. Spengler, *Metall* 1958, *12*, 105.
20. Sunden, N. *Z. Elektrochem.* 1953, *57*, 100.
21. Strachan, J.F.; Harris, N.L. *J. Inst. Metals* 1956-57, *85*, 17.

COMPONENTS:	ORIGINAL MEASUREMENTS:
(1) Indium; In; [7440-74-6] (2) Mercury; Hg; [7439-97-6]	Kozin, L.F.; Tananaeva, N.N. *Zh. Neorg. Khim.* 1961, *6*, 909-12.
VARIABLES:	PREPARED BY:
Temperature: (-38)-150°C	C. Guminski; Z. Galus

EXPERIMENTAL VALUES:

Liquidus temperatures of the In-Hg system were abstracted from the phase diagram by the compilers:

t/°C	at % In	t/°C	at % In	t/°C	at % In	t/°C	at % In
-38.0	0.1	-14.4	14.00	-20.7	45.00	66.0	77.50
-37.1	0.25	-14.5	15.00	-19.2	47.50	79.0	80.00
-36.0	0.30	-14.7	16.00	-18.9	48.00	80.0	80.25
-35.1	0.50	-14.9	17.00	-18.65	49.00	90.0	82.50
-33.2	1.00	-15.7	19.00	-18.6	50.00	101.0	85.00
-28.0	2.50	-16.0	20.00	-18.65	51.00	103.0	85.50
-26.6	3.00	-18.2	22.50	-19.2	52.00	106.0	86.00
-24.5	4.00	-20.4	25.00	-20.5	55.00	108.0	87.50
-22.5	5.00	-24.0	27.50	-26.00	60.00	114.0	88.00
-18.4	7.50	-29.2	30.00	-26.4	61.50	123.0	90.00
-16.9	9.00	-32.6	32.00	-26.0	63.00	134.0	94.60
-16.0	10.00	-31.4	35.00	+25.0	70.00	150.0	97.50
-14.83	12.00	-27.6	37.50	+37.0	72.50		
-14.5	13.00	-24.9	40.00	53.0	75.00		

AUXILIARY INFORMATION

METHOD/APPARATUS/PROCEDURE:	SOURCE AND PURITY OF MATERIALS:
Separate alloys were prepared for each composition from pure metals; the containment tube was probably glass. The samples were covered with a thin layer of glycerol. The samples were cooled and heating curves were determined. Calibrated thermometers were used for the heating curves, but a Pt, Pt-Rh thermocouple was used to record the differential heating curves.	Mercury was purified by treatment with HNO_3-$Hg_2(NO_3)_2$, then distilled twice in vacuum. Indium was 99.999% pure.
	ESTIMATED ERROR: Soly: nothing specified. Temp: precision ± 0.1 K.
	REFERENCES:

COMPONENTS:	ORIGINAL MEASUREMENTS:
(1) Indium; In; [7440-74-6] (2) Mercury; Hg; [7439-97-6]	Spicer, W.M.; Banick, C.J. *J. Am. Chem. Soc.* 1953, *75*, 2268-2269.
VARIABLES: Temperature: 10-151°C	PREPARED BY: C. Guminski; Z. Galus

EXPERIMENTAL VALUES:

The liquidus temperatures of indium-rich amalgams were reported:

t/°C	mass % In	at % In[a]
151.3	97.46	98.53
135.1	90.12	94.09
121.7	84.07	90.22
108.2	79.30	87.00
94.2	74.70	83.77
78.1	69.84	80.19
59.2	64.89	76.36
37.6	60.01	72.40
10.3	54.92	68.05

[a] by compilers.

AUXILIARY INFORMATION

METHOD/APPARATUS/PROCEDURE:	SOURCE AND PURITY OF MATERIALS:
Indium-rich alloys were made by adding the desired amount of Hg to the previously analyzed alloy in a test tube. The amalgams were protected from oxidation with mineral oil. The amalgam was analyzed gravimetrically by dissolving the In in conc. HCl then weighing the Hg residue. Cooling curves were determined by inserting the glass-clad copper-constantan thermocouple into the amalgam and reading the temperature with a precision potentiometer. Down to 60°C the samples were cooled in a tube furnace, then at lower temperatures the sample tube was placed in water jacket and the latter was cooled with various solutions to obtain cooling curves.	Indium was 99.97% pure. Mercury was purified by spraying through a column of dilute HNO_3, washed, dried, then distilled under vacuum. ESTIMATED ERROR: Soly: nothing specified. Temp: precision $\pm$ 0.02 K. REFERENCES:

COMPONENTS:	ORIGINAL MEASUREMENTS:
(1) Indium; In; [7440-74-6] (2) Mercury; Hg; [7439-97-6]	Ito, H.; Ogawa, E.; Yanagase, T. *Nippon Kinzoku Gakkaishi* 1951, *15B*, 382-4.
VARIABLES: Temperature	PREPARED BY: C. Guminski; Z. Galus

EXPERIMENTAL VALUES:

The melting point of $InHg_5$, at 16.6 at % In, and that of InHg, at 50.0 at % In, were determined to be -17 and -23°C, respectively.

AUXILIARY INFORMATION

METHOD/APPARATUS/PROCEDURE:	SOURCE AND PURITY OF MATERIALS:
The amalgams were prepared by mixing weighed amounts of the metals in a sealed glass tube. The freezing points were determined by thermal analysis. The temperatures were measured with a copper-constantan thermocouple which was carefully calibrated by comparison with a calibrated Pt-PtRh thermocouple and with a mercury thermometer.	Indium was electrolytic material from zinc-fusion residue which was obtained from Hikoshima Refinery. Mercury was purified by vacuum distillation.
	ESTIMATED ERROR: Soly: nothing specified. Temp: nothing specified; precision no better than ± 0.5 K (compilers).
	REFERENCES:

COMPONENTS:	ORIGINAL MEASUREMENTS:
(1) Indium; In; [7440-74-6] (2) Mercury; Hg; [7439-97-6]	Sunden, N. Z. Elektrochem. 1953, 57, 100-2.

VARIABLES:	PREPARED BY:
One temperature: 20°C	C. Guminski; Z. Galus

EXPERIMENTAL VALUES:

Solubility of indium in mercury at 20.0°C was reported to be 55 mass %. The atom % solubility calculated by the compilers is 68 at %.

AUXILIARY INFORMATION

METHOD/APPARATUS/PROCEDURE:	SOURCE AND PURITY OF MATERIALS:
The amalgams were prepared by dissolution of indium in mercury under a nitrogen atmosphere. Potentials of the cell, Hg, Hg_2Cl_2, NaCl $\mid$ $In(ClO_4)_3$ $\mid$ In(Hg) were measured. The plot of EMF against the logarithm of amalgam concentration showed a breakpoint at saturation.	Indium was 99.97% pure from Indium Corp. of America. Mercury was distilled. Other chemicals were analytically pure from Merck, or they were recrystallized before use.
	ESTIMATED ERROR: Soly: nothing specified; precision probably no better than ± 1% (compilers). Temp: nothing specified.
	REFERENCES:

COMPONENTS:	ORIGINAL MEASUREMENTS:
(1) Indium; In; [7440-74-6] (2) Mercury; Hg; [7439-97-6]	Kozin, L.F. *Tr. Inst. Khim. Nauk Akad. Nauk Kaz. SSR* <u>1962</u>, *9*, 71-80.
VARIABLES: Temperature: 20-80°C	PREPARED BY: C. Guminski; Z. Galus

EXPERIMENTAL VALUES:

Solubility of indium in mercury:

$t/°C$	Soly/at %
20	68.0
50	73.3
80	80.0

Similar measurements at -1.5 to 14°C gave unreliable results, probably because of slow equilibration at lower temperatures.

AUXILIARY INFORMATION

METHOD/APPARATUS/PROCEDURE:	SOURCE AND PURITY OF MATERIALS:
Amalgams were prepared by dissolution of various amounts of indium in mercury. Potentials of the cell, In(Hg) | 0.1 mol dm^{-3} In(ClO$_4$)$_3$, 0.9 mol dm^{-3} NaClO$_4$ | NaCl, Hg$_2$Cl$_2$, Hg were determined. The solutions were protected from oxidation with a stream of pure nitrogen. The plot of EMF against the logarithm of the amalgams concentration showed a breakpoint at saturation.	The salts were twice recrystallized. Mercury was purified chemically, then twice distilled. Indium was 99.999% pure.
	ESTIMATED ERROR: Soly: nothing specified. Temp: precision ± 0.2 K.
	REFERENCES:

COMPONENTS:	ORIGINAL MEASUREMENTS:
(1) Indium; In; [7440-74-6] (2) Mercury; Hg; [7439-97-6]	Eggert, G.L. *Trans. ASM* 1962, *55*, 891-97.
VARIABLES: Temperature: (-36)-141°C	PREPARED BY: C. Guminski; Z. Galus

EXPERIMENTAL VALUES:

The author determined the complete phase diagram and reported numerical values only for the eutectics at -37.4 and -30.8°C (at 34.0 and 61.7 at %, respectively), and for the congruent melting points at -14.6 and -19.2°C (at 16.7 and 50.0 at % In, respectively). The following data points were read from the phase diagram by the compilers:

Soly/at %	t/°C	Soly/at %	t/°C	Soly/at %	t/°C	Soly/at %	t/°C
1.0	-35.6	22.8	-19.0	44.7	-21.0	70.0	26.9
3.0	-29.5	24.0	-20.7	47.6	-22.7	72.8	38.8
5.0	-25.7	24.8	-21.3	52.4	-20.0	79.8	70.0
8.0	-22.0	27.4	-25.8	54.0	-21.8	82.7	90.0
10.0	-18.0	29.0	-28.6	57.3	-23.6	83.2	93.5
12.0	-16.7	31.3	-32.6	58.7	-25.0	84.3	105.2
15.0	-14.8	35.0	-34.4	59.2	-27.9	86.8	119.0
17.4	-14.7	37.0	-29.7	60.0	-28.9	88.8	132.4
18.4	-14.7	39.0	-27.1	63.5	-17.1	93.0	141.2
20.0	-16.2	42.6	-22.7	66.8	+5.0		

AUXILIARY INFORMATION

METHOD/APPARATUS/PROCEDURE:	SOURCE AND PURITY OF MATERIALS:
Weighed quantities of the metals were mixed at room temperature in glass tubes. The latter were inserted inside a larger glass tube jacket and the space between tubes was packed with Cu wool. The assembly was immersed in a mixture of dry-ice and trichloroethylene to obtain cooling curves; temperatures were determined with a calibrated copper-constantan thermocouple inserted into the liquid amalgam and the data were recorded on a strip-chart recorder. Precise thermopotentials at occurrences on the cooling curves were measured with a precision potentiometer. Low temperature microscopy was observed on a microscope stage upon repeated melting and freezing.	Indium was 99.98% pure from Indium Corp. of America. Mercury purity was 99.999_5%.
	ESTIMATED ERROR: Soly: nothing specified. Temp: precision ± 0.3 K.
	REFERENCES:

COMPONENTS:	ORIGINAL MEASUREMENTS:
(1) Indium; In; [7440-74-6] (2) Mercury; Hg; [7439-97-6]	Chiaranzelli, R.V.; Brown, O.L.I. J. Chem. Eng. Data 1962, 7, 477-78.
VARIABLES: Temperature: (-37)-11°C	PREPARED BY: C. Guminski; Z. Galus

EXPERIMENTAL VALUES:

The liquidus temperatures of the saturated indium amalgams were determined:

mass % Hg	at % In[a]	t/°C	mass % Hg	at % In[a]	t/°C	mass % Hg	at % In[a]	t/°C
99.95	0.087	-36.7	96.00	6.79	-17.4	82.35	27.24	-22.0
99.90	0.17	-35.3	94.63	9.02	-16.0	79.02	31.69	-28.3
99.80	0.35	-34.0	93.21	11.30	-12.3	76.89	34.43	-32.0
99.56	0.77	-31.6	91.85	13.42	-12.0	75.32	36.41	-30.0
99.21	1.37	-29.4	91.25	14.35	-12.4	69.22	43.72	-19.1
98.90	1.91	-26.0	90.70	15.19	-12.8	66.15	47.21	-17.0
98.73	2.25	-24.0	90.40	15.75	-14.1	63.66	43.74	-16.6
98.70	2.30	-25.3	90.07	16.15	-13.0	58.91	54.93	-19.2
98.60	2.46	-24.5	88.98	16.33	-13.9	54.03	59.79	-25.0
98.37	2.81	-23.5	87.54	19.92	-14.6	51.53	62.17	-27.6
98.03	3.39	-24.7	87.06	20.54	-16.0	49.99	63.61	-28.6
97.49	4.30	-19.7	85.82	22.40	-17.0	48.02	65.41	-3.5
96.97	5.18	-21.0	84.50	24.27	-18.0	46.03	67.20	+10.8

[a] by compilers.

AUXILIARY INFORMATION

METHOD/APPARATUS/PROCEDURE:	SOURCE AND PURITY OF MATERIALS:
Weighed portions of the metals were mixed in Pyrex test tubes, and generally heated and cooled while stirring for several cycles. Some of the alloys were covered with mineral oil, but no oxidation was noticeable on unprotected samples. Heating and cooling curves were observed with a calibrated, glass-sheathed copper-constantan thermocouple.	Mercury was 99.9999% pure. Indium was 99.9995% pure.
	ESTIMATED ERROR: Soly: precision better than ± 0.7%. Temp: precision ± 0.02 K.
	REFERENCES:

COMPONENTS:	ORIGINAL MEASUREMENTS:
(1) Indium; In; [7440-74-6] (2) Mercury; Hg; [7439-97-6]	Jangg, G. Z. Metallk. 1962, 53, 612-14.

VARIABLES:	PREPARED BY:
Temperature: (-37)-(-13)°C	C. Guminski; Z. Galus

EXPERIMENTAL VALUES:

The data were presented as a phase diagram. The following numerical liquidus data were reported:

t/°C	at % In
-13.2	16.6
-36.8	33.3
-18.5	50.0
-30.0	62.7

The results at the higher temperatures show excellent agreement with (1).

AUXILIARY INFORMATION

METHOD/APPARATUS/PROCEDURE:	SOURCE AND PURITY OF MATERIALS:
Appropriate amounts of both metals were melted in a closed glass container, and cooling curves were recorded with calibrated alcohol and mercury thermometers. Samples of amalgams were analyzed by unspecified method.	Indium was specified as being of high purity. Mercury was treated with H_2SO_4 then triply-distilled under vacuum.
	ESTIMATED ERROR: Nothing specified.
	REFERENCES: 1. Spicer, W.M.; Banick, C.J. J. Am. Chem. Soc. 1953, 75, 2268.

COMPONENTS:	ORIGINAL MEASUREMENTS:
(1) Indium; In; [7440-74-6] (2) Mercury; Hg; [7439-97-6]	Coles, B.R.; Merriam, M.F.; Fisk, Z. *J. Less-Common Met.* 1963, 5, 41-48.
VARIABLES: Temperature: (-37)-143°C	**PREPARED BY:** C. Guminski; Z. Galus

EXPERIMENTAL VALUES:

The phase diagram for the In-Hg system was presented, and numerical values were reported only for the congruently melting, peritectic and eutectic points. These points were as follows:

at % In	14.2±0.2[a]	34.7±0.2[b]	50.0±0.2[a]	61.5±0.4[b]	86.6[c]
$t/°C$	-14.2±0.2	-37.2±0.2	-19.3±0.2	31.0±0.2	108±1

Other liquidus points were read from the phase diagram by the compilers:

at % In	$t/°C$	at % In	$t/°C$	at % In	$t/°C$	at % In	$t/°C$
0.9	-34.9	18.2	-16.2	58.7	-25.3	89.3	118.5
1.8	-32.3	20.0	-18.5	60.8	-28.9	91.4	125.9
2.8	-28.5	25.1	-23.6	62.3	-27.9	92.9	132.3
5.9	-21.0	29.7	-29.5	64.6	-11.0	93.6	134.3
7.4	-18.7	33.0	-35.1	67.6	+13.8	94.1	136.4
9.6	-16.0	36.2	-35.1	80.5	81.5	95.9	142.8
11.1	-15.5	40.5	-27.7	84.2	98.0		
12.4	-15.2	44.1	-22.6	86.9	109.7		
16.5	-15.1	54.2	-20.2	88.3	114.3		

[a] Congruent melting point.
[b] Eutectic point.
[c] Peritectic point.

AUXILIARY INFORMATION

METHOD/APPARATUS/PROCEDURE:	SOURCE AND PURITY OF MATERIALS:
Desired quantities of each metal to yield 50-100 g of a given amalgam were melted and stirred in an alumina crucible exposed to air. Temperature of the melt was determined with a calibrated, glass-sheathed, copper-constantan thermocouple which was inserted into the alloy during the determination of the heating and cooling curves. X-ray diffraction data, using CuKα radiation, were obtained to identify crystal phases.	Indium from Indium Corp. of America was better than 99.999% pure. Mercury, "Vacumetal" from Metal Salts Corp., was better than 99.999% pure.
	ESTIMATED ERROR: Soly: precision ± 1%. Temp: precision ± 0.2 K.
	REFERENCES:

COMPONENTS:	ORIGINAL MEASUREMENTS:
(1) Indium; In; [7440-74-6] (2) Mercury; Hg; [7439-97-6]	Cusack, N.; Kendall, P.; Fielder, M. *Phil. Mag.*, Ser. 8, <u>1964</u>, *10*, 871-82.
VARIABLES:	PREPARED BY:
Temperature: (-37)-(-16)°C	C. Guminski; Z. Galus

EXPERIMENTAL VALUES:

The data were presented only as the liquidus curve for the In-Hg system. The data points were read off the curve by the compilers:

Soly/at %	t/°C	Soly/at %	t/°C	Soly/at %	t/°C
2.3	-31	34.7	-37	57.0	-20
6.0	-21	37.3	-30	61.0	-25
14.0	-15.5	40.5	-26	63.0	-30
22.0	-20	47.0	-20.5	64.0	-20
31.5	-30	51.3	-18.5		

AUXILIARY INFORMATION

METHOD/APPARATUS/PROCEDURE:	SOURCE AND PURITY OF MATERIALS:
Amalgams were prepared, presumably, by weighing desired quantities of each metal with subsequent mixing and alloying in vacuo. The freezing points were obtained from cooling curves.	Indium from BDH and from L. Light and Co. was 99.999% pure. Mercury was purified by multiple distillation and had only 10^{-4} mass % of impurities.
	ESTIMATED ERROR: Soly: nothing specified. Temp: nothing specified.
	REFERENCES:

COMPONENTS:	ORIGINAL MEASUREMENTS:
(1) Indium; In; [7440-74-6] (2) Mercury; Hg; [7439-97-6]	Morawietz, W. *Chem. Ing. Tech.* 1964 *36*, 638-45.
VARIABLES: Temperature	PREPARED BY: C. Guminski; Z. Galus

EXPERIMENTAL VALUES:

The results were presented as a phase diagram. The indium solubility at room temperature was reported to be 120 parts In/100 parts Hg by mass. The corresponding atomic % solubility calculated by the compilers is 67.7 at %.

AUXILIARY INFORMATION

METHOD/APPARATUS/PROCEDURE:	SOURCE AND PURITY OF MATERIALS:
The alloys were obtained by electro-reduction, and thermal analysis curves were recorded. Detailed description of the method was not specified.	Indium was stated as being of high purity. Mercury purity not specified.
	ESTIMATED ERROR: Nothing specified.
	REFERENCES:

COMPONENTS:	ORIGINAL MEASUREMENTS:

(1) Indium; In; [7440-74-6]
(2) Mercury; Hg; [7439-97-6]
	1. Kozin, L.F.; Sudakov, V.A.
	   Izv. Akad. Nauk Kaz. SSR, Ser. Khim.
	   1970, No. 1, 50-5.
	2. Same authors.
	   Izv. Akad. Nauk SSSR, Metally
	   1970, No. 5, 197-201.

VARIABLES:	PREPARED BY:
Temperature: (-37)-140°C	C. Guminski; Z. Galus

EXPERIMENTAL VALUES:

The data were presented graphically as a partial phase diagram in (1). The complete phase diagram was presented in (2), and numerical data were presented for the congruently melting, eutectic, and peritectic points. The experimental liquidus points were read off the curve in (1) by the compilers. The phase diagram from (2) is presented in the critical evaluation, Fig. 1.

t/°C	at % In	Ref.
-15.0	14.8	[2]
-37.5	35.0	[2]
-18.6	50.0	[2]
-31.5	63.0	[2]
97.0	84.8	[1]
105	85.0	[2]
105.2	87.2	[1]
118.7	89.5	[1]
127	93.0	[1]
139.6	95.6	[1]

AUXILIARY INFORMATION

METHOD/APPARATUS/PROCEDURE:	SOURCE AND PURITY OF MATERIALS:
Details of the procedure were not described in (1), but it was probably identical to that in (2). The amalgams were prepared by precisely weighing the metals in an atmosphere of dry carbon dioxide, then the samples were sealed in glass tubes. The melting points were obtained from cooling curves; the temperatures were determined with Pt, Pt-Rh calibrated thermocouples.	Indium was 99.999% pure. Mercury was specified as "R-0".
	ESTIMATED ERROR:
	Soly: precision ± 0.01% in (2). Temp: precision ± 0.5 K in (2).
	REFERENCES:

COMPONENTS:	ORIGINAL MEASUREMENTS:
(1) Indium; In; [7440-74-6] (2) Mercury; Hg; [7439-97-6]	Franck, G. *Tech.-Wiss. Abh. Osram Gesel.* 1973, *11*, 101-105. *Z. Naturforsch.*, A 1971, *26*, 150-3.
VARIABLES:	PREPARED BY:
Temperature: 80-130°C	C. Guminski; Z. Galus

EXPERIMENTAL VALUES:

The data were reported graphically as a liquidus curve. The following points on the liquidus were read off by the compilers:

$t/°C$	at % Hg	at % In
130	7.5	92.5
120	9.7	90.3
110	12.5	87.5
100	15.2	84.8
90	18.0	82.0
80	19.6	80.4

AUXILIARY INFORMATION

METHOD/APPARATUS/PROCEDURE:	SOURCE AND PURITY OF MATERIALS:
Method of preparation of alloys was not described. The alloy, in the form of cubes approximately 1 mm^3, was vacuum-sealed in a Supracil silica cuvette. The vapor pressure of the alloys was determined as a function of temperature by measuring the Hg 2537 Å resonance line absorption, and comparing that for the alloy vapor against that of pure Hg to eliminate the effect of Doppler line broadening in the absorption. The freezing point of the alloy was determined as the breakpoint in the relationship of the optical absorption as a function of temperature.	Mercury was specified as being of high purity. Indium purity was not specified.
	ESTIMATED ERROR: Temp: nothing specified. Composition: precision $\pm$ 0.3% (compilers).
	REFERENCES:

COMPONENTS:	EVALUATOR:
(1) Thallium; Tl; [7440-28-0] (2) Mercury; Hg; [7439-97-6]	C. Guminski; Z. Galus Department of Chemistry University of Warsaw Warsaw, Poland July, 1985

CRITICAL EVALUATION:

Tammann (1) reported the first solubility study in the Tl-Hg system; he observed that the addition of 0.469 at % Tl into mercury depressed the freezing point of Hg by 0.81 K. The fact that thallium has a high solubility in mercury near room temperature was indicated by an early potentiometric study by Spencer (2) who reported a solubility of 41.5 at % at 291 K. Sucheni (3) also reported an early potentiometric study at 273 and 310 K, and he observed that amalgams which contained more than 43 at % Tl are heterogeneous at 310 K; his solubility of 28 at % at 273 K is too high as compared to later works.

Kurnakov and Pushin (4) were the first to report a phase diagram for this system. However, their thermoanalytical determination of the liquidus in the range of 8 to 40 at % Tl did not agree with later works by other more accurate measurements. Pushin (5) subsequently redetermined and corrected the liquidus in the range of 19.1 to 39.5 at % Tl. The measurements of Pavlovich (6) were in agreement with (4) in the range of 0-8 and 40-100 at % Tl, but the former author showed that the maximum in the liquidus occurred at 29 at % Tl and 288 K, as compared to 33.33 at % Tl found by (4). Roos (7) also determined the phase diagram for this system from a detailed study which took into account the effect of impurities; he found the first eutectic at 214.2 K at 8.56 at % and the second at 273.78 K at 40.0 at % Tl. Roos found that the coordinates for the maximum in the liquidus were 28.6 at % Tl at 287.6 K. Richards and Daniels (8) and Richards and Smyth (9) applied thermal analysis and potentiometry to confirm the results of Roos; however, Richards et al. found slightly higher solubilities at lower temperatures and slightly lower solubilities at higher temperatures as compared to Roos. Kozin (10) employed potentiometric measurements at 298 to 353 K and found that the solubility of thallium increased from 42.6 to 53.2 at % in this temperature range; these results were in agreement with the earlier measurements. Claire and Rey (11) verified parts of the Tl-Hg phase diagram in the thallium-rich region. Moser (12), without presenting experimental detail, reported the eutectics at 213.2 and 272.4 K at 8.5 and 40.0 at % Tl, respectively; Siede (13) also found the first eutectic at 8.5 at % Tl, but at 214.8 K. Resistivity measurements performed by Schulz and Spiegler (14) confirmed the melting temperature of Tl_2Hg_5 at 287.7 K.

Without presenting details of his density measurements of Tl amalgams, Kanda (15) reported a solubility of 42 at % Tl at 296 K. Strachan and Harris (16) reported only that the solubility is higher than 13.1 at % at room temperature. Kozin's (17) predicted solubility of 34.6 at % at 298 K is too low. Zebreva and coworkers (18) determined a solubility of 44.0 at % at 298 K by thermometric titration; this value is slightly too high.

Richter and Pistorius investigated the effect of pressure on the congruently melting point (Tl_2Hg_5) (19) and on the eutectic points (20), and these authors observed that the above temperatures increased nearly linearly with increasing pressure up to approximately 30 kbar. Based on these measurements, liquidus lines for the Tl-Hg system were presented for the pressure range of 0 to 50 kbar.

The phase diagram for the Tl-Hg system is shown in Fig. 1 (21).

(Continued next page)

COMPONENTS:	EVALUATOR:
(1) Thallium; Tl; [7440-28-0] (2) Mercury; Hg; [7439-97-6]	C. Guminski; Z. Galus Department of Chemistry University of Warsaw Warsaw, Poland July, 1985

CRITICAL EVALUATION: (continued)

Recommended Solubility of Thallium in Mercury

T/K	Soly/at %	Reference
214	8.5[a]	6,7,12,13,20
245	12	7
274	19	5,6,8
274	40[a]	7,20
288	28.6	5,6,7,14,19
293	42[b]	2,8,15
298	42.7[b]	10,18,15
323	47[c]	4,8,10
373	56[c]	4,6
473	76[c]	4,7,11
573	99[c]	9,11

[a] eutectic point.
[b] average value of data from cited references.
[c] interpolated from data of cited references.

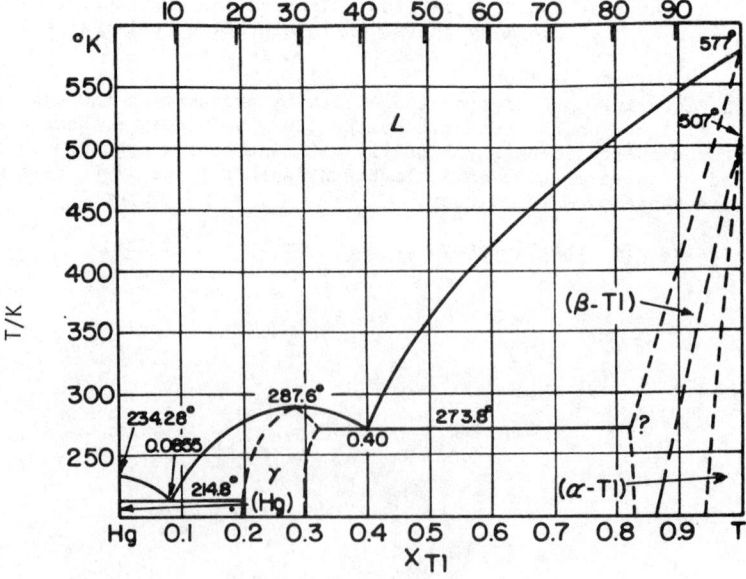

Fig. 1. Hg-Tl system (21)

(Continued next page)

COMPONENTS:	EVALUATOR:
(1) Thallium; Tl; [7440-28-0] (2) Mercury; Hg; [7439-97-6]	C. Guminski; Z. Galus Department of Chemistry University of Warsaw Warsaw, Poland July, 1985

CRITICAL EVALUATION: (continued)

References

1. Tammann, G. *Z. Phys. Chem.* 1889, *3*, 443.
2. Spencer, J. F. *Z. Elektrochem.* 1905, *11*, 681.
3. Sucheni, A. *Z. Elektrochem.* 1906, *12*, 726.
4. Kurnakov, N.S.; Pushin, N.A. *Zh. Russ. Fiz. Khim. Obshch., Ser. Khim.* 1901, *33*, 565; *Z. Anorg. Chem.* 1902, *30*, 86.
5. Pushin, N.A. *Bull. Soc. Chim.*, Belgrade, 1949, *14*, 101.
6. Pavlovich, P. *Zh. Russ. Fiz. Khim. Obshch., Ser Khim.* 1915, *47*, 29.
7. Roos, G.D. *Z. Anorg. Chem.* 1916, *94*, 358.
8. Richards, T.W.; Daniels, F. *J. Am. Chem. Soc.* 1919, *41*, 1732.
9. Richards, T.W.; Smyth, C.P. *J. Am. Chem. Soc.* 1922, *44*, 524.
10. Kozin, L.F. *Tr. Inst. Khim. Nauk Akad. Nauk Kaz. SSR* 1962, *9*, 71, 81, 93.
11. Claire, Y.; Rey, J. *J. Less-Common Metals* 1980, *70*, 33.
12. Moser, H. *Phys. Z.* 1936, *37*, 885.
13. Siede, B. *Metall* 1963, *17*, 1031.
14. Schulz, L.G.; Spiegler, P. *Trans. Metall. Soc. AIME* 1959, *215*, 87.
15. Kanda, F.A. *U.S. At. Energ. Comm. Rep.*, NYO-2731-4, 1967.
16. Strachan, J.F.; Harris, N.L. *J. Inst. Metals* 1956-57, *85*, 17.
17. Kozin, L.F. *Fiziko-Khimicheskie Osnovy Amalgamnoi Metallurgii*, Nauka, Alma-Ata, 1964.
18. Zebreva, A.I.; Filippova, L.M.; Omarova, N.D.; Gayfullin, A.Sh. *Izv. Vyssh. Ucheb. Zaved., Khim. Khim. Tekhnol.* 1976, *19*, 1043.
19. Richter, P.W.; Pistorius, C.W.F.T. *J. Less-Common Metals* 1972, *29*, 217.
20. Richter, P.W.; Pistorius, C.W.F.T. *Acta Met.* 1973, *21*, 391.
21. Hultgren, R.; Desai, P.D.; Hawkins, D.T.; Gleiser, M.; Kelley, K.K. *Selected Values of the Thermodynamic Properties of Binary Alloys*, Am. Soc. Met., Metals Park, OH, 1973, p. 990.

COMPONENTS:	ORIGINAL MEASUREMENTS:
(1) Thallium; Tl; [7440-28-0] (2) Mercury; Hg; [7439-97-6]	Tammann, G. Z. Phys. Chem. 1889, 3, 443-9.

VARIABLES:	PREPARED BY:
Temperature: 234 K	C. Guminski; Z. Galus

EXPERIMENTAL VALUES:

Depression of the melting point of mercury, ΔT, was determined after addition of small amounts of thallium:

$\Delta T/K$	mass % Tl	at % Tl[a]
0.01	0.034	0.034
0.18	0.079	0.078
0.30	0.143	0.141
0.35	0.226	0.222
0.62	0.395	0.388
0.81	0.480	0.469

[a] by compilers.

AUXILIARY INFORMATION

METHOD/APPARATUS/PROCEDURE:	SOURCE AND PURITY OF MATERIALS:
The melting temperature were determined, but absolute values not given. Details of experiment not specified, therefore, compilers assume that $\Delta T/K$ in the above table is based on the melting point of Hg of 234.28 K (1).	Nothing specified.

	ESTIMATED ERROR:
	Soly: nothing specified. Temp: precision ± 0.05 K.

	REFERENCES:
	1. Hultgren, R.; Desai, P.D.; Hawkins, D.T.; Gleiser, M.; Kelley, K.K. *Selected Values of the Thermodynamic Properties of Binary Alloys*, Am. Soc. Met., Metals Park, OH, 1973, p. 990.

COMPONENTS:	ORIGINAL MEASUREMENTS:
(1) Thallium; Tl; [7440-28-0] (2) Mercury; Hg; [7439-97-6]	Pavlovich, P. *Zh. Russ. Fiz. Khim. Obshch. Ser. Khim.* <u>1915</u>, *47*, 29-46.
VARIABLES:	PREPARED BY:
Temperature: (-60)-297°C	C. Guminski; Z. Galus

EXPERIMENTAL VALUES:

Crystallization temperatures of saturated thallium amalgams were reported for two series of measurements.

Series I

t/°C	at % Tl	t/°C	at % Tl	t/°C	at % Tl
-40	1.0	15	28.8	68	50.7
-48	4.9	15	29.1	116	58.9
-60	8.0	14	31.7	155	68.4
-16	14.6	13.5	32.8	221	82.8
+ 1	18.1	12	33.8	261	90.8
4	19.9	7	38.0	276	95.0
8	21.0	6	39.2	285	97.0
11	23.5	29	44.2	297	99.0
14	25.0				

Series II

t/°C	at % Tl	t/°C	at % Tl
13.5	25.8	13.9	32.4
14.6	27.6	13.5	33.0
14.8	28.7	12.9	33.7
14.8	29.7	12.0	34.7
14.6	30.7	11.0	35.6
14.3	31.6	9.8	36.5

AUXILIARY INFORMATION

METHOD/APPARATUS/PROCEDURE:	SOURCE AND PURITY OF MATERIALS:
The amalgams were obtained by mixing the two metals, with heating if needed, and the cooling curves were recorded with the use of thermoelement. For 0-20% Tl, the heating curves also were recorded. The alloys were protected against oxidation with vaseline.	Pure thallium from Kahlbaum. Mercury purity not specified.
	ESTIMATED ERROR: Soly: precision ± 1%. Temp: precision ± 1 K.
	REFERENCES:

COMPONENTS:	ORIGINAL MEASUREMENTS:
(1) Thallium; Tl; [7440-28-0] (2) Mercury; Hg; [7439-97-6]	Roos, G.D. Z. Anorg. Chem. 1916, 94, 358-70.
VARIABLES: Temperature: (-59)-261°C	PREPARED BY: C. Guminski; Z. Galus

EXPERIMENTAL VALUES:

Crystallization temperatures of thallium amalgams determined in four series:

I. Kahlbaum Tl under CO_2 atmosphere

$t/°C$	at % Tl	$t/°C$	at % Tl	$t/°C$	at % Tl	$t/°C$	at % Tl
-43.4	2.43	-28.4	12.06	14.40	18.90	5.00	37.90
-46.5	4.2	-10.0	14.9	14.22	29.90	0.62	40.0
-47.0	5.4	11.5	23.0	14.14	30.20	75.5	50.68
-53.0	7.0	13.20	25.1	13.66	31.20	138.0	62.65
-59.0	8.56	13.40	25.9	12.75	32.50	183.5	72.24
-51.0	9.1	14.05	27.10	11.95	33.33	222.0	81.54
-45.8	9.8	14.37	28.10	9.90	35.00	261.5	90.31
-38.4	10.5						

II. Kahlbaum Tl under petroleum

$t/°C$	at % Tl	$t/°C$	at % Tl
14.0	41.25	12.85	32.3
2.4	40.5	13.75	31.2
4.40	38.2	14.30	29.8
8.50	36.0	14.45	28.4
10.90	34.1	14.30	27.6

III. Thöl Tl in CO_2 atmosphere

$t/°C$	at % Tl	$t/°C$	at % Tl
9.5	35.2	14.0	29.9
11.8	33.33	14.18	29.1
12.6	32.1	14.25	28.3
13.2	31.45	14.10	27.7
13.58	30.8		

IV. Electrolytic Tl in CO_2 atmosphere

$t/°C$	at % Tl	$t/°C$	at % Tl
13.40	31.70	14.40	27.36
14.41	29.50	13.95	26.25
14.53	28.56	12.55	24.18

Author found that Thöl Tl contained small amounts of Pb, resulting in decreased M.P. for Tl_2Hg_5. Therefore, results with Kahlbaum and electrolytic Tl were recommended.

AUXILIARY INFORMATION

METHOD/APPARATUS/PROCEDURE:	SOURCE AND PURITY OF MATERIALS:
Weighed quantities of the metals were mixed and cooling curves were determined with either a mercury thermometer or thermocouples. The amalgams were protected against oxidation with either petroleum or pure, dry CO_2.	Pure thallium from A. Thöl and from Kahlbaum, and electrolytically prepared by the authors.
	ESTIMATED ERROR: Soly: nothing specified. Temp: precision ± 0.01 K.
	REFERENCES:

COMPONENTS:	ORIGINAL MEASUREMENTS:
(1) Thallium; Tl; [7440-28-0] (2) Mercury; Hg; [7439-97-6]	Richards, T.W.; Daniels, F. *J. Am. Chem. Soc.* 1919, *41*, 1732-68.

VARIABLES:	PREPARED BY:
Temperature: (-6.5)-40°C	C. Guminski; Z. Galus

EXPERIMENTAL VALUES:

Freezing points of amalgams determined thermometrically:

	Series I				Series II	
$t/°C$	mass % Tl	at % Tl		$t/°C$	mass % Tl	at % Tl
+1.6	42.8	42.3		9.2	36.5	36.0
5.3	38.8	38.3		11.7	34.4	33.9
12.0	34.0	33.5		14.1	31.5	31.1
13.9	31.7	31.3		14.8	29.0	28.6
14.9	29.1	28.7		13.2	25.4	25.0
14.3	26.4	26.0		11.5	23.8	23.4
12.3	24.2	23.8		4.0	20.0	19.7
3.0	19.5	19.2				

	Series III	
$t/°C$	mass % Tl	at % Tl
+0.9	40.90	40.47
5.9	38.83	38.37
9.5	37.19	36.71
12.8	32.63	32.31
14.3	27.60	27.24
5.7	20.63	20.31
-0.9	18.27	17.97
-6.5	16.92	16.65

Freezing points determined from EMF measurements

$t/°C$	mass % Tl	at % Tl
20.00	43.3	42.8
30.00	44.5	44.0
40.00	45.8	45.3

AUXILIARY INFORMATION

METHOD/APPARATUS/PROCEDURE:	SOURCE AND PURITY OF MATERIALS:
Amalgams were prepared by mixing weighed amounts of Hg and Tl in a closed tube which contained acid of known concentration. The acid neutralized any Tl_2O on the metal, and the net Tl was determined by back-titration of the acid with standard alkali. The clean amalgam was removed from the tube under a H_2 atmosphere and used for the various measurements. The freezing points in Series I and II were made on small amounts of concentrated amalgams contained in a small glass bulb with a thermometer placed in the bulb; the freezing point was determined by plunging the bulb in cold water. Series III was determined on larger amounts of amalgam with a Beckmann freezing point apparatus. In the EMF method, the potential of the cell, $Tl(Hg)_x	Tl_2SO_4	Tl(Hg)_y$, was determined at a fixed temperature at increasing Tl concentration. At the saturation point the EMF attained constant reading.	Thallium from various sources was transformed into Tl_2SO_4, the latter recrystallized more than 3 times after contact with very pure, electrolytic Tl, then the pure Tl was prepared by electrolysis of the sulfate solution which also contained $(NH_4)_2C_2O_4$. Mercury was purified with H_2SO_4, then with $Hg_2(NO_3)_2$-HNO_3 mixture, then vacuum distilled and finally distilled under hydrogen.
	ESTIMATED ERROR: Composition: precision better than ± 0.3%. Temp: precision of thermal analysis better than ± 0.1 K; EMF: ± 0.01 K.
	REFERENCES:

COMPONENTS:	ORIGINAL MEASUREMENTS:
(1) Thallium; Tl; [7440-28-0] (2) Mercury; Hg; [7439-97-6]	Richards, T.W.; Smyth, C.P. J. Am. Chem. Soc. 1922, 44, 524-45.
VARIABLES: Temperature: 231-300°C	PREPARED BY: C. Guminski; Z. Galus

EXPERIMENTAL VALUES:

Freezing points of thallium amalgams were presented graphically; the mass % data were read from the graph by the compilers and recalculated to at %:

	Hg				Hg	
$t/°C$	mass %	at %	$t/°C$	mass %	at %
299.5	1.0	1.0	276	7.8	7.7
295.5	2.3	2.3	272	8.8	8.7
292	3.2	3.1	264.5	10.5	10.3
289.5	3.6	3.5	257.5	12.0	11.8
287.5	4.2	4.1	252.3	13.5	13.3
283.0	5.4	5.3	246.5	15.0	14.8
277.5	6.9	6.8	238.5	16.5	16.2
272.5	7.7	7.6	231.5	18.2	17.9
257.5	11.3	11.1			
244.5	14.5	14.3			
232.0	17.5	17.2			

AUXILIARY INFORMATION

METHOD/APPARATUS/PROCEDURE:	SOURCE AND PURITY OF MATERIALS:
The amalgams were prepared by mixing weighed quantities of the metals in an earthenware dish; the mixture was covered with a layer of paraffin, and the amalgam formed by gently heating the dish. Cooling curves were determined in a large glass tube with a thermometer inserted into the amalgam. The amalgams were analyzed by decomposing with standardized acid and back titration of the acid with standard alkali.	Crude Tl was purified by treatment with dil. H_2SO_4, filtered, and TlCl precipitated from the filtrate with dil. HCl. The TlCl was converted to the sulfate and recrystallized at least twice. Tl was electrolytically prepared as a sponge from the aqueous sulfate solution, then fused and filtered through a capillary as bright metal. Hg was purified with $Hg_2(NO_3)_2$, then distilled.
	ESTIMATED ERROR: Temp: precision $\pm$ 0.1 K. Precision of chemical analysis: $\pm$ 0.2%.
	REFERENCES:

COMPONENTS:	ORIGINAL MEASUREMENTS:
(1) Thallium; Tl; [7440-28-0] (2) Mercury; Hg; [7439-97-6]	Pushin, N.A. *Bull. Soc. Chim.*, Belgrade, 1949, *14*, 101-3.
VARIABLES: Temperature: 2.6-15°C	PREPARED BY: C. Guminski; Z. Galus

EXPERIMENTAL VALUES:

Crystallization temperatures of thallium amalgams were reported:

t/°C	at % Hg	t/°C	at % Hg
2.6	60.5	14.5	71.4
4.3	61.4	14.4	72.2
6.9	62.8	14.3	72.8
8.8	64.0	14.0	73.6
10.2	65.0	13.4	74.6
11.7	66.2	12.5	75.5
12.8	67.5	11.5	76.4
13.8	68.8	9.1	77.9
14.2	69.8	7.8	78.6
14.4	70.3	5.2	79.7
14.5	71.0	2.6	80.9

AUXILIARY INFORMATION

METHOD/APPARATUS/PROCEDURE:	SOURCE AND PURITY OF MATERIALS:
Thermal analysis was utilized to determine the crystallization temperatures, but experimental details were not given. The method was probably similar to, or an improved version of, that in (1).	Nothing specified.
	ESTIMATED ERROR: Soly: nothing specified. Temp: nothing specified; probably ± 0.1 K (compilers).
	REFERENCES: 1. Kurnakov, N.S.; Pushin, N.A. *Z. Anorg. Chem.* 1902, *30*, 86.

COMPONENTS:	ORIGINAL MEASUREMENTS:
(1) Thallium; Tl; [7440-28-0] (2) Mercury; Hg; [7439-97-6]	Schulz, L.G.; Spiegler, P. *Trans. Metall. Soc. AIME* 1959, *215*, 87-90.
VARIABLES: One temperature	PREPARED BY: C. Guminski; Z. Galus

EXPERIMENTAL VALUES:

The melting point of the congruently melting compound, Tl_2Hg_5, was confirmed to be 14.5°C. The solubility of Tl at this temperature, therefore, is 28.6 at % (compilers).

AUXILIARY INFORMATION

METHOD/APPARATUS/PROCEDURE:	SOURCE AND PURITY OF MATERIALS:
The alloys of composition, 28.6 ± 0.2 at % Tl, were prepared either by directly mixing weighed amounts of the metals in the measurement cell or by premixing the metals then loading the amalgam into the cell under vacuum. The cell consisted of two Teflon-cup reservoirs connected at the end of a capillary tube in which were placed the thin electrodes. Of several metals used for the electrodes, nickel gave the most uniform results. The specific resistivity of the amalgam was obtained by comparing the resistance of the amalgam against that of pure Hg in the same cell. The melting point of Tl_2Hg_5 was obtained by measuring the resistance of the liquid amalgam as the temperature was decreased from 24°C to lower temperatures. There was a linear decrease in resistance with decreasing temperature down to 16°C, then at temperatures below 14.5°C the resistance remained constant.	Mercury of "triply distilled quality" from Bethlehem Apparatus Co., Inc. Thallium purity not specified.
	ESTIMATED ERROR: Precision of chemical analysis: ± 1%. Temp: precision ± 0.2 K.
	REFERENCES:

COMPONENTS:	ORIGINAL MEASUREMENTS:
(1) Thallium; Tl; [7440-28-0] (2) Mercury; Hg; [7439-97-6]	Kozin, L.F. *Tr. Inst. Khim. Nauk. Akad. Nauk Kaz. SSR* <u>1962</u>, *9*, 71-80.
VARIABLES: Temperature: 25-80°C	PREPARED BY: C. Guminski; Z. Galus

EXPERIMENTAL VALUES:

Solubility of thallium in mercury:

$t/°C$	Soly/at %
25	42.6
40	46.7
60	49.8
80	53.2

Measurements at 5 and 15°C also were made, but results were practically identical to that at 25°C.

AUXILIARY INFORMATION

METHOD/APPARATUS/PROCEDURE:	SOURCE AND PURITY OF MATERIALS:
The amalgams were obtained by dissolution of thallium in mercury. The potentials of the cell, $Tl(Hg)_x	TlClO_4$ (0.1 mol dm^{-3}) + $NaClO_4$ (0.9 mol dm^{-3})$| NaCl, Hg_2Cl_2, Hg$ were determined. Amalgams were protected from oxidation by passing pure nitrogen over the cell. The saturation point corresponded to any inflection in the curve relating cell EMF to log of Tl concentration.	Salts were recrystallized twice. Mercury was purified chemically and double distilled. Thallium was 99.999% pure.
	ESTIMATED ERROR: Soly: nothing specified. Temp: precision $\pm$ 0.2 K.
	REFERENCES:

COMPONENTS:	ORIGINAL MEASUREMENTS:
(1) Thallium; Tl; [7440-28-0] (2) Mercury; Hg; [7439-97-6]	Richter, P.W.; Pistorius, C.W.F.T. *J. Less-Common Metals* 1972, 29, 217-19.
VARIABLES: Pressure	PREPARED BY: C. Guminski; Z. Galus

EXPERIMENTAL VALUES:

Melting point of Tl_2Hg_5 (28.6 at % Tl) was presented graphically as a function of pressure. Experimental points were fitted by equation,

$$t/°C = 13.7 + 3.44 \; P$$

where P is in kbar. Standard deviation was 1.3°C. The data points were read from the curve by the compilers:

P/kbar	T/K
0	286.9±0.5[a]
2.8	294.9
3.9	298.5
5.4	303.5
7.0	309.4
8.6	314.4
10.4	319.7
11.4	324.3
12.6	329.6
13.2	331.1
22.6	363.7
27.4	381.1
28.6	385.5
30.0	389.9
30.9	393.7
32.8	399.0

[a] numerical value is given for atmospheric pressure only.

AUXILIARY INFORMATION

METHOD/APPARATUS/PROCEDURE:	SOURCE AND PURITY OF MATERIALS:
The metals were weighed and thoroughly mixed at room temperature. Tl_2Hg_5 obtained was stored at 273 K under nitrogen. Pressure was generated in a piston-cylinder apparatus. The melting points were observed by means of differential thermal analysis; Chromel-Alumel thermocouples were used. The samples were contained in stainless steel or aluminum capsules with no evidence of contamination.	99.999% pure Tl from Koch-Light. Triply-distilled mercury from Johnson-Matthey & Co.
	ESTIMATED ERROR: Temp: precision ± 1 K. Pressure: precision ± 0.5 kbar.
	REFERENCES:

COMPONENTS:	ORIGINAL MEASUREMENTS:
(1) Thallium; Tl; [7440-28-0] (2) Mercury; Hg; [7439-97-6]	Richter, P.W.; Pistorius, C.W.F.T. *Acta Met.* 1973, *21*, 391-94.
VARIABLES: Pressure	PREPARED BY: C. Guminski; Z. Galus

EXPERIMENTAL VALUES:

The pressure dependence of the eutectic temperatures was determined and fitted to the equations, where P is in kbar and t in °C:

(I) $t/°C = 60.0 + 4.09 P + 0.0132 P^2$ for eutectic at 8.5 at % Tl

(II) $t/°C = 0.9 + 3.65 P$ for eutectic at 40 at % Tl.

The authors found eutectic temperatures at 1 bar to be $-60 \pm 1°C$ and $0.9 \pm 0.5°C$, respectively, at 8.5 and 40 at % Tl. There was only a very slight pressure dependence in eq. (I). The published pressure dependence of the melting points of Hg (1), Hg_5Tl_2 (2), and Tl (3) were used with the eutectic data to construct liquidus curves at various pressures, as shown in the figure. In the construction of liquidus lines it was assumed that the eutectic composition was independent of pressure.

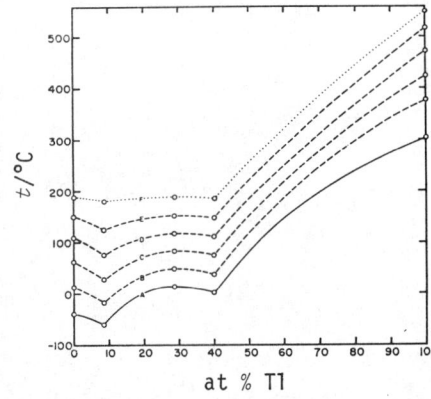

Liquidus lines in the system Hg-Tl at various pressures.
- A: atmospheric pressure;
- B: 10 kbar;
- C: 20 kbar;
- D: 30 kbar;
- E: 40 kbar;
- F: Extrapolated to 50 kbar.

AUXILIARY INFORMATION

METHOD/APPARATUS/PROCEDURE:	SOURCE AND PURITY OF MATERIALS:
Thallium and mercury in the eutectic compositions were mixed at room temperature, then stored under N_2. Samples for measurements were contained in stainless steel capsules, with no evidence of contamination. In order to prevent leakage, the pressure plate was first cooled to well below the eutectic points before pressure was applied by a piston to seal a capsule in situ. Heating and cooling rates in the differential thermal analyses were in the range of 0.4-1.1°C/sec, and temperatures were measured with a Chromel-Alumel thermocouple.	Tl: 99.999% pure from Koch-Light. Hg: triple distilled from Johnson-Matthey Co. ESTIMATED ERROR: Temp: precision $\pm$ 1 K. Pressure: precision $\pm$ 0.5 kbar. REFERENCES: 1. Klement, W.; Jayaraman, A.; Kennedy, G.C. *Phys. Rev.* 1963, *131*, 1. 2. Richter, P.W.; Pistorius, C.W.F.T. *J. Less-Common Metals* 1972, *29*, 217. 3. Adler, P.N.; Margolin, H. *Trans. Met. Soc. AIME* 1964, *230*, 1048.

COMPONENTS:	ORIGINAL MEASUREMENTS:
(1) Thallium; Tl; [7440-28-0] (2) Mercury; Hg; [7439-97-6]	Zebreva, A.I.; Filippova, L.M.; Omarova, N.D.; Gayfullin, A.Sh. *Izv. Vyssh. Ucheb. Zaved., Khim. Khim. Tekhnol.* 1976, *19*, 1043-6.
VARIABLES:	PREPARED BY:
One temperature: 25°C	C. Guminski; Z. Galus

EXPERIMENTAL VALUES:

Solubility of thallium in mercury at 25°C was reported to be 44.0 at %.

AUXILIARY INFORMATION

METHOD/APPARATUS/PROCEDURE:	SOURCE AND PURITY OF MATERIALS:
The heterogeneous thallium amalgam was prepared by mixing weighed amounts of the metals. Heat effects (Q) were recorded when subsequent portions of mercury were added. The inflection point on a plot of Q vs. amalgam concentration corresponds to the solubility of thallium in mercury.	"Pure" metals were used.
	ESTIMATED ERROR: Soly: accuracy no better than a few percent (compilers). Temp: not specified.
	REFERENCES:

COMPONENTS:	ORIGINAL MEASUREMENTS:
(1) Thallium; Tl; [7440-28-0] (2) Mercury; Hg; [7439-97-6]	Claire, Y.; Rey, J. *J. Less-Common Metals* 1980, *70*, 33-8.
VARIABLES: Temperature: 279-556 K	PREPARED BY: C. Guminski; Z. Galus

EXPERIMENTAL VALUES:

Liquidus points in the Tl-Hg system were determined:

T/K	Soly/at %	T/K	Soly/at %
279	22.5	491	79.59
282	23.2	504	83.16
282	36.1	517	86.85
279	37.2	534	90.83
273.5	40.5	551	94.62
457	72.52	556	96.82

Partial molar enthalpy and integral enthalpy of mixing are presented for various temperatures and concentrations.

AUXILIARY INFORMATION

METHOD/APPARATUS/PROCEDURE:	SOURCE AND PURITY OF MATERIALS:
The amalgams were prepared by mixing weighed portions of the metals which were contained in evacuated glass ampules. The liquidus were obtained mostly by differential thermal analysis by slowly heating the ampules followed by slow cooling. The liquidus also was determined by microcalorimetry by plotting the enthalpy of mixing against the composition at a fixed temperature; the breakpoint in the curve corresponded to the liquidus temperature, or other phase changes.	Both metals were of 99.999% purity.
	ESTIMATED ERROR: Soly: nothing specified. Temp: precision probably better than $\pm$ 1 K (compilers).
	REFERENCES:

COMPONENTS:	EVALUATOR:
(1) Carbon; C; [7440-44-0] (2) Mercury; Hg; [7439-97-6]	C. Guminski; Z. Galus Department of Chemistry University of Warsaw Warsaw, Poland July, 1985

CRITICAL EVALUATION:

There is general agreement that carbon is insoluble in mercury. However, when mercury was boiled in a carbon crucible, traces of graphite were precipitated upon cooling (1); this suggested that carbon may have a very low solubility provided that there was no mechanical fragmentation of carbon from the crucible during the experiment. On the other hand, no corrosion of carbon was observed when mercury was circulated over carbon at 719 K for 30 days (2). Because of its high melting point, the solubility of carbon in Hg should be extremely low.

Solid HgC_2 may be prepared by reaction of C_2H_2 with certain Hg compounds, but the carbide is not formed by direct contact of the elements (3).

References

1. Ruff, O.; Bergdahl, B. *Z. Anorg. Chem.* 1919, *106*, 91.
2. Nejedlik, J.F.; Vargo, E.J. *Electrochem. Technol.* 1965, *3*, 250.
3. Frad, W.A. *U.S. At. Ener. Comm. Rep.*, IS-722, 1963, p. 21.

COMPONENTS:	EVALUATOR:
(1) Silicon; Si; [7440-21-3] (2) Mercury; Hg; [7439-97-6]	C. Guminski; Z. Galus Department of Chemistry University of Warsaw Warsaw, Poland July, 1985

CRITICAL EVALUATION:

Silicon is not attacked by mercury at room and elevated temperatures (1). Strachan and Harris (2) stated that the solubility of silicon is lower than 7×10^{-3} at % at room temperature. Calculations of solubility according to equations given by Kozin give extremely low values: 7.4×10^{-46} (3) and 2.0×10^{-25} at % (4) at 298 K. However, assuming that the corrosiveness of Si is proportional to its solubility in Hg, one may surmise, after the work of Nejedlik and Vargo (5), that the solubility of Si in Hg at 719 K should be of similar order of magnitude as the solubility of vanadium in Hg at the same temperature, i.e., 10^{-5} at %.

Further experimental work is needed in this system.

References

1. Winkler, J. *J. Prakt. Chem.* 1864, *91*, 193.
2. Strachan, J.F.; Harris, N.L. *J. Inst. Metals* 1956-57, *85*, 17.
3. Kozin, L.F. *Tr. Inst. Khim. Nauk Akad. Nauk Kaz. SSR* 1962, *9*, 101.
4. Kozin, L.F. *Fiziko-Khimicheskie Osnovy Amalgamnoi Metallurgii*, Nauka, Alma-Ata, 1964.
5. Nejedlik, J.F.; Vargo, E.J. *Electrochem. Technol.* 1965, *3*, 250.

COMPONENTS:	EVALUATOR:
(1) Germanium; Ge; [7440-56-4] (2) Mercury; Hg; [7439-97-6]	C. Guminski; Z. Galus Department of Chemistry University of Warsaw Warsaw, Poland July, 1985

CRITICAL EVALUATION:

Germanium has a low solubility in mercury. Edwards (1), without giving details of his electrical resistivity measurements of germanium amalgams, reported that the solubility at 573 K is at least 0.074 at %. Strachan and Harris (2) stated that the solubility should be lower than 3×10^{-3} at % at room temperature. Stepanova and Zakharov (3,4) showed that germanium may be electrolytically introduced into mercury with the formation of supersaturated amalgams; from oxidation currents of the amalgams the solubility of germanium at 298 K was estimated by these authors to be 2.7×10^{-4} at %. This value is too high, and is rejected, as compared to more precise measurements discussed below; the error in this solubility value is connected with the graphical procedure for the solubility determination. Moreover, Karpinski and Kublik (5) showed that under experimental conditions similar to those of (3,4) some of the germanium crystals may be oxidized, thus resulting in significantly overstated values for the solubility.

Karpinski and Kublik (5) reported on an exhaustive electroanalytical study of the formation and dissolution of germanium amalgams. These authors determined the solubility at 298.2 K to be 3.0×10^{-7} at %. Gladyshev and Tember (6), by employing radioactive ^{71}Ge, found that the solubility at 293 K is 1×10^{-5} at %. In an earlier reference (7) attributed to the latter authors, the solubility at 298 K was reported to be lower than 3×10^{-6} at %; details of the experimental procedure for this radioactive isotope work were not presented. Gladyshev and coworkers reported additional polarographic measurements of germanium amalgams, as follows: 1.4×10^{-5} (8) and 1.5×10^{-5} at % (9) at 293 K, and 1×10^{-5} at % (10) at 298 K. These values may be overstated because of too short drop-times of the mercury electrode during the polarographic measurements.

Kozin's estimated solubilities of 1.3×10^{-18} (11) and 1.1×10^{-12} at % (12) at 298 K are clearly too low.

Sarieva et al (13) performed polarographic studies at 293 to 353 K and these authors reported only the upper limits of the germanium solubility in this temperature range; the solubility limits at 293 and 353 K were 4.3×10^{-5} and 9.2×10^{-4}, respectively.

The saturated germanium amalgam is in equilibrium with solid germanium (5).

The tentative solubility of germanium in mercury at 298 K is 3×10^{-7} at % (5).

References

1. Edwards, T.J. *Phil. Mag., Ser. 7*, 1926, 2, 15.
2. Strachan, J.F.; Harris, N.L. *J. Inst. Metals* 1956-57, 85, 17.
3. Stepanova, O.S.; Zakharov, M.S. *Elektrokhimiya* 1966, 2, 777.
4. Stepanova, O.S.; Zakharov, M.S. *Izv. Tomsk. Politekh. Inst.* 1966, 151, 21.
5. Karpiński, Z.J.; Kublik, Z. *J. Electroanal. Chem. Interfacial Electrochem.* 1977, 81, 53.
6. Gladyshev, V.P.; Tember, G.A. *Izv. Akad. Nauk Kaz. SSR, Ser. Khim.* 1972, No. 2, 14.
7. Tember, G.A.; Gladyshev, V.P., cited by M.T. Kozlovsky, A.I. Zebreva, V.P. Gladyshev, *Amalgamy i Ikh Primenenie*, Nauka, Alma-Ata, 1971, p. 20.
8. Gladyshev, V.P.; Syroeshkina, T.V.; Sarieva, L.S. *Izv. Vyssh. Ucheb. Zaved., Khim. Khim. Tekhnol.* 1980, 23, 936.
9. Gladyshev, V.P.; Syroeshkina, T.V.; Sarieva, L.S. *Zh. Anal. Khim.* 1979, 34, 296.
10. Gladyshev, V.P.; Kovaleva, S.V.; Sarieva, L.S. *Zh. Anal. Khim.* 1982, 37, 1762.
11. Kozin, L.F. *Tr. Inst. Khim. Nauk Akad. Nauk Kaz. SSR* 1962, 9, 101.
12. Kozin, L.F. *Fiziko-Khimicheskie Osnovy Amalgamnoi Metallurgii*, Nauka, Alma-Ata, 1964.
13. Sarieva, L.S.; Kovaleva, S.V.; Gladyshev, V.P. *Zh. Fiz. Khim.* 1984, 58, 502.

COMPONENTS:	ORIGINAL MEASUREMENTS:
(1) Germanium; Ge; [7440-56-4] (2) Mercury; Hg; [7439-97-6]	Gladyshev, V.P.; Tember, G.A. *Izv. Akad. Nauk Kaz. SSR., Ser Khim.* <u>1972</u>, No. 2, 14-21.
VARIABLES:	PREPARED BY:
One temperature: 293 K	C. Guminski; Z. Galus

EXPERIMENTAL VALUES:

Solubility of germanium in mercury at 293 K was reported to be 0.005 ± 0.002 mg/10 cm^3 Hg. The corresponding mass % and atomic % solubilities calculated by the compilers are 3×10^{-6} mass % and 1×10^{-5} at %.

AUXILIARY INFORMATION

METHOD/APPARATUS/PROCEDURE:	SOURCE AND PURITY OF MATERIALS:
Amalgams were obtained by electroreduction, on a Hg cathode, of Ge(IV) in 0.5 mol dm^{-3} H$_2$SO$_4$; the Ge also contained radioactive ^{71}Ge, and oxygen was eliminated from the solution by a stream of hydrogen. The amalgam was then transferred into another cell for solubility measurements. Based on radioactivity measurements, a set of kinetic curves of aging of the amalgam was recorded. It was assumed that Ge crystals from the amalgam should cover the Hg surface while the bulk of the amalgam was a saturated solution. Independently of initial Ge content, the final level of radioactivity of the homogeneous phase was always the same after 16 h; this suggests that the level measured corresponds to the saturated amalgam of germanium.	Mercury was purified by electrolysis then distilled from quartz apparatus. GeO$_2$ was of "semiconductor" purity. Water was distilled in a quartz apparatus. H$_2$SO$_4$ was purified by electrolysis. ESTIMATED ERROR: Soly: precision $\pm$ 40%. Temp: nothing specified. REFERENCES:

COMPONENTS:	ORIGINAL MEASUREMENTS:
(1) Germanium; Ge; [7440-56-4] (2) Mercury; Hg; [7439-97-6]	Karpinski, Z.J.; Kublik, Z. J. Electroanal. Chem. Interfacial Electrochem. 1977, 81, 53-66.
VARIABLES:	PREPARED BY:
One temperature: 25°C	C. Guminski; Z. Galus

EXPERIMENTAL VALUES:

Solubility of germanium in mercury at 25.0°C was reported to be $(2.0 \pm 0.5) \times 10^{-7}$ mol dm^{-3}. The corresponding atomic % solubility calculated by the compilers is 3.1×10^{-7} at %.

AUXILIARY INFORMATION

METHOD/APPARATUS/PROCEDURE:	SOURCE AND PURITY OF MATERIALS:
Germanium amalgams were obtained electrolytically on the hanging mercury-drop electrode from solutions of Ge(IV) of concn. 10^{-6}–10^{-5} mol dm^{-3} in a phosphate buffer at pH = 7.8. Chronoamperometric measurements: initially, reduction at -1.75 V vs. saturated Hg_2SO_4 electrode followed by pause of 15-60 seconds, then oxidation at -0.75 V. Measurements made at different Ge(IV) concentrations and the oxidation current, i_3, at -0.75 V was plotted against the time, t_2, of applied potential, -1.25 V, at which no Ge(IV) reduction current flowed. For $t_2 < 10$ min, i_3 systematically decreased with increase in t_2; for $t_2 > 10$ min i_3 was independent of t_2, indicating saturation equilibrium. Solubility was calculated from the determined diffusion coefficient of Ge in Hg and the time during oxidation.	Supporting electrolytes were prepared with analytical reagents (Ciech) and triply-distilled water, then purified with charcoal and electrolyzed at -1.7 V. GeO_2 was 99.999% pure from Fluka. Hg purified with acidified $Hg_2(NO_3)_2$ solution then distilled under reduced pressure.
	ESTIMATED ERROR: Soly: precision $\pm$ 25%. Temp: precision $\pm$ 0.2 K.
	REFERENCES:

COMPONENTS:	ORIGINAL MEASUREMENTS:
(1) Germanium; Ge; [7440-56-4] (2) Mercury; Hg; [7439-97-6]	1. Gladyshev, V.P.; Syroeshkina, L.S.; Sarieva, L.S. *Izv. Vyssh. Ucheb. Zaved., Khim. Khim. Tekhnol.* 1980, *23*, 936-9. 2. Same authors. *Zh. Anal. Khim.* 1979, *34*, 296-9. 3. Gladyshev, V.P.; Kovaleva, S.V.; Sarieva, L.S. *Zh. Anal. Khim.* 1982, *37*, 1762-6.
VARIABLES:	PREPARED BY:
Temperature: 293-298 K	C. Guminski; Z. Galus

EXPERIMENTAL VALUES:

Solubility of germanium in mercury:

T/K	Soly/mass %	Soly/at %[a]	Reference
293	$(5.0 \pm 0.5) \times 10^{-6}$	$(1.4 \pm 0.1) \times 10^{-5}$	[1]
293	5.5×10^{-6}	1.5×10^{-5}	[2]
298	3×10^{-6}	1×10^{-5}	[3]

[a] by compilers

It appears that these results may be too high because the mercury drop-times during the polarographic measurements may have been too short to reach equilibrium in the amalgam (compilers).

AUXILIARY INFORMATION

METHOD/APPARATUS/PROCEDURE:	SOURCE AND PURITY OF MATERIALS:
Ge(II) was reduced on the dropping-Hg electrode by polarography; Ge(II) concentrations were $1 \times 10^{-5} - 1 \times 10^{-2}$ mol dm^{-3}. The electrolyte was 1-10 mol dm^{-3} HCl + 0.5 mol dm^{-3} Na$_2$H$_2$PO$_2$. Argon was passed for 15 min. through the solutions to remove oxygen. The electrode process proceeded with 100% current efficiency. Stationary concentrations of germanium amalgams at various Ge(II) concentrations in the solutions were calculated coulometrically; the inflection point in the plot of peak current vs. logarithm of Ge(II) concentration indicated the saturation concentration.	GeO$_2$ and HCl were of high purity. NaH$_2$PO$_4$, chemically pure, was recrystallized. Hg was specified as "R-O" grade.
	ESTIMATED ERROR:
	Soly: precision $\pm$ 10% in (1); nothing specified in (2) and (3). Temp: nothing specified.
	REFERENCES:

COMPONENTS:	EVALUATOR:
(1) Tin; Sn; [7440-31-5] (2) Mercury; Hg; [7439-97-6]	C. Guminski; Z. Galus Department of Chemistry University of Warsaw Warsaw, Poland July, 1985

CRITICAL EVALUATION:

Tammann (1) reported the first study on the phase relationship in the Sn-Hg system. This author found that the melting point, m.p., of Hg is elevated by the addition of small amounts of Sn; the elevation of the m.p. was 2.4 K at 0.474 at % Sn. Tammann reported that the m.p. of Hg was 244 K, as compared to the true m.p. of 234.13 K; it is the opinion of the evaluators that Tammann inadvertently misstated the m.p., and that his experimental value was 234 K.

Heycock and Neville (2) studied this system in the tin-rich region and found the continuous decrease in the m.p. with addition of up to 9.29 at % Hg; the m.p. was 486.21 K at 9.29 at % Hg.

The first extensive phase diagram studies of the Sn-Hg system were reported by Pushin (3) and by Van Heteren (4). Both authors used thermal analysis to determine the crystallization temperatures over the complete composition range, and there was excellent agreement in the liquidus temperatures. The liquidus curve in Hansen's phase diagram (5) is based mainly on these data. More recent determinations of the liquidus by thermal analysis (6-11) and by EMF measurements of concentration cells (9,12,13) confirm the validity of the liquidus curve obtained by Pushin and by Van Heteren. However, Hansen's phase diagram has been revised by Hultgren et al. (14) because of the more recent determinations (11) of the compositions in the solid phases in the Sn-rich region.

The solubility of tin also has been determined at selected temperatures by chemical analyses of the equilibrium liquid phase. Gouy (15) reported the first determination of the solubility of tin near room temperature. Van Heteren (4) and Haring and White (16) obtained a solubility of 1.21 and 1.263 at %, respectively, at 298 K, while Joyner (17) found a solubility of 1.24 at % Sn at 298.6 K. Bennett and Lewis (18) found the solubility at 303 and 313 K to be 1.43 and 1.76 at %, respectively. Filippova and coworkers (19,20) determined the solubility of 1.29 at % at 298 K by calorimetry.

The solubility of gray and white tin in mercury was determined by Van Lent (21); in the temperature range of 239.6 to 273.2 K it was found that the solubility difference between these two forms of tin may be as high as 10%. The author suggests that some of the discrepancies in the previously reported solubilities in this low temperature range may be attributed to the difference in solubility between the two forms of tin.

Strachan and Harris (22) determined the solubility of 0.256 at % at room temperature. Campbell and Carter (23) reported that the solubility of tin increased from 0.28 to 3.65 at % in the temperature range of 303 to 343 K, while Shalaevskaya and coworkers (24) found that the solubility increased from 2.59 to 3.86 at % in the range of 295 to 333 K. Kozin (25) estimated a solubility of 17.02 at % at 298 K. The values from (24,25) are rejected because the solubilities are either too high or too low.

The Sn-Hg phase diagram is shown in Fig. 1 (14).

(continued next page)

COMPONENTS:	EVALUATOR:
(1) Tin; Sn; [7440-31-5] (2) Mercury; Hg; [7439-97-6]	C. Guminski; Z. Galus Department of Chemistry University of Warsaw Warsaw, Poland July, 1985

CRITICAL EVALUATION: (continued)

Recommended (r) and tentative values of the solubility of tin in mercury:

T/K	Soly/at %	Reference
238	0.23 (gray Sn)[a]	[21]
238	0.26 (white Sn)[a]	[21]
253	0.35 (gray Sn)	[21]
253	0.38 (white Sn) (r)[b]	[4,9,21]
263	0.47 (gray Sn)	[21]
263	0.49 (white Sn) (r)	[9,21]
266	0.54 (white Sn)	[21]
273	0.66 (r)	[9,13,21]
293	1.05 (r)[b]	[4,9,12,13,17]
298.2	1.26 (r)	[9,12,13,16,17,19,20]
323	2.4 (r)[b]	[9,12,13,18]
373	30 (r)	[3,4,7-10]
473	84 (r)	[3,4,7,8,10]

[a] Extrapolated value from data of cited references.
[b] Interpolated value from data of cited references.

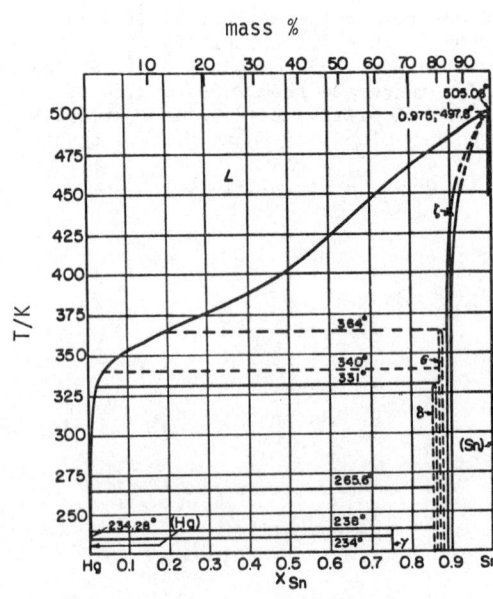

Fig. 1. Sn-Hg system (14)

COMPONENTS:	EVALUATOR:
(1) Tin; Sn; [7440-31-5] (2) Mercury; Hg; [7439-97-6]	C. Guminski; Z. Galus Department of Chemistry University of Warsaw Warsaw, Poland July, 1985

CRITICAL EVALUATION:

References

1. Tammann, G. *Z. Phys. Chem.* <u>1889</u>, *3*, 441.
2. Heycock, C.T.; Neville, F.H. *J. Chem. Soc.* <u>1890</u>, 376.
3. Pushin, N.A. *Zh. Russ. Fiz. Khim. Obshch., Ser. Khim.* <u>1902</u>, *34*, 856; *Z. Anorg. Chem.* <u>1903</u>, *36*, 201.
4. Van Heteren, W.J. *Z. Anorg. Chem.* <u>1904</u>, *42*, 129.
5. Hansen, M.; Anderko, K. *Constitution of Binary Alloys*, 2nd ed., McGraw-Hill Book Co., New York, <u>1958</u>, pp. 837-39.
6. Honda, K.; Ishigaki, T. T. *Sci. Rep. Tohoku Univer.* <u>1925</u>, *14*, 219.
7. Gayler, M.L.V. *J. Inst. Metals* <u>1937</u>, *60*, 379.
7a. Prytherch, W.E. *cited by ref. (7)*.
8. Taylor, D.F.; Burns, C.L. *J. Res. Nat. Bur. Stand.* <u>1963</u>, *67A*, 55.
9. Petot-Ervas, G.; Caillet, M.; Desrè, P. *C.R. Acad. Sci., Ser. 2*, <u>1967</u>, *264*, 490.
10. Yan-Sho-Syan, G.V.; Semibratova, N.M.; Nosek, M.V. *Tr. Inst. Khim. Nauk Akad. Nauk Kaz. SSR* <u>1969</u>, *24*, 120.
11. Predel, B.; Rothacker, D. *Acta Met.* <u>1969</u>, *17*, 783.
12. Bonnier, E.; Desrè, P.; Petot-Ervas, G. *C.R. Acad. Sci., Ser. 2*, <u>1962</u>, *255*, 2432.
13. Petot-Ervas, G.; Desrè, P.; Bonnier, E. *Bull. Soc. Chim. Fr.* <u>1967</u>, 1261.
14. Hultgren, R.; Desai, P.D.; Hawkins, D.T.; Gleiser, M.; Kelley, K.K. *Selected Values of the Thermodynamic Properties of Binary Alloys*, Am. Soc. Metals, Metals Park, OH <u>1973</u>, pp. 978-85.
15. Gouy, M. *J. phys.* <u>1895</u>, *4*, 320.
16. Haring, M.M.; White, J.C. *Trans. Electrochem. Soc.* <u>1938</u>, *73*, 211.
17. Joyner, R.A. *J. Chem. Soc.* <u>1911</u>, 195.
18. Bennett, J.A.R.; Lewis, J.B. *J. Chim. Phys.* <u>1958</u>, *55*, 83; *Am. Inst. Chem. Eng. J.* <u>1958</u>, *4*, 418.
19. Filippova, L.M.; Zebreva, A.I.; Zhumakanov, V.Z. *Izv. Vyssh. Ucheb. Zaved., Khim. Khim. Tekhnol.* <u>1982</u>, *25*, 827.
20. Filippova, L.M.; Zebreva, A.I.; Zhumakanov, V.Z. *Ukr. Khim. Zh.* <u>1981</u>, *47*, 473.
21. Van Lent, P.H. *Acta Met.* <u>1961</u>, *9*, 125.
22. Strachan, J.F.; Harris, N.L. *J. Inst. Metals* <u>1956-57</u>, *85*, 17.
23. Campbell, A.N.; Carter, H.D. *Trans. Faraday Soc.* <u>1933</u>, *29*, 1295.
24. Shalaevskaya, V.N.; Igolinskii, V.A.; Kataev, G.A. *Dep. VINITI* <u>1975</u>, 588-75; abstracted in *Zh. Fiz. Khim.* <u>1975</u>, *49*, 1587.
25. Kozin, L.F. *Fiziko-Khimicheskie Osnovy Amalgamnoi Metallurgii*, Nauka, Alma-Ata, <u>1964</u>.

COMPONENTS:	ORIGINAL MEASUREMENTS:
(1) Tin; Sn; [7440-31-5] (2) Mercury; Hg; [7439-97-6]	Tammann, G. *Z. Phys. Chem.* 1889, *3*, 441-9.
VARIABLES: Temperature	PREPARED BY: C. Guminski; Z. Galus

EXPERIMENTAL VALUES:

Elevation of the melting point of Hg, $\Delta T/K$, upon addition of tin:

$\Delta T/K$	mass %	at %[a]
0.60	0.063	0.106
1.1	0.148	0.250
2.1	0.219	0.369
2.4	0.281	0.474

[a] by compilers.

Solubilities at Sn content higher than 0.25 at % are erroneous (compilers).

The melting point of Hg was reported to be 244 instead of 234 K, but it is the opinion of the compilers that the former value was a typographical error in the original publication.

AUXILIARY INFORMATION

METHOD/APPARATUS/PROCEDURE:	SOURCE AND PURITY OF MATERIALS:
The melting temperatures were determined. No further details were given.	Nothing specified.
	ESTIMATED ERROR: Soly: nothing specified. Temp: precision better than $\pm$ 0.1 K.
	REFERENCES:

COMPONENTS:	ORIGINAL MEASUREMENTS:
(1) Tin; Sn; [7440-31-5] (2) Mercury; Hg; [7439-97-6]	Heycock, C.T.; Neville, F.H. *J. Chem. Soc.* 1890, *57*, 376-93.
VARIABLES:	PREPARED BY:
Temperature: 213-231°C	C. Guminski; Z. Galus

EXPERIMENTAL VALUES:

Crystallization temperatures of tin amalgams:

$t/°C$	at Hg/100 at Sn	at % Sn
231.4	0.0911	99.91
231.2	0.1809	99.82
230.89	0.3127	99.69
230.22	0.5889	99.41
229.05	1.078	98.93
227.53	1.7256	98.30
225.05	2.772	97.30
223.07	3.886	96.25
219.39	6.141	94.21
214.62	9.21	91.57
213.06	10.24	90.71

AUXILIARY INFORMATION

METHOD/APPARATUS/PROCEDURE:	SOURCE AND PURITY OF MATERIALS:
The experiments were performed in heavy iron blocks, and the amalgams were protected from the atmosphere by a surface layer of paraffin. The melting temperatures were determined with calibrated mercury thermometers.	Nothing specified.
	ESTIMATED ERROR: Soly: nothing specified. Temp: precision probably ± 0.05 K (compilers).
	REFERENCES:

COMPONENTS:	ORIGINAL MEASUREMENTS:
(1) Tin; Sn; [7440-31-5] (2) Mercury; Hg; [7439-97-6]	Pushin, N.A. Zh. Russ. Fiz. Khim. Obshch., Ser. Khim. 1902, 34, 856-78. Z. Anorg. Chem. 1903, 36, 201-54.
VARIABLES:	PREPARED BY:
Temperature: 25-229°C	C. Guminski; Z. Galus

EXPERIMENTAL VALUES:

Crystallization temperature of tin amalgams:

t/°C	at % Hg	t/°C	at % Hg	t/°C	at % Hg
229.4	0.7	170.5	30.9	101.5	69.2
227	1.7	166	33.2	98	71.5
224	3.0	159.2	36.2	97	73.3
221	4.8	152	40.0	93.5	74.6
218.2	6.3	140.5	45.7	88.7	80.0
215.5	7.8	132.5	50.0	81.5	87.4
211.7	10.0	122.7	54.6		
207.5	12.1	117.5	58.2		
199.7	16.2	114	60.1		
192.5	20.0	108	63.8		
185.2	23.5	105	66.7		
180	26.4	102	68.2		

AUXILIARY INFORMATION

METHOD/APPARATUS/PROCEDURE:	SOURCE AND PURITY OF MATERIALS:
The amalgams were obtained by mixing and heating the metals together. The experiments were carried out under paraffin or vaseline, and the cooling curves were recorded.	Nothing specified.
	ESTIMATED ERROR: Soly: nothing specified. Temp: precision ± 0.5 K.
	REFERENCES:

COMPONENTS:	ORIGINAL MEASUREMENTS:
(1) Tin; Sn; [7440-31-5] (2) Mercury; Hg; [7439-97-6]	Van Heteren, W.J. Z. Anorg. Chem. 1904, 42, 129-73.
VARIABLES: Temperature: (-37)-212°C	PREPARED BY: C. Guminski; Z. Galus

EXPERIMENTAL VALUES:

Liquidus temperatures of the Sn-Hg system:

t/°C	at % Sn	t/°C	at % Sn
-37.7	0.05	99.0	28.96
-36.8	0.1	102.4	31.87
-35.6	0.2	103.4	32.46
-34.35	0.3	107.4	35.33
65.2	5.17	115.2	40.27
79.7	10.79	133.4	49.99
88.4	18.11	155.2	61.44
90.0	20.37	173.0	70.31
94.0	24.53	183.7	76.62
95.4	25.23	198.55	82.84
98.75	28.45	211.6	89.95

The following solubilities of Sn in Hg were also reported:

t/°C	Soly/at %
-18.8	0.36
0	0.59
15	0.97
25	1.21

AUXILIARY INFORMATION

METHOD/APPARATUS/PROCEDURE:	SOURCE AND PURITY OF MATERIALS:
The amalgams were prepared from weighed amounts of the metals in CO_2 atmosphere. The liquid fraction of the amalgam was filtered into a separate glass tube and covered with paraffin or ricin oil. The amalgams were heated and cooling curves were recorded with the use of a recording thermometer at the higher temperatures, and with a toluol thermometer at the lower temperatures. In the solubility measurements, the amalgams were filtered through a chamois skin. The samples were weighed and analyzed; tin was probably determined gravimetrically as SnO_2 (compilers).	Mercury was twice-distilled under vacuum. Tin from Bankazinn contained traces of lead; it was melted, washed and dried before use.
	ESTIMATED ERROR: Soly: nothing specified. Temp: precision better than ± 0.5 K (compilers).
	REFERENCES:

COMPONENTS:	ORIGINAL MEASUREMENTS:
(1) Tin; Sn; [7440-31-5] (2) Mercury; Hg; [7439-97-6]	Joyner, R.A. *J. Chem. Soc.* 1911, 195-205.
VARIABLES: Temperature: 14-163°C	PREPARED BY: C. Guminski; Z. Galus

EXPERIMENTAL VALUES:

The solubility of tin in mercury:

$t/°C$	Soly/at % Sn
14.0	1.05
25.4	1.24
63.2	4.04
90.0	18.0
163.0	66.7

AUXILIARY INFORMATION

METHOD/APPARATUS/PROCEDURE:	SOURCE AND PURITY OF MATERIALS:
Amalgam was prepared by combining Sn filings with Hg in sealed tubes containing H_2, then heating the tubes in a thermostat. Liquid phase was pipetted through glass-wool filter, and weighed sample was analyzed gravimetrically for Sn as the oxide.	Commercial Sn was dissolved in HCl, and the crystallized $SnCl_2$ was treated with HNO_3 to be converted to metastannic acid. The latter was dried and reduced to Sn with coal gas or H_2. The finely divided Sn was then fused under KCN and cast into bars. Mercury purity not specified.
	ESTIMATED ERROR: Soly: precision probably no better than $\pm$ 0.5% (compilers). Temp: precision better than $\pm$ 0.1 K (compilers).
	REFERENCES:

COMPONENTS:	ORIGINAL MEASUREMENTS:
(1) Tin; Sn; [7440-31-5] (2) Mercury; Hg; [7439-97-6]	Honda, K.; Ishigaka, T. *Sci. Rep. Tohoku Univ.* 1925, *14*, 219-32.
VARIABLES: Temperature: 508 K	PREPARED BY: C. Guminski; Z. Galus

EXPERIMENTAL VALUES:

Depression of freezing point of tin amalgam containing 1 at % of mercury was determined to be 3.04 K. The melting temperature of pure tin was assumed to be 505.0 K.

AUXILIARY INFORMATION

METHOD/APPARATUS/PROCEDURE:	SOURCE AND PURITY OF MATERIALS:
The usual method of thermal analysis was used. The alloys to be tested were melted in an alundum tube. The melts were protected from oxidation with a thick layer of asbestos wool which was covered with paraffin or vaseline. Temperatures were measured with a copper-constantan thermocouple.	Metals probably were extra pure grade from Merck (compilers).
	ESTIMATED ERROR: Soly: nothing specified. Temp: precision better than $\pm$ 0.05 K.
	REFERENCES:

COMPONENTS:	ORIGINAL MEASUREMENTS:
(1) Tin; Sn; [7440-31-5] (2) Mercury; Hg; [7439-97-6]	Gayler, M.L.V. J. Inst. Metals 1937, 60, 379-406.
VARIABLES: Temperature: 75-230°C	PREPARED BY: C. Guminski; A. Galus

EXPERIMENTAL VALUES:

Gayler presented a phase diagram based mainly on the unpublished data of Prytherch (1) and of Van Heteren (2), with three points from the author's own measurements. The mass % liquidus data points were read from the curve and converted to atomic % by the compilers.

$t/°C$	mass %	at %	$t/°C$	mass %	at %	$t/°C$	mass %	at %
75	5.2	8.5	124	35.4	48.1	222	94.0	96.4
85	10.0	15.8	174	60.0	71.7	228	96.5	97.9
90	13.0	20.2	181	70.0	79.8	229	97.6	98.6
94	17.2	26.0	199	75.8	84.1	230	99.0	99.4
107	25.3	36.4	210	83.2	89.3	93-104[a]	20	30
108	27.2	38.7	215	88.0	92.5	102-112[a]	30	42
						151-157[a]	50	63

[a] Data of Gayler.

AUXILIARY INFORMATION

METHOD/APPARATUS/PROCEDURE:	SOURCE AND PURITY OF MATERIALS:
The amalgams were prepared from the pure metals. The alloys were placed in silica tubes and sealed in an atmosphere of dry hydrogen. The cooling and heating curves were recorded with the use of thermocouples. Prytherch's method is not specified in detail, but he also applied thermal analysis.	Chemically pure tin contained a trace of iron. Mercury was chemically purified and redistilled.
	ESTIMATED ERROR: Nothing specified.

	REFERENCES:
	1. Prytherch, W.E.
	   Unpublished work cited by Gayler in this paper.
	2. Van Heteren, W.J.
	   Z. Anorg. Chem. 1904, 42, 129-73.

COMPONENTS:	ORIGINAL MEASUREMENTS:
(1) Tin; Sn; [7440-31-5] (2) Mercury; Hg; [7439-97-6]	Haring, M.M.; White, J.C. *Trans. Electrochem. Soc.* 1938, *73*, 211-21.
VARIABLES:	PREPARED BY:
One temperature: 25°C	C. Guminski; Z. Galus

EXPERIMENTAL VALUES:

The solubility of tin in mercury at 25°C was reported to be 1.263 at %.

AUXILIARY INFORMATION

METHOD/APPARATUS/PROCEDURE:	SOURCE AND PURITY OF MATERIALS:
Known quantities of tin and mercury were placed in a flask with a few milliliters of 0.06 mol dm^{-3} HCl; the latter solvent was used to remove the oxide film on the tin. The flask was heated in a beaker of boiling water with shaking. The hot amalgam was then rapidly passed through two capillaries into an air-free cell through a special funnel. The double filtration in the capillaries removed any solid amalgam and traces of oxide. A known quantity of amalgam was dissolved in conc. HNO$_3$, evaporated to dryness, heated to drive off the Hg, then ignited to constant weight to determine the Sn as the oxide.	Mercury was sprayed through a column of 1.0 mol dm^{-3} HNO$_3$ and Hg$_2$(NO$_3$)$_2$, then dried, and twice distilled. Tin was prepared by electrolysis of stannous chloride in hydrochloric acid.
	ESTIMATED ERROR: Soly: precision ± 0.2%. Temp: precision ± 0.02 K.
	REFERENCES:

COMPONENTS:	ORIGINAL MEASUREMENTS:
(1) Tin; Sn; [7440-31-5] (2) Mercury; Hg; [7439-97-6]	Bennett, J.A.R.; Lewis, J.B. J. Chim. Phys. 1958, 55, 83-7. Am. Inst. Chem. Eng. J. 1958, 4, 418-22.
VARIABLES:	PREPARED BY:
Temperature: 30-40°C	C. Guminski; Z. Galus

EXPERIMENTAL VALUES:

The solubility of tin in mercury at 30 and 40°C was reported to be 0.85 and 1.05 mass %.

The corresponding atomic % solubilities calculated by the compilers are 1.43 and 1.76 at %, respectively.

AUXILIARY INFORMATION

METHOD/APPARATUS/PROCEDURE:	SOURCE AND PURITY OF MATERIALS:
The Sn amalgams were prepared by dissolution of rotating Sn cylinders in Hg. The dissolution vessel was mounted inside a glove box filled with pure argon. The amalgam samples were analyzed by distilling out the Hg at 573 K in a nitrogen atmosphere. The sample and the residue were weighed for analysis.	Metal purities were 99.99%.
	ESTIMATED ERROR: Soly: nothing specified. Temp: precision $\pm$ 0.2 K.
	REFERENCES:

COMPONENTS:	ORIGINAL MEASUREMENTS:
(1) Tin; Sn; [7440-31-5] (2) Mercury; Hg; [7439-97-6]	Van Lent, P.H. *Acta Met.* 1961, *9*, 125-28.
VARIABLES: Temperature: (-34)-0°C	PREPARED BY: C. Guminski; Z. Galus

EXPERIMENTAL VALUES:

Solubility of gray and white tin in mercury:

t/°C	Sn (gray) Soly/at %	Sn (white) Soly/at %
-33.6	0.243 ± 0.001	0.269 ± 0.002
-21.6	0.344	0.369 ± 0.002
-10.6	0.467 ± 0.004	0.492 ± 0.002
-6.55	0.566 ± 0.002	0.560 ± 0.001
0.00	0.659 ± 0.003	0.656

AUXILIARY INFORMATION

METHOD/APPARATUS/PROCEDURE:	SOURCE AND PURITY OF MATERIALS:
Amalgams were prepared by adding Sn powder to Hg which was contained in a stoppered tube. The gray Sn amalgam was prepared at -40°C, then stored at -20°C for 12 hours. The white Sn amalgam was prepared and stored at room temperature. The equilibrations were made by suspending the amalgam tubes in a dewar tube which contained various salt-water eutectic mixtures. The tubes were vigorously agitated for 8 hours, then 40 g of the amalgam solution was removed, and the Sn was determined gravimetrically as SnO_2.	Mercury was purified by air oxidation of impurities and vacuum distilled. Tin purity not specified.
	ESTIMATED ERROR: Soly: precision better than ± 1%. Temp: precision better than ± 0.1 K.
	REFERENCES:

COMPONENTS:	ORIGINAL MEASUREMENTS:
(1) Tin; Sn; [7440-31-5] (2) Mercury; Hg; [7439-97-6]	Taylor, D.F.; Burns, C.L. J. Res. Nat. Bur. Stand. 1963, 67A, 55-70.
VARIABLES: Temperature: 99-230°C	PREPARED BY: C. Guminski; Z. Galus

EXPERIMENTAL VALUES:

Liquidus temperatures of mercury-tin alloys:

$t/°C$	at % Sn	$t/°C$	at % Sn
231.9	100.0	204.0	87.11
230.1	98.81	203.2	85.72
222.9	96.97	199.5	84.29
219.3	95.74	197.5	82.04
218.4	93.83	191.6	79.85
216.0	92.53	176.1	71.72
214.4	91.22	157.5	62.83
212.8	89.86	139.0	52.98
208.3	88.53	118.9	42.01
208.4	88.50	99.0	29.70

AUXILIARY INFORMATION

METHOD/APPARATUS/PROCEDURE:	SOURCE AND PURITY OF MATERIALS:
Weighed amounts of Sn and Hg were sealed in Pyrex tubes provided with reentrant thermocouple wells. Before sealing, the tubes were repeatedly evacuated and flushed with dry H_2 and finally sealed with a residual H_2 pressure of 2-5 torr. The alloys were homogenized by heating to 250°C and holding for at least 1 hour, then quenched in water at 20-25°C. Heating and cooling curves were recorded as soon as possible after annealing by measuring the temperature of the alloy, and the differential temperature of the alloy vs. pure Hg. A minimum of six heating-cooling runs were made on each composition. Tin-rich Alloys Constant temperature diffusion followed by serial sectioning and analyses were carried out to identify the various phases and their compositions in high Sn alloys (max. $t/°C$ = 110). Hg analysis was by modification of that of Crawford and Larson (1): known weight of sample was heated in vacuum at 500°C and Hg determined by the weight loss. X-ray diffraction studies on these alloys also reported.	Refined Hg from N.B.S. contained <1.1 mg/kg metallic impurity. Baker and Adamson reagent grade tin sticks and tin from Consolidated Mining and Smelting Company of Canada Limited were used. Analyses of tin specimens were given.
	ESTIMATED ERROR: Soly: nothing specified. Temp: precision ± 0.5 K.
	REFERENCES: 1. Crawford, W.H.; Larson, J.H. J. Dental Research 1955, 34, 313.

COMPONENTS:	ORIGINAL MEASUREMENTS:
(1) Tin; Sn; [7440-31-5] (2) Mercury; Hg; [7439-97-6]	Petot-Ervas, G.; Caillet, M.; Desrè, P. *C.R. Acad. Sci., Ser. 2* 1967, *264*, 490-3.
VARIABLES:	PREPARED BY:
Temperature: (-35)-192°C	C. Guminski; Z. Galus

EXPERIMENTAL VALUES:

Solubility of tin in mercury at various temperatures; data in first four columns by EMF measurements, and last two columns by thermal analysis:

$t/°C$	Soly/at %	$t/°C$	Soly/at %	$t/°C$	Soly/at %
54	2.5	-35.4	0.16	79	9
61	3.0	-28.4	0.29	147	57
67.5	4.0	-17.9	0.41	192	80
70	5.0	-8.4	0.52		
78	8.0	1.1	0.65		
85	15	16.5[a]	0.97±0.02		
92	20	26[a]	1.27±0.02		
103	30	30[a]	1.40±0.03		
108.5	35	40[a]	1.88±0.02		
113.5	40	50[a]	2.59±0.04		
123	45	60[a]	3.34±0.02		
129	50	72[a]	5.6±0.5		
142.5	55				

[a] Previously published in refs. (1) and (2).

AUXILIARY INFORMATION

METHOD/APPARATUS/PROCEDURE:	SOURCE AND PURITY OF MATERIALS:
Measurements of the EMF of the cell, $Sn|Sn(II)|Sn(Hg)$, were performed in an argon atmosphere. At temperatures below 200°C the electrolytes were $SnCl_2$-NH_4Cl and $SnCl_2$-LiCl in water or glycerine. At temperatures above 200°C EMF measurements were made by using the molten electrolyte, $SnCl_2$-$ZnCl_2$-KCl-LiCl. Solubility corresponds to breakpoint of EMF vs. log (concentration). Method of thermal analysis is not described in detail.	Not specified.
	ESTIMATED ERROR: Soly: precision better than ± 2%. Temp: precision ± 0.1 K for EMF measurements.
	REFERENCES: 1. Bonnier, E.; Desrè, P.; Petot-Ervas, G. *C.R. Acad. Sci., Ser. 2* 1962, *255*, 2432-4. 2. Petot-Ervas, G.; Desrè, P.; Bonnier, E. *Bull. Soc. Chim. Fr.* 1967, 1261-4.

COMPONENTS:	ORIGINAL MEASUREMENTS:
(1) Tin; Sn; [7440-31-5] (2) Mercury; Hg; [7439-97-6]	Yan-Sho-Syan, G.V.; Semibratova, N.M.; Nosek, M.V. *Tr. Inst. Khim. Nauk Akad. Nauk Kaz. SSR* 1969, *24*, 120-7.

VARIABLES:	PREPARED BY:
Temperature: 70-215°C	C. Guminski; Z. Galus

EXPERIMENTAL VALUES:

Liquidus temperatures of tin-mercury alloys:

$t/°C$	Soly/at %
70	10.0
84	20.0
102	30.0
115	40.0
130	50.0
150	60.0
173	70.0
182	75.0
193	80.0
201	85.0
215	90.0

AUXILIARY INFORMATION

METHOD/APPARATUS/PROCEDURE:	SOURCE AND PURITY OF MATERIALS:
Thermal analysis was used in the determination of liquidus temperatures. The procedure was probably the same as described in (1).	Mercury was chemically purified and then twice-distilled under vacuum. Tin purity was 99.9998%.
	ESTIMATED ERROR: Soly: nothing specified. Temp: precision ± 1 K.
	REFERENCES: 1. Yan-Sho-Syan, G.V.; Nosek, M.V.; Semibratova, N.M.; Shalamov, A.E. *Tr. Inst. Khim. Nauk Akad. Nauk Kaz. SSR* 1967, *15*, 139-49.

COMPONENTS:	ORIGINAL MEASUREMENTS:
(1) Tin; Sn; [7440-31-5] (2) Mercury; Hg; [7439-97-6]	Predel, B.; Rothacker, D. *Acta Met.* 1969, *17*, 783-91.
VARIABLES: Temperature: 209-230°C	PREPARED BY: C. Guminski; Z. Galus

EXPERIMENTAL VALUES:

The authors present a revised phase diagram for the composition range of 87.5-100 at %
Sn. The solubilities were read from the liquidus line by the compilers:

t/°C	Soly/at %	t/°C	Soly/at %
230.3	99.3	219.7	94.5
228.4	98.8	218.6	93.7
226.6	98.1	217.6	93.4
224.7	97.5	215.4	91.9
221.8	96.4	213.6	91.5
221.6	96.0	212.1	90.6
221.2	95.5	211.4	90.1
220.9	95.1	210.0	89.1
222.8	97.0	209.5	88.5

AUXILIARY INFORMATION

METHOD/APPARATUS/PROCEDURE:	SOURCE AND PURITY OF MATERIALS:
The amalgams were prepared from weighed amounts of the metals, then differential thermal analysis curves were recorded to determine the liquidus points.	Tin was 99.999% pure. Mercury was 99.9995% pure.
	ESTIMATED ERROR: Nothing specified.
	REFERENCES:

COMPONENTS:	ORIGINAL MEASUREMENTS:
(1) Tin; Sn; [7440-31-5] (2) Mercury; Hg; [7439-97-6]	Filippova, L.M.; Zebreva, A.I.; Zhumakanov, V.Z. *Ukr. Khim. Zh.* 1981, *47*, 473-6. *Izv. Vyssh. Ucheb. Zaved., Khim. Khim. Tekhnol.* 1982, *25*, 827-9.
VARIABLES:	PREPARED BY:
One temperature: 298 K	C. Guminski; Z. Galus

EXPERIMENTAL VALUES:

The solubility of tin in mercury at 298 K was reported to be 0.87 ± 0.06 mol dm^{-3}. The corresponding atomic % solubility calculated by the compilers is 1.29 at %.

AUXILIARY INFORMATION

METHOD/APPARATUS/PROCEDURE:	SOURCE AND PURITY OF MATERIALS:
The heterogeneous amalgam was prepared by mixing weighed amounts of the metals. Heat effect, Q, was measured directly during thermometric titration when subsequent portions of mercury were added to the amalgam. The inflection point on the plot of Q vs. amalgam concentration of tin corresponded to the solubility of tin in mercury. Experiments were performed in argon atmosphere.	Source and purity of Sn and Hg not specified. Argon was of "A" class purity.
	ESTIMATED ERROR: Soly: precision $\pm$ 7%. Temp: not specified.
	REFERENCES:

COMPONENTS:	EVALUATOR:
(1) Lead; Pb; [7439-92-1] (2) Mercury; Hg; [7439-97-6]	C. Guminski; Z. Galus Department of Chemistry University of Warsaw Warsaw, Poland July, 1985

CRITICAL EVALUATION:

Tammann (1) was the first to report on the solubility of lead in mercury by determining the freezing point upon addition of small quantities of lead to mercury. At a lead concentration of 0.347 at % he found that the melting point of mercury was elevated by 1.30 K.

Pushin (2) and Jänecke (3) determined the crystallization temperatures of lead amalgams over nearly the complete composition range with good agreement. The major portion of the liquidus for the phase diagram (4) of this system, Fig. 1, is based upon the data of these authors. Yan-Sho-Syan and coworkers (5) performed exhaustive thermographic experiments and their liquidus line in the composition range of 0-65 at % Pb differs significantly from that of the previous (2-4) and some subsequent results.

A number of workers have reported solubility determinations over narrow composition ranges, especially for those near room temperature. Thompson (6) employed a filtration method to obtain a solubility in the temperature range of 293-342 K, and the interpolation of his data yields a solubility of 1.63 at % at 298 K. This solubility is in good agreement with the carefully determined value of 1.65 at % which was obtained by Haring et al. (7) from EMF measurements. Filippova et al. (8,9), from thermometric titration, also determined a solubility of 1.65 at % at 298 K. These three determinations at 298 K are considered to be the most accurate at this temperature. The solubility of 1.42 at % determined by EMF measurements (10) is considered too low by the evaluators.

Gouy (11) reported a lead solubility of 1.3 at % at 288-291 K, while Jangg and Kirchmayr (12) determined a solubility of 1.35 at % at 288 K. The latter value is in good agreement with the extrapolated data of (6). Moshkevich and Ravdel (13) determined the solubility of lead in the Hg-rich region, at 237-323 K, by observing the decrease in weight of a lead disc which was rotated in a known quantity of mercury. These authors' results were in good agreement with the acceptable solubilities reported by other workers (6-9,12).

There have been other reports of the solubility in the Hg-rich region, but these solubilities are rejected because they are either too low (1.05 at % at room temperature (14), 0.99 at % at 293 K (15), 1.16 at % at 291 K (16), and a set of points on the liquidus line shifted down to 303 K in the 64-95 at % range (17)), or too high (1.9 at % in the temperature range of 273-302 K (18) and higher than 1.00 at % at 273 K (19)). Kozin's estimate (20) of the 298 K solubility of lead in mercury, 26.9 at %, is clearly too high.

Heycock and Neville (21) determined the solubility in the Pb-rich region by observing the freezing point depression of lead by addition of up to 6.08 at % of mercury. Ishigaki and Honda (22) similarly determined the freezing point depression of lead upon addition of 1.0 and 2.0 at % Hg. The results of the measurements by both groups of authors agreed with the data of Pushin (2), and Jänecke (3).

As shown in Fig. 1 (4), the saturated amalgams are in equilibrium with either Pb or Pb_2Hg. However, the phase diagram is not yet completely clear.

(continued next page)

COMPONENTS:	EVALUATOR:
(1) Lead; Pb; [7439-92-1] (2) Mercury; Hg; [7439-97-6]	C. Guminski; Z. Galus Department of Chemistry University of Warsaw Warsaw, Poland July, 1985

CRITICAL EVALUATION: (continued)

Recommended (r) and tentative values of the solubility of lead:

T/K	Soly/at %	Reference
237	0.44	[13]
258	0.73	[13]
273	0.96	[13]
293.2	1.47 (r)[b]	[6,13]
298.2	1.63 (r)[a]	[7-9,13]
323	2.7 (r)[a]	[6,13]
373	13[a]	[2,3,5]
473	63[a]	[2,5]
573	93 (r)[b]	[2,3,5,21]

[a] mean value from cited references.
[b] Interpolated value from data of cited references.

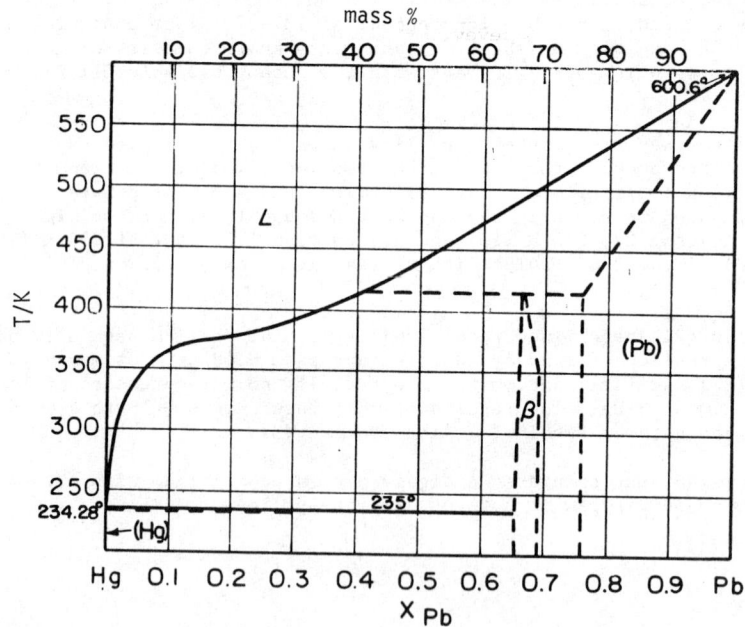

Fig. 1. The Pb-Hg system (4)

(continued next page)

COMPONENTS:	EVALUATOR:
(1) Lead; Pb; [7439-92-1] (2) Mercury; Hg; [7439-97-6]	C. Guminski; Z. Galus Department of Chemistry University of Warsaw Warsaw, Poland July, 1985

CRITICAL EVALUATION: (continued)

References

1. Tammann, G. *Z. Phys. Chem.* 1889, *3*, 441.
2. Pushin, N. *Zh. Russ. Fiz. Khim. Obshch., Ser. Khim.* 1902, *34*, 856; *Z. Anorg. Chem.* 1903, *36*, 201.
3. Jänecke, E. *Z. Phys. Chem.* 1907, *60*, 400.
4. Hultgren, R.; Desai, P.D.; Hawkins, D.T.; Gleiser, M.; Kelley, K.K. *Selected Values of the Thermodynamic Properties of Binary Alloys*, Am. Soc. Metals, Metals Park, OH 1973, p. 971.
5. Yan-Sho-Syan, G.V.; Nosek, M.V.; Semibratova, N.M.; Shalamov, A.E. *Tr. Inst. Khim. Nauk Akad. Nauk Kaz. SSR* 1967, *15*, 139.
6. Thompson, H.E. *J. Phys. Chem.* 1935, *39*, 655.
7. Haring, M.M.; Hatfield, M.R.; Zapponi, P.P. *Trans. Electrochem. Soc.* 1939, *75*, 473.
8. Filippova, L.M.; Gayfullin, A.Sh.; Zebreva. A.I. *Prikl. Teor. Khim.*, Alma-Ata 1974, No. 5, 76.
9. Filippova, L.M.; Zebreva, A.I.; Korobkina, N.P. *Ukr. Khim. Zh.* 1978, *44*, 791.
10. Hoyt, C.S.; Stegman, G. *J. Phys. Chem.* 1934, *38*, 753.
11. Gouy, M. *J. Phys.* 1895, *4*, 320.
12. Jangg, G.; Kirchmayr, H. *Z. Chem.* 1963, *3*, 47.
13. Moshkevich, A.S.; Ravdel, A.A. *Zh. Prikl. Khim.* 1970, *43*, 71.
14. Strachan, J.F.; Harris, N.L. *J. Inst. Metals* 1956-57, *85*, 17.
15. Nigmatullina, A.A.; Zebreva, A.I. *Izv. Akad. Nauk Kaz. SSR, Ser. Khim.* 1965, *15*, No. 2, 20.
16. Spencer, J.F. *Z. Elektrochem.* 1905, *11*, 683.
17. Fay, H.; North, E. *Am. Chem. J.* 1901, *25*, 216.
18. Babinski, J.J.; cited by G. Timofeyev, *Z. Phys. Chem.* 1912, *78*, 304.
19. Richards, T.W.; Garrod-Thomas, R.N. *Z. Phys. Chem.* 1910, *72*, 165.
20. Kozin, L.F. *Fiziko-Khimicheskie Osnovy Amalgamnoi Metallurgii*, Nauka, Alma-Ata, 1964.
21. Heycock, C.T.; Neville, F.H. *J. Chem. Soc.* 1892, *61*, 888.
22. Honda, K.; Ishigaki, T. *Sci. Rep. Tohoku Univer.* 1925, *14*, 219.

COMPONENTS:	ORIGINAL MEASUREMENTS:
(1) Lead; Pb; [7439-92-1] (2) Mercury; Hg; [7439-97-6]	Tammann, G. *Z. Phys. Chem.* 1889, *3*, 441-9.
VARIABLES: Temperature	PREPARED BY: C. Guminski; Z. Galus

EXPERIMENTAL VALUES:

Elevation of the melting point of mercury, $\Delta T/K$, upon addition of small amounts of lead:

$\Delta T/K$	g Pb/100 g Hg	at % Pb[a]
−0.02	0.015	0.015
+0.027	0.070	0.068
+0.37	0.172	0.166
+0.89	0.247	0.239
+1.24	0.333	0.322
+1.30	0.359	0.347

[a] by compilers.

The melting point of Hg was reported to be 244 instead of 234 K, but it is the opinion of the compilers that the former value was a typographical error in the original publication.

AUXILIARY INFORMATION

METHOD/APPARATUS/PROCEDURE:	SOURCE AND PURITY OF MATERIALS:
The melting temperatures of the amalgams were determined. No further details were given.	Nothing specified.
	ESTIMATED ERROR: Soly: nothing specified. Temp: precision ± 0.05 K.
	REFERENCES:

COMPONENTS:	ORIGINAL MEASUREMENTS:
(1) Lead; Pb; [7439-92-1] (2) Mercury; Hg; [7439-97-6]	Heycock, T.C.; Neville, F.H. *J. Chem. Soc.* 1892, *61*, 888-914.
VARIABLES:	PREPARED BY:
Temperature: 304-323°C	C. Guminski; Z. Galus

EXPERIMENTAL VALUES:

Freezing points of lead amalgams:

t/°C	atom Hg/100 at Pb	at % Hg[a]
323.89	0.729	0.724
315.48	3.29	3.18
304.69	6.74	6.08

[a] by compilers.

AUXILIARY INFORMATION

METHOD/APPARATUS/PROCEDURE:	SOURCE AND PURITY OF MATERIALS:
Weighed quantities of the metals were placed in a hard glass tube then evacuated prior to sealing. The tube was heated to a red heat and well shaken. Temperatures of crystallization were measured with calibrated thermometers.	Not specified.
	ESTIMATED ERROR: Soly: nothing specified. Temp: precision probably better than $\pm$ 0.05 K (compilers).
	REFERENCES:

COMPONENTS:	ORIGINAL MEASUREMENTS:
(1) Lead; Pb; [7439-92-1] (2) Mercury; Hg; [7439-97-6]	Pushin, N.A. *Zh. Russ. Fiz. Khim. Obshch., Ser. Khim.* 1902, *34*, 856-78. *Z. Anorg. Chem.* 1903, *36*, 201-54.
VARIABLES: Temperature: 23-318°C	PREPARED BY: C. Guminski; Z. Galus

EXPERIMENTAL VALUES:

Crystallization temperatures of lead amalgams:

t/°C	at % Hg	t/°C	at % Hg	t/°C	at % Hg
318.5	2.6	189.5	40.6	116.75	70.8
305.25	6.3	179	44.1	110.5	75.0
288	11.0	∼174	46.2	104.5	79.9
267.5	16.6	162.5	50.0	101	83.0
247	22.6	155.5	52.6	96.75	86.4
241	24.5	149.5	54.7	90.75	89.7
232	27.0	137	60.0	∼84	92.7
222.75	29.9	129.5	63.5	∼71	95.0
212	33.33	123.5	66.7	∼50	96.7
204	35.8	120.2	68.4	<23	98.2
191.5	39.8				

AUXILIARY INFORMATION

METHOD/APPARATUS/PROCEDURE:	SOURCE AND PURITY OF MATERIALS:
The amalgams were obtained by mixing the metals, followed by heating. The cooling curves were recorded. The experiments were carried out under paraffin or vaseline.	Nothing specified.
	ESTIMATED ERROR: Soly: nothing specified. Temp: precision ± 0.5 K.
	REFERENCES:

COMPONENTS:	ORIGINAL MEASUREMENTS:
(1) Lead; Pb; [7439-92-1] (2) Mercury; Hg; [7439-97-6]	Jänecke, E.J. Z. Phys. Chem. 1907, 60, 399-412.
VARIABLES: Temperature: 106-307°C	PREPARED BY: C. Guminski; Z. Galus

EXPERIMENTAL VALUES:

Temperatures of crystallization:

$t/°C$	at % Hg
307	5
293	9
278	13
264	16.5
252	20
236	25
224	28.5
210	33.5
161	50
124	66.5
106	80

AUXILIARY INFORMATION

METHOD/APPARATUS/PROCEDURE:	SOURCE AND PURITY OF MATERIALS:
The cooling of the amalgams was measured with mercury thermometers or thermo-elements, and microscopic observations were carried out in parallel.	Not given.
	ESTIMATED ERROR: Soly: nothing specified. Temp: precision $\pm$ 1 K (compilers).
	REFERENCES:

COMPONENTS:	ORIGINAL MEASUREMENTS:
(1) Lead; Pb; [7439-92-1] (2) Mercury; Hg; [7439-97-6]	Honda, K.; Ishigaki, T. *Sci. Rep. Tohoku Univer.* <u>1925</u>, *14*, 219-232.
VARIABLES:	PREPARED BY:
Temperature	C. Guminski; Z. Galus

EXPERIMENTAL VALUES:

Depression of freezing point of lead with 1.0 and 2.0 at % of mercury was reported to be 3.38 and 6.88 K, respectively. The melting point of lead was assumed to be 600.6 K.

AUXILIARY INFORMATION

METHOD/APPARATUS/PROCEDURE:	SOURCE AND PURITY OF MATERIALS:
The usual method of thermal analysis was used. The alloys to be tested were melted in an alundum tube. The melts were protected from oxidation with a thick layer of asbestos wool, over which paraffin or vaseline was poured. Temperatures were measured with a copper-constantan thermocouple.	Nothing specified, but probably extra pure metals from Merck were used (compilers).
	ESTIMATED ERROR:
	Soly: nothing specified. Temp: precision probably ± 0.05 K (compilers).
	REFERENCES:

COMPONENTS:	ORIGINAL MEASUREMENTS:
(1) Lead; Pb; [7439-92-1] (2) Mercury; Hg; [7439-97-6]	Hoyt, C.S.; Stegman, G. *J. Phys. Chem.* 1934, *38*, 753-9.
VARIABLES: One temperature: 298 K	PREPARED BY: C. Guminski; Z. Galus

EXPERIMENTAL VALUES:

Solubility of lead in mercury at 298.16 K was reported to be 1.42 at %.

AUXILIARY INFORMATION

METHOD/APPARATUS/PROCEDURE:	SOURCE AND PURITY OF MATERIALS:
The amalgam was prepared by adding predetermined amounts of Pb to a known amount of Hg. The mixture was homogenized by warming and agitating in the separatory funnel in which the amalgam was prepared. EMF's of the cell $Pb(Hg)_{sat} \mid PbSO_4 \mid ZnSO_4 \mid PbSO_4 \mid xPb(Hg)$ were measured. All operations were performed in hydrogen atmosphere.	Mercury was purified with concentrated H_2SO_4 and then distilled 3 times under reduced pressure. Lead was Kahlbaum's "for analysis."
	ESTIMATED ERROR: Soly: precision better than 1%. Temp: precision $\pm$ 0.02 K.
	REFERENCES:

COMPONENTS:	ORIGINAL MEASUREMENTS:
(1) Lead; Pb; [7439-92-1] (2) Mercury; Hg; [7439-97-6]	Thompson, H.E., Jr. J. Phys. Chem. 1935, 39, 655-64
VARIABLES: Temperature: 19-69°C	PREPARED BY: C. Guminski; Z. Galus

EXPERIMENTAL VALUES:

The solubility of lead in mercury:

t/°C	Soly/at %
19.7	1.469
30.7	1.811
39.9	2.203
47.4	2.588
48.2	2.631
60.6	3.438
69.2	4.279

AUXILIARY INFORMATION

METHOD/APPARATUS/PROCEDURE:	SOURCE AND PURITY OF MATERIALS:
Both metals were sealed in Pyrex glass tubes at a pressure of about 0.01 mm. Tubes were placed in the thermostat and shaken for several hours to saturate the mercury with lead. Then the homogeneous amalgam was filtered off and analyzed for the content of lead. The analysis consisted in the vaporization of the mercury from the amalgam and weighing the residue as the amount of metal dissolved.	High purity lead from the U.S. Bureau of Standards. Spectrographic analysis showed that only calcium was present in quantities more than a trace. Mercury was purified with 6 mol dm^{-3} nitric acid and then triply distilled.
	ESTIMATED ERROR: Soly: precision better than 0.015%. Temp: precision $\pm$ 0.1 K.
	REFERENCES:

COMPONENTS:	ORIGINAL MEASUREMENTS:
(1) Lead; Pb; [7439-92-1] (2) Mercury; Hg; [7439-97-6]	Haring, M.M.; Hatfield, M.R.; Zapponi, P.T. *Trans. Electrochem. Soc.* 1939, *75*, 473-84.
VARIABLES:	PREPARED BY:
One temperature: 25°C	C. Guminski; Z. Galus

EXPERIMENTAL VALUES:

The solubility of lead in mercury at 25°C was reported to be 1.650 at %.

AUXILIARY INFORMATION

METHOD/APPARATUS/PROCEDURE:	SOURCE AND PURITY OF MATERIALS:
Amalgams were prepared electrolytically and the homogeneous amalgams were separated by filtration. The EMF of the amalgam cells enabled the determination of the activity of lead in the saturated amalgams of various concentrations. The standard potential of the lead electrode also was determined. The cell was of the type: (Pt),H_2 | (1 atm) | $HClO_4$ (xm) | $HClO_4$ (xm) + $Pb(ClO_4)_2$ (ym) | Pb(Hg) where m is the concentration in mol kg^{-1}.	All materials were of reagent grade. Mercury was purified with dilute nitric acid and mercurous nitrate and then distilled.
	ESTIMATED ERROR: EMF's: precision ± 0.05 mV. Temp: precision ± 0.02 K.
	REFERENCES:

COMPONENTS:	ORIGINAL MEASUREMENTS:
(1) Lead; Pb; [7439-92-1] (2) Mercury; Hg; [7439-97-6]	Jangg, G.; Kirchmayr, H. *Z. Chem.* 1963, *3*, 47-56.
VARIABLES: One temperature: 15°C	PREPARED BY: C. Guminski; Z. Galus

EXPERIMENTAL VALUES:

The solubility of lead in mercury at 15°C was reported to be about 0.90 mol dm^{-3}. The corresponding atomic % solubility calculated by the compilers is 1.35 at %.

AUXILIARY INFORMATION

METHOD/APPARATUS/PROCEDURE:	SOURCE AND PURITY OF MATERIALS:
The amalgams were obtained by electrolysis. Potential of the lead amalgam was measured against the constant potential electrode in the cell, Pb(Hg)|Pb(CH$_3$COO)$_2$||KCl|Hg$_2$Cl$_2$, Hg The concentration of Pb(CH$_3$COO)$_2$ was 0.01, 0.1 or 1.0 mol dm^{-3} with addition of 5 x 10^{-3} mol dm^{-3} of CH$_3$COOH. The concentration of the saturated amalgam was evaluated from the break in the curve relating potential to the logarithm of the amalgam concentration. The experiments were performed in an inert gas atmosphere.	Nothing specified.
	ESTIMATED ERROR: Soly: precision ± 10% or better. Temp: nothing specified.
	REFERENCES:

COMPONENTS:	ORIGINAL MEASUREMENTS:
(1) Lead; Pb; [7439-92-1] (2) Mercury; Hg; [7439-97-6]	Moshkevich, A.S.; Rav'del, A.A. *Zh. Prikl. Khim.* 1970, *43*, 71-5.
VARIABLES:	PREPARED BY:
Temperature: (-36)-50°C	C. Guminski; Z. Galus

EXPERIMENTAL VALUES:

The solubility of lead in mercury:

t/°C	Soly/mass %	Soly/at %[a]
-36	0.45	0.44
-15	0.75	0.73
0	0.99	0.96
15	1.35	1.31
25	1.62	1.58
50	2.78	2.69

[a] by compilers.

AUXILIARY INFORMATION

METHOD/APPARATUS/PROCEDURE:	SOURCE AND PURITY OF MATERIALS:
A lead disk was rotated in a known volume of mercury at a precisely controlled rate. The concentration of dissolved lead in mercury was determined on the basis of a change of weight of the lead disk. To protect the amalgam against oxidation, it was covered by a layer of glycerine or acetone.	Pure lead used was analyzed by spectral analysis. Hg purity not specified.
	ESTIMATED ERROR: Soly: precision ± 1-2%. Temp: precision ± 0.3 K.
	REFERENCES:

COMPONENTS:	ORIGINAL MEASUREMENTS:
(1) Lead; Pb; [7439-92-1] (2) Mercury; Hg; [7439-97-6]	Filippova, L.M.; Zebreva, A.I.; Korobkina, N.P. *Ukr. Khim. Zh.* 1978, *44*, 791-3.
VARIABLES:	PREPARED BY:
Temperature: 25-40°C	C. Guminski; Z. Galus

EXPERIMENTAL VALUES:

Solubility of lead in mercury at 25 and 40°C was reported to be 1.1 and 1.4 mol dm^{-3}. The corresponding atomic % solubilities calculated by the compilers are 1.65 and 2.1 at %, respectively.

The same result at 25°C was also reported in (1).

AUXILIARY INFORMATION

METHOD/APPARATUS/PROCEDURE:	SOURCE AND PURITY OF MATERIALS:
The heterogeneous lead amalgam was obtained by dissolution of lead in mercury. Heats of dilution (Q) of the amalgams of various compositions (heterogeneous and homogenous) were measured upon addition of mercury. A break in the curve of Q vs. N_{Pb} corresponds to the composition of the saturated amalgam. All operations were carried out in an argon atmosphere.	Not given.
	ESTIMATED ERROR: Soly: precision no better than 1%. Temp: nothing specified.
	REFERENCES: 1. Filippova, L.M.; Gayfullin, A.Sh.; Zebreva, A.I. *Prikl. Teor. Khim.*, Alma-Ata 1974, No. 5, 76-82.

COMPONENTS:	ORIGINAL MEASUREMENTS:
(1) Lead; Pb; [7439-92-1] (2) Mercury; Hg; [7439-97-6]	Yan-Sho-Syan, G.V.; Nosek, M.V.; Semibratova, N.M.; Shalamov, A.E. *Tr. Inst. Khim. Nauk Akad. Nauk Kaz. SSR* 1967, *15*, 139-49.
VARIABLES: Temperature: 323-590 K	PREPARED BY: C. Guminski; Z. Galus

EXPERIMENTAL VALUES:

Liquidus temperatures of the Pb-Hg system determined on amalgams with different pretreatment:

	T/K				T/K	
at % Pb	Freshly Prepared and Quenched	Soaked at 623 K for 37 hrs and Quenched		at % Pb	Freshly Prepared and Quenched	Soaked at 623 K for 37 hrs and Quenched
2.5	323	–		50.0	451	445
5.0	365	–		52.5	454	–
7.5	373	370		55.0	458	457
10.0	377	371		57.5	461	–
12.5	388	371		60.0	468	471
15.0	391	378		62.5	474	475
17.5	394	378		65.0	482	–
20.0	394	384		67.5	488	–
22.5	396	–		70.0	494	–
25.0	399	388		72.5	501	–
27.5	404	–		75.0	509	–
30.0	411	398		77.5	519	–
32.5	413	–		82.5	532	–
35.0	–	410		85.0	541	–
37.5	420	–		87.5	553	546
40.0	423	418		90.0	565	551
42.5	433	–		92.5	570	–
45.0	438	429		95.0	583	–
47.5	444	–		97.5	590	–

AUXILIARY INFORMATION

METHOD/APPARATUS/PROCEDURE:	SOURCE AND PURITY OF MATERIALS:
The alloys were prepared by mixing weighed amounts of lead and mercury. The mixtures were placed in glass tubes and sealed for the different pretreatments of the amalgams. Thermographic analysis was performed to determine the liquidus temperatures. Comments The results for the freshly quenched amalgams are erroneous because of segregation of some fractions of the alloys (compilers). The authors did not specify the temperature from which the freshly prepared samples were quenched.	Mercury was purified chemically and electrochemically, then twice distilled under vacuum. Lead was 99.999% pure with regard to 17 metallic impurities.
	ESTIMATED ERROR: Soly: nothing specified. Temp: precision ± 3 K.
	REFERENCES:

COMPONENTS:	EVALUATOR:
(1) Arsenic; As; [7440-38-2] (2) Mercury; Hg; [7439-97-6]	C. Guminski; Z. Galus Department of Chemistry University of Warsaw Warsaw, Poland July, 1985

CRITICAL EVALUATION:

The solubility of arsenic in mercury was speculated to be very low by Tammann and Hinnüber (1). Kozin estimated solubilities of 2.8×10^{-13} (2) and 1.6×10^{-9} at % (3) at 298 K. Gladyshev (4) reported on arsenic solubility of 1.6×10^{-9} at % at room temperature, a value identical to Kozin's second estimated solubility (3), but because no details of the experimental determination were presented for ref. (4) it is difficult to assess the validity of this result. Nevertheless, the data of Refs. (2-4) confirm that of (1). Strachan and Harris (5) reported a solubility determination of 0.646 at % at room temperature, but this value is much too high; the error in this determination is attributed to evaporation losses of arsenic during the analysis.

Kamenev and coworkers (6) reported that the saturated amalgam of arsenic should be in equilibrium with As_2Hg_3; however, the solubility could not be estimated from the experiments performed by these authors.

It is clear that further solubility measurements are needed in this system.

References

1. Tammann, G.; Hinnüber, J. *Z. Anorg. Chem.* 1927, *160*, 249.
2. Kozin, L.F. *Tr. Inst. Khim. Nauk Akad. Nauk Kaz. SSR* 1962, *9*, 101.
3. Kozin, L.F. *Fiziko-Khimicheskie Osnovy Amalgamnoi Metallurgii*, Nauka, Alma-Ata, 1964.
4. Gladyshev, V.P.; cited by Kozin, L.F.; Nigmetova, R.Sh.; Dergacheva, M.B. *Termodinamika Binarnykh Amalgamnykh Sistem*, Nauka, Alma-Ata, 1977, p. 268.
5. Strachan, J.F.; Harris, N.L. *J. Inst. Metals* 1956-57, *85*, 17.
6. Kamenev, A.I.; Mustafa, I.; Agasyan, P.K. *Zh. Anal. Khim.* 1984, *39*, 1242.

COMPONENTS:	EVALUATOR:
(1) Antimony; Sb; [7440-36-0] (2) Mercury; Hg; [7439-97-6]	C. Guminski; Z. Galus Department of Chemistry University of Warsaw Warsaw, Poland July, 1985

CRITICAL EVALUATION:

The solubility of antimony in mercury near room temperature has been shown to be low. Tammann and Hinnüber (1) determined a solubility of 4.8×10^{-5} at % at 291 K by EMF measurements, whereas Strachan and Harris (2) reported a solubility of 3.3×10^{-2} at %. These values are too low and too high, respectively, when compared to more reliable measurements which have been reported subsequently. At 293 K the following solubilities have been reported: 3.5×10^{-4} at % by Levitskaya and Zebreva (3), 3.6×10^{-4} at % by Zebreva and Kozlovskii (4), 1.1×10^{-3} at % by Zaichko and Zakharov (5), and 9×10^{-4} at % by Lange and Bukhman (6). In refs. (4-6) voltammetry was used to determine the solubility of Sb by anodic oxidation of the amalgams of various concentrations, while in refs. (3) and (4) the determinations were made potentiometrically on the amalgam concentration cells. Verplaetse and coworkers (7) determined the solubility of Sb in Hg by cyclic and stripping voltammetry at 298 K and reported a value of 1.27×10^{-3} at %; this solubility is in good agreement with those reported above (5,6). Zaichko and Zakharov (8) also determined the antimony solubility by voltammetry, presumably at room temperature, and reported a value of 1×10^{-3} at %. Liebl (9) reported a solubility of 3.8×10^{-3} at % at room temperature, but no details of the coulometric method were described; the latter solubility is tenfold higher than that reported by Zebreva and Kozlovskii (4).

Zakharova and coworkers (10) determined the antimony solubility, probably at 298 K, by chronoamperometric oxidation of the amalgam, and reported a value of 1.0×10^{-3} at %. At 293 K Bukhman and Dragavtseva (11) reported a solubility of 6.8×10^{-4} at %. Ignateva and Dubova (12), without presenting experimental details and presumably at room temperature, reported a solubility of $6.6-7.0 \times 10^{-4}$ at %. Kozin's (13) estimated solubility of 5×10^{-5} at % at 298 K is much too low. Toibaev (14) stated that the saturated antimony amalgam at 293 K should contain less than 9×10^{-4} at % antimony; the solubility measurements reported above appear to confirm the latter statement.

Jangg and coworkers (15,16) determined the solubility of antimony at high temperatures and showed that the saturated amalgam is in equilibrium with pure antimony; they also showed that there is complete miscibility at temperatures above 904 K. The extrapolation of the high temperature solubilities to 298 K yields a solubility near 10^{-3} at %. The high temperature measurements of Jangg and coworkers showed a tendency for the antimony to supersaturate; if this tendency extends to room temperature the lower values of the solubility would probably be more reliable, as reported by other workers discussed above.

The homogeneous amalgam is in equilibrium with pure Sb. However, as shown (17) on the inset in Fig. 1 there appears to be a break in the solubility curve near 473 K; the break suggests the peritectic formation of a compound, although this compound was not detected. The formation of Hg_3Sb_2 was reported by Ugai and Gordin (18).

Tentative values of the antimony solubility in mercury:

T/K	Soly/at %	Reference
293	4×10^{-4}	[3,4]
298	5×10^{-4} [a]	[3]
323	1.5×10^{-3} [a]	[3,6]
373	2×10^{-2}	[16]
473	0.12	[16]
573	0.7 [b]	[15,16]
673	13 [b]	[15,16]
773	54 [a]	[15]
873	91	[15]

[a] Interpolated value from cited references.

[b] Mean value from data of cited references.

(continued next page)

COMPONENTS:	EVALUATOR:
(1) Antimony; Sb; [7440-36-0] (2) Mercury; Hg; [7439-97-6]	C. Guminski; Z. Galus Department of Chemistry University of Warsaw Warsaw, Poland July, 1985

CRITICAL EVALUATION: (continued)

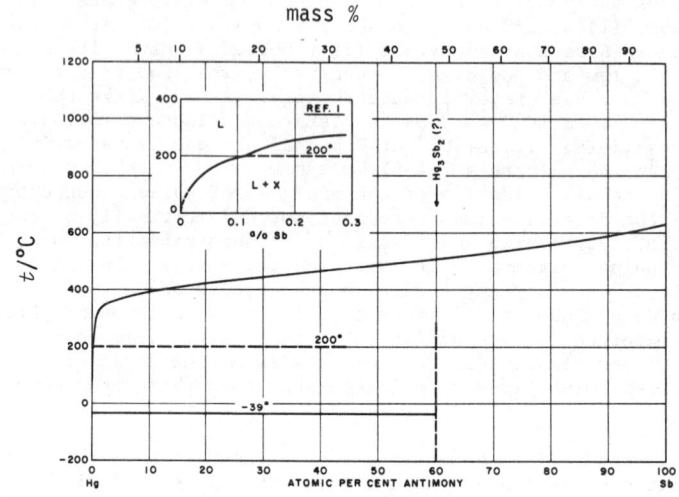

Fig. 1. Hg-Sb System (17).

References

1. Tammann, G.; Hinnüber, J. *Z. Anorg. Chem.* 1927, *160*, 249.
2. Strachan, J.F.; Harris, N.L. *J. Inst. Metals* 1956-57, *85*, 17.
3. Levitskaya, S.A.; Zebreva, A.I. *Elektrokhimiya* 1966, *2*, 92.
4. Zebreva, A.I.; Kozlovskii, M.T. *Collect. Czech. Chem. Commun.* 1960, *25*, 3188.
5. Zaichko, L.F.; Zakharov, M.S. *Zh. Anal. Khim.* 1966, *21*, 65.
6. Lange, A.A.; Bukhman, S.P. *Elektrokhimiya* 1974, *10*, 391.
7. Verplaetse, H.; Donche, H.; Temmerman, E.; Verbeek, F. *J. Electroanal. Chem. Interfacial Electrochem.* 1978, *93*, 213.
8. Zaichko, L.F.; Zakharov, M.S. *Izv. Tomsk. Politekhn. Inst.* 1971, *174*, 59.
9. Liebl, G.; cited by H. Spengler, *Metall.* 1958, *12*, 105.
10. Zakharova, E.A.; Volkova, V.N.; Polyakova, T.P. *Dep. ONIITEKhim*, 214-81, 1981.
11. Bukhman, S.P.; Dragavtseva, N.A. *Izv. Akad. Nauk Kaz. SSR, Ser. Khim.* 1970, No. 5, 23.
12. Ignateva, L.A.; Dubova, N.M. *Mater. Konfer. Molodykh Uchen. Tomsk. Gos. Univer.*, 1973, Tomsk, 1974, No. 2, 173; *C.A.* 1977, *87*, 122896p.
13. Kozin, L.F. *Fiziko Khimicheskie Osnovy Amalgamnoi Metallurgii*, Nauka, Alma-Ata, 1964.
14. Toibaev, B.K. *Tr. Inst. Khim. Nauk Akad. Nauk Kaz. SSR* 1969, *32*, 35.
15. Jangg, G.; Lihl, F.; Legler, E. *Z. Metallk.* 1962, *53*, 313.
16. Jangg, G.; Palman, H. *Z. Metallk.* 1963, *54*, 364.
17. Shunk, F. *Constitution of Binary Alloys, Second Supplement*, McGraw-Hill, N.Y., 1969, p. 433.
18. Ugai, Ya.A.; Gordin, V.L. *Zh. Neorg. Khim.* 1962, *7*, 703.

COMPONENTS:	ORIGINAL MEASUREMENTS:
(1) Antimony; Sb; [7440-36-0] (2) Mercury; Hg; [7439-97-6]	Lange, A.A.; Bukhman, S.P. *Elektrokhimiya* 1974, *10*, 391-5.
VARIABLES: Temperature: 20-80°C	PREPARED BY: C. Guminski; Z. Galus

EXPERIMENTAL VALUES:

Solubility of antimony in mercury:

t/°C	Soly/at %
20	0.9×10^{-3}
40	1.75×10^{-3}
60	2.55×10^{-3}
80	3.4×10^{-3}

The enthalpy of solution of Sb at saturation, calculated from the $(T/K)^{-1}$ dependence of the solubility, was 21.1 kJ mol^{-1}.

AUXILIARY INFORMATION

METHOD/APPARATUS/PROCEDURE:	SOURCE AND PURITY OF MATERIALS:
Amalgams prepared by electrolysis of $Sb_2(SO_4)_3$ solutions in 1-3 mol dm^{-3} H_2SO_4 at a mercury cathode. Sb content of amalgam determined by difference in Sb(III) concentration before and after electrolysis; Sb(III) concentration determined by bromate titration. Limiting anodic currents (i.e., limiting diffusion currents, i_d (compilers)) were measured for amalgams of varying Sb content. A plot of i_d vs. Sb content gave a sharp break at the saturation value of Sb content. A second break in the i_d vs. Sb content curve was observed for supersaturated amalgams and was attributed by the authors to the oxidation of elemental Sb in a two-phase amalgam.	Nothing specified.
	ESTIMATED ERROR: Soly: precision of method probably around 10% (compilers). Temp: nothing specified.
	REFERENCES:

COMPONENTS:	ORIGINAL MEASUREMENTS:
(1) Antimony; Sb; [7440-36-0] (2) Mercury; Hg; [7439-97-6]	Bukhman, S.P.; Dragavtseva. N.A. *Izv. Akad. Nauk Kaz. SSR, Ser. Khim.* <u>1970</u>, *20*, No. 5, 23-31.
VARIABLES: Temperature: 20°C	PREPARED BY: C. Guminski; Z. Galus

EXPERIMENTAL VALUES:

Solubility of antimony in mercury at 20°C was reported to be 6.8×10^{-4} at %.

AUXILIARY INFORMATION

METHOD/APPARATUS/PROCEDURE:	SOURCE AND PURITY OF MATERIALS:
The amalgam was prepared by electrolysis and then was aged for one hour. The antimony content was determined by the "bromate method". Polarization curves (i vs. E) of the amalgam oxidation were recorded to determine the potential of the limiting current. In other experiments the potentiostatic curves (i vs. t) were recorded at the potentials of the limiting current (0.3 V vs. NHE). There was a breakpoint in the curve when the amalgam became saturated with antimony.	Nothing specified.
	ESTIMATED ERROR: Soly: not specified; precision no better than $\pm$ 20% (compilers). Temp: nothing specified.
	REFERENCES:

COMPONENTS:	ORIGINAL MEASUREMENTS:
(1) Antimony, Sb; [7440-36-0] (2) Mercury; Hg; [7439-97-6]	Jangg, G.; Lihl, F.; Legler, E. Z. Metallk. 1962, 53, 313-16.
VARIABLES: Temperature: 573-904 K	PREPARED BY: C. Guminski; Z. Galus

EXPERIMENTAL VALUES:

Liquidus temperatures of the antimony-mercury system:

T/K	Soly/at %
573.2	0.8
655.2	6.4
673.2	11.9
683.2	15.5
713.2	28.6
738.2	39.7
758.2	50.0
766.2	51.9
783.2	59.5
801.2	67.5
833.2	79.5
868.2	91.0
903.7	100

Antimony and mercury did not form any compound over the complete composition range, but a single eutectic was observed on the Hg-rich side; the eutectic temperature was within $\pm$ 0.1 K from the freezing point of Hg.

AUXILIARY INFORMATION

METHOD/APPARATUS/PROCEDURE:	SOURCE AND PURITY OF MATERIALS:
The liquidus temperature was determined thermographically from cooling and heating curves of the amalgams which were sealed in an ampule of Supremaxglas. The undercooling of the melt was minimized by a strong mechanical vibration of the sample on a vibrating table. The liquidus temperature was determined from the breakpoint in the temperature versus time plot.	Nothing specified.
	ESTIMATED ERROR: Soly: nothing specified. Temp: precision $\pm$ 5 K.
	REFERENCES:

COMPONENTS:	ORIGINAL MEASUREMENTS:
(1) Antimony; Sb; [7440-36-0] (2) Mercury; Hg; [7439-97-6]	Zebreva, A.I.; Kozlovskii, M.T. *Collect. Czech. Chem. Commun.* 1960, 25, 3188-94.
VARIABLES: Temperature: 20°C	PREPARED BY: C. Guminski; Z. Galus

EXPERIMENTAL VALUES:

The solubility of antimony in mercury at 20°C was reported to be $(2.4 \pm 0.2) \times 10^{-4}$ mol dm^{-3} from potentiometric measurements and 2.6×10^{-4} mol dm^{-3} from polarographic measurements. The respective atomic % solubilities calculated by the compilers are 3.5×10^{-4} and 3.7×10^{-4} at %.

AUXILIARY INFORMATION

METHOD/APPARATUS/PROCEDURE:	SOURCE AND PURITY OF MATERIALS:
The antimony amalgam was prepared by electrolysis of $Sb_2(SO_4)_3$ on the mercury cathode. The solubility was determined by polarography and potentiometry. In the former method the limiting current was linearly dependent on the concentration only up to the saturation point of the amalgam. In the case of potentiometry the potential of the amalgam electrode was linearly dependent on the logarithm of the antimony content for homogeneous solution in mercury. At saturation an inflection was observed in the curve of the latter relationship.	Mercury was chemically purified with $Hg_2(NO_3)_2$ then distilled under vacuum. Other chemicals were chemically pure.
	ESTIMATED ERROR: Soly: nothing specified, but may be greater than $\pm$ 10% (compilers). Temp: nothing specified.
	REFERENCES:

COMPONENTS:	ORIGINAL MEASUREMENTS:
(1) Antimony; Sb; [7440-36-0] (2) Mercury; Hg; [7439-97-6]	Jangg, G.; Palman, H. *Z. Metallk.* 1963, *54*, 364-9.

VARIABLES:	PREPARED BY:
Temperature: 96-453°C	C. Guminski; Z. Galus

EXPERIMENTAL VALUES:

The mass % solubility of antimony in mercury was presented graphically as a function of temperature. The data points were read off the curve and the solubilities converted to atomic % by the compilers.

$t/°C$	Soly/mass %	Soly/at %	$t/°C$	Soly/mass %	Soly/at %
96	0.012	0.020	272	0.19	0.31
130	0.020	0.033	300	0.38	0.62
150	0.026	0.043	310	0.84	1.3
190	0.054	0.089	333	1.2	2.0
200	0.074	0.12	340	2.0	3.2
210	0.080	0.13	350	3.0	4.8
240	0.091	0.15	375	6.5	10.3
250	0.12	0.19	400	9.2	14.3
260	0.13	0.21	425	13	20
			453	24	34

The saturated amalgam was reported to be in equilibrium with pure antimony.

AUXILIARY INFORMATION

METHOD/APPARATUS/PROCEDURE:	SOURCE AND PURITY OF MATERIALS:
The heterogeneous amalgam was introduced into a specially constructed apparatus made of refractory chromium steel. Such steel apparatus could be used because the solubility of iron in mercury is very low and the Cr(III)-oxide film inhibits the wetting of the steel by mercury. After twelve hours of equilibration at the experimental temperature the amalgam was filtered through a sintered-iron frit under purified nitrogen pressure. Usually 3- to 4-fold filtration was necessary. The metal content of the filtered, saturated amalgam was then determined by an unspecified method. For experiments carried out below 320°C the amalgam was equilibrated in a glass vessel.	Nothing specified.
	ESTIMATED ERROR: Soly: precision ± 5%. Temp: precision ± 2 K.
	REFERENCES:

COMPONENTS:	ORIGINAL MEASUREMENTS:
(1) Antimony; Sb; [7440-36-0] (2) Mercury; Hg; [7439-97-6]	Levitskaya, S.A.; Zebreva, A.I. *Elektrokhimiya* 1966, *2*, 92-6.
VARIABLES: Temperature: 20-80°C	PREPARED BY: C. Guminski; Z. Galus

EXPERIMENTAL VALUES:

Solubility of antimony in mercury:

$t/°C$	Soly/mol dm^{-3}	Soly/at %^{a}
20	2.40×10^{-4}	3.6×10^{-4}
40	8.24×10^{-4}	1.22×10^{-3}
60	1.70×10^{-3}	2.52×10^{-3}
80	2.76×10^{-3}	4.08×10^{-3}

^{a}by compilers.

AUXILIARY INFORMATION

METHOD/APPARATUS/PROCEDURE:

The amalgam was prepared by electro-reduction of Sb(III) at the mercury cathode. EMF were determined on the cell,

$$\text{Sb(Hg)} \left| \begin{array}{l} \text{Sb}_2(\text{SO}_4)_3 \ (10^{-3} \text{ mol dm}^{-3}) + \\ \text{KNaC}_4\text{H}_4\text{O}_6 \ (0.075 \text{ mol dm}^{-3}) \\ + \text{H}_2\text{SO}_4 \ (1 \text{ mol dm}^{-3}) \end{array} \right| \text{Sb(Hg)}_x$$

The EMF varied linearly with the logarithm of the amalgam concentration up to the solubility limit. Beyond the latter the EMF remained virtually constant.

SOURCE AND PURITY OF MATERIALS:

Nothing specified.

ESTIMATED ERROR:

Soly: nothing specified; precision may be no better than $\pm$ 15% (compilers).

Temp: nothing specified.

REFERENCES:

COMPONENTS:	ORIGINAL MEASUREMENTS:
(1) Antimony; Sb; [7440-36-0] (2) Mercury; Hg; [7439-97-6]	Verplaetse, H.; Donche, H.; Tammermann, E.; Verbeek, F. *J. Electroanal. Chem. Interfacial Electrochem.* 1978, *93*, 213-19.
VARIABLES:	PREPARED BY:
One temperature: 25°C	C. Guminski; Z. Galus

EXPERIMENTAL VALUES:

The solubility of antimony in mercury at 25°C was reported to be 1.27×10^{-3} at %.

The enthalpy of solution of Sb in Hg was reported to be 16.7 kJ mol^{-1}.

AUXILIARY INFORMATION

METHOD/APPARATUS/PROCEDURE:	SOURCE AND PURITY OF MATERIALS:
Antimony amalgam was prepared by the electroreduction of Sb(III) on the hanging-mercury and sitting-mercury drop electrodes. In the case of voltammetric oxidation of Sb from the heterogeneous amalgam, the shape of the peak current was changed. The charge corresponding to the oxidation curve where this deformation was just detectable was used to calculated the solubility of this metal in mercury. To ensure equilibrium in the amalgam the oxidation process was carried out some time after the preparation of the amalgam.	Mercury was purified by distillation. It was then anodically dissolved and cathodically deposited in 0.5 mol dm^{-3} HNO$_3$. All solutions were prepared with analytical grade reagents and double-distilled water.
	ESTIMATED ERROR: Soly: precision $\pm$ 4%. Temp: nothing specified.
	REFERENCES:

COMPONENTS:	EVALUATOR:
(1) Bismuth; Bi; [7440-69-9] (2) Mercury; Hg; [7439-97-6]	C. Guminski; Z. Galus Department of Chemistry University of Warsaw Warsaw, Poland July, 1985

CRITICAL EVALUATION:

Tammann (1) reported on the first study of the Bi-Hg system by determining the solidification temperatures upon addition of small amounts of bismuth to mercury. He found that the melting point of mercury was depressed by 0.30 K at a bismuth concentration of 0.217 at %.

The liquidus has been determined over wide concentration ranges by several workers. Pushin (2) reported the first extensive study of this system by thermoanalysis over the range of 1.4 to 97.3 at % Bi; however, Pushin's bismuth solubility at concentrations below 5 at % is too high by comparison with later measurements. Petot-Ervas et al. (3,4) determined the liquidus in the range of 0.1 to 30 at % Bi by measuring the EMF of concentration cells and from 30 to 90 at % Bi by thermoanalysis. Nosek and Yan-Sho-Syan (5) used thermoanalysis to determine the solubility of bismuth over a temperature range of 269 to 533 K, but the solubilities reported by these authors are lower than those of (3). Predel and Rothacker (6) redetermined the Bi-Hg phase diagram, but the solubilities of bismuth determined by these authors in the middle range of the amalgam composition lie between those of (3,4) and of (5). It has been shown (4-6) that the equilibrium solid phase in this system is bismuth. In the opinion of the evaluators, the data of Petot-Ervas et al. (3,4) are the preferred solubilities.

The solubility of bismuth was determined over narrower temperature ranges by the following authors with satisfactory agreement with those of (3,4): Dergacheva and Kozin (7) employed EMF measurements to determine the solubilities between 298 and 348 K; Kozin and Nigmetova (8) also used the same technique with satisfactory results; Schenk et al. (9) employed thermoanalysis over the temperature range of 303 to 373 K; Heycock and Neville (10) reported four points in the Bi-rich region.

Single determinations of the solubility of bismuth near room temperature have been reported by several authors (11-14).

The reported solubilities of 0.84 at % at room temperature (15) and of 0.82 at % at 298 K (16) are too low and are rejected. Kozin's (17) estimated solubility of 2.8 at % at 298 K is too high. Campbell and Kartzmark (18) reported that they exactly confirmed the results of Pushin (2), but no data were presented by these authors.

The phase diagram for this system is shown in Fig. 1 (19).

Recommended (r) and tentative values of the solubility of bismuth in mercury:

T/K	Soly/at %	Reference
234.1	0.072	[4]
243	0.15	[4]
253	0.26[a]	[4]
263	0.36	[4]
273	0.6[a]	[4]
293	1.1	[3,4,12]
298	1.3[a]	[3,4]
323	3.7[b]	[3,4]
373	22	[3,4]
473	70 (r)	[2-4]

[a] Interpolated value from cited references.
[b] Mean value from data of cited references.

(Continued next page)

COMPONENTS:	EVALUATOR:
(1) Bismuth; Bi; [7440-69-9] (2) Mercury; Hg; [7439-97-6]	C. Guminski; Z. Galus Department of Chemistry University of Warsaw Warsaw, Poland July, 1985

CRITICAL EVALUATION: (continued)

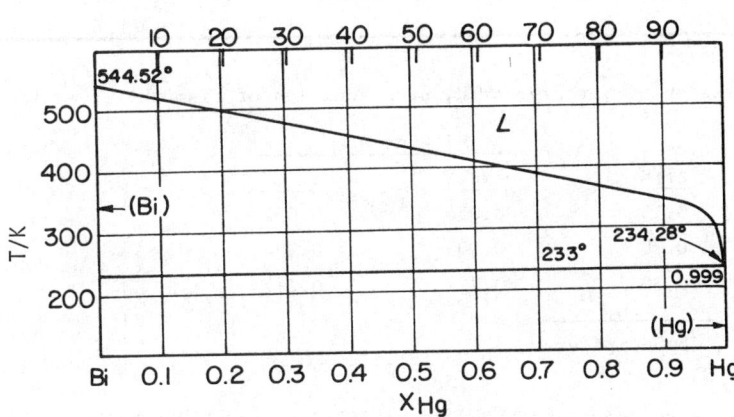

Fig. 1. The Bi-Hg system (19).

References

1. Tammann, G. *Z. Phys. Chem.* 1889, *3*, 444.
2. Pushin, N.A. *Zh. Russ. Fiz. Khim. Obshch., Ser. Khim.* 1902, *34*, 856; *Z. Anorg. Chem.* 1903, *36*, 201.
3. Petot-Ervas, G.; Desrè, P.; Bonnier, E. *C.R. Acad. Sci., Ser. 2* 1965, *261*, 3406.
4. Petot-Ervas, G.; Allibert, M.; Petot, C.; Desrè, P.; Bonnier, E. *Bull. Soc. Chim. Fr.* 1969, 1477.
5. Nosek, M.V.; Yan-Sho-Syan, G.V. *Izv. Akad. Nauk Kaz. SSR, Ser. Khim.* 1965, No. 4, 26.
6. Predel, B.; Rothacker, D. *J. Less-Common Met.* 1966, *10*, 392.
7. Dergacheva, M.B.; Kozin, L.F. *Zh. Fiz. Khim.* 1977, *51*, 417.
8. Kozin, L.F.; Nigmetova, R.Sh. *Zh. Prikl. Khim.* 1967, *40*, 1914.
9. Schenck, H.; Steinmetz, E.; Frohberg, M.G. *Arch. Eisenhüttenw.* 1963, *34*, 561.
10. Heycock, C.T.; Neville, F.H. *J. Chem. Soc.* 1892, *61*, 888.
11. Gouy, M. *J. Phys.* 1895, *4*, 320.
12. Nigmatulina, A.A.; Zebreva, A.I. *Izv. Akad. Nauk Kaz. SSR, Ser. Khim.* 1964, No. 4, 18.
13. Filippova, L.M.; Zhumakanov, V.Z.; Zebreva, A.I. *Izv. Vyssh. Ucheb. Zaved., Khim. Khim. Tekhnol.* 1978, *21*, 1450.
14. Same authors, ibid 1980, *23*, 204.
15. Strachan, J.F.; Harris, N.L. *J. Inst. Metals* 1956-57, *85*, 17.
16. Smith, D.L. *U.S. At. Ener. Comm. Rep.*, IS-T-544, 1972.
17. Kozin, L.F. *Fiziko-Khimicheskie Osnovy Amalgamnoi Metallurgii*, Nauka, Alma-Ata 1964.
18. Campbell, A.N.; Kartzmark, E.M. *Can. J. Chem.* 1965, *43*, 1924.
19. Hultgren, R.; Desai, P.D.; Hawkins, D.T.; Gleiser, M.; Kelley, K.K. *Selected Values of the Thermodynamic Properties of Binary Alloys*, Am. Soc. Metals, Metal Park, OH 1973, p. 412.

COMPONENTS:	ORIGINAL MEASUREMENTS:
(1) Bismuth; Bi; [7440-69-9] (2) Mercury; Hg; [7439-97-6]	Tammann, G. Z. Phys. Chem. 1889, 3, 441-9.
VARIABLES: Temperature: -39°C	PREPARED BY: C. Guminski; Z. Galus

EXPERIMENTAL VALUES:

Melting point depression of mercury, $\Delta T/K$, upon addition of bismuth:

	Bi Content	
$\Delta T/K$	mass %	at %[a]
0.15	0.054	0.052
0.30	0.109	0.104
0.30	0.227	0.217

[a] by compilers

The melting point of mercury was reported to be 244 instead of 234 K, but it is the opinion of the compilers that the former value was a typographical error in the original publication.

AUXILIARY INFORMATION

METHOD/APPARATUS/PROCEDURE:	SOURCE AND PURITY OF MATERIALS:
The melting points were determined thermometrically. No further details were given.	Nothing specified.
	ESTIMATED ERROR: Soly: nothing specified. Temp: precision better than ± 0.1 K.
	REFERENCES:

COMPONENTS:	ORIGINAL MEASUREMENTS:
(1) Bismuth; Bi; [7440-69-9] (2) Mercury; Hg; [7439-97-6]	Heycock, C.T.; Neville, F.H. J. Chem. Soc. 1892, 888-914.
VARIABLES: Temperature: 258-267°C	PREPARED BY: C. Guminski; Z. Galus

EXPERIMENTAL VALUES:

Freezing point of Bi-Hg amalgams:

t/°C	at. Hg/100 at. Bi	at % Hg[a]
266.65	–	0
266.17	0.225	0.224
264.65	0.911	0.903
259.77	3.27	3.17
257.80	4.29	4.11

[a] by compilers

AUXILIARY INFORMATION

METHOD/APPARATUS/PROCEDURE:	SOURCE AND PURITY OF MATERIALS:
The amalgams were prepared by thoroughly mixing weighed quantities of the metals at red heat after they had been sealed in evacuated hard-glass tubes. Freezing points of the amalgams were determined with carefully calibrated thermometers.	Nothing specified.
	ESTIMATED ERROR: Soly: nothing specified. Temp: precision no better than ± 0.05 K.
	REFERENCES:

COMPONENTS:	ORIGINAL MEASUREMENTS:
(1) Bismuth; Bi; [7440-69-9] (2) Mercury; Hg; [7439-97-6]	Pushin, N.A. Zh. Russ. Fiz. Khim., Obshch., Ser. Khim. 1902, 34, 856-904. Z. Anorg. Chem. 1903, 36, 201-54.
VARIABLES: Temperature: 18-262°C	PREPARED BY: C. Guminski; Z. Galus

EXPERIMENTAL VALUES:

Freezing points of bismuth amalgams:

t/°C	at % Hg	t/°C	at % Hg	t/°C	at % Hg
261.7	2.7	189.5	36.3	104.5	73.0
254	6.2	182.0	40.0	98.0	76.2
245	10.0	169.5	45.0	90.0	79.4
240.5	12.1	156.7	50.0	81.7	83.7
233	15.7	142.7	56.0	68	89.4
224	20.0	133.7	60.0	56	93.3
219.2	22.3	125.0	64.2	~44	95.8
213.2	25.0	117.2	66.7	~32	97.5
205	28.9	113.0	68.6	~18	98.6
195.7	33.3				

AUXILIARY INFORMATION

METHOD/APPARATUS/PROCEDURE:	SOURCE AND PURITY OF MATERIALS:
The amalgams were prepared by heating and mixing appropriate weights of each metal. Cooling curves were determined with the amalgams protected from oxidation by a surface film of paraffin or vaseline.	Nothing specified.
	ESTIMATED ERROR: Soly: nothing specified. Temp: precision ± 0.5 K.
	REFERENCES:

COMPONENTS:	ORIGINAL MEASUREMENTS:
(1) Bismuth; Bi; [7440-69-9] (2) Mercury; Hg; [7439-97-6]	Schenk, H.; Steinmetz, E.; Frohberg, M.G. *Arch. Eisenhüttenw.* 1963, *34*, 562-63.
VARIABLES: Temperature: 18-100°C	PREPARED BY: C. Guminski; Z. Galus

EXPERIMENTAL VALUES:

The solubility of bismuth in mercury was reported graphically as a plot of the logarithm of solubility versus $1/(T/K)$. The data points were read from the curve by the compilers.

$t/°C$	Soly/at %
18	0.46[a]
30	1.15
40	2.2
49	3.3
60	5.6
69	8.7
80	11.0
90	16.2
100	23.5

[a] From EMF measurement; the value is erroneous (compilers).

AUXILIARY INFORMATION

METHOD/APPARATUS/PROCEDURE:	SOURCE AND PURITY OF MATERIALS:
Bismuth particles were introduced into the mercury phase under argon atmosphere in small glass container. The container was placed in a thermostated bath. The amalgams were filtered through glasswool filter. The filtrate was analyzed by a complexometric method with Titriplex (from Merck). To test for saturation, the filtrations were made after various times from the moment of mixing of the metals.	Bismuth and mercury were chemically pure grade.
	ESTIMATED ERROR: Soly: nothing specified. Temp: nothing specified.
	REFERENCES:

COMPONENTS:	ORIGINAL MEASUREMENTS:
(1) Bismuth; Bi; [7440-69-9] (2) Mercury; Hg; [7439-97-6]	Nigmatullina, A.A.; Zebreva, A.I. *Izv. Akad. Nauk Kaz. SSR, Ser. Khim.* <u>1964</u>, *14*, No. 4, 18-22.
VARIABLES: Temperature: 20°C	PREPARED BY: C. Guminski; Z. Galus

EXPERIMENTAL VALUES:

The solubility of bismuth in mercury at 20°C was reported to be 1.07 at %.

AUXILIARY INFORMATION

METHOD/APPARATUS/PROCEDURE:	SOURCE AND PURITY OF MATERIALS:
The amalgams were prepared by electrolysis and were used as the electrodes in a concentration cell. The concentration of one electrode was kept constant while the Bi concentration in the other amalgam electrode was varied. The curve of EMF vs. logarithm of the ratio of Bi concentration in the electrodes exhibited a breakpoint at amalgam saturation.	Nothing specified.
	ESTIMATED ERROR: Soly: precision no better than several percent. Temp: precision ± 0.1 K.
	REFERENCES:

COMPONENTS:	ORIGINAL MEASUREMENTS:
(1) Bismuth; Bi; [7440-69-9] (2) Mercury; Hg; [7439-97-6]	Nosek, M.V.; Yan-Sho-Syan, G.V. *Izv. Akad. Nauk Kaz. SSR, Ser. Khim.* <u>1965</u>, *15*, No. 4, 26-32.
VARIABLES: Temperature: (-4)-265°C	PREPARED BY: C. Guminski; Z. Galus

EXPERIMENTAL VALUES:

The data were presented graphically as a phase diagram; the experimental liquidus points were read from the curve by the compilers.

t/°C	Soly/at %	t/°C	Soly/at %
-4	1.00	196	57.49
46	2.65	205	60.16
62	5.04	209	65.09
81	7.63	210	69.82
85	10.24	213	67.56
102	14.78	227	75.04
109	20.05	228	77.41
110	17.50	233	80.16
121	25.20	241	84.98
134	27.51	242	87.56
137	30.16	248	89.91
144	35.06	260	94.95
153	37.50	265	97.34
160	40.05		
172	45.02		
177	47.52		
188	55.00		
175	50.13		

AUXILIARY INFORMATION

METHOD/APPARATUS/PROCEDURE:	SOURCE AND PURITY OF MATERIALS:
The liquidus was determined by thermal analyses. For each composition, the alloy was heated to 573 K then cooled at a rate of 1-3 K per minute. A pyrometer of the Kurnakov-type was used for the thermal analyses.	Mercury was purified by chemical and electrochemical methods, then distilled twice under reduced pressure. Bismuth was 99.998% pure.
	ESTIMATED ERROR: Soly: nothing specified. Temp: precision ± 2 K.
	REFERENCES:

Bismuth

COMPONENTS:	ORIGINAL MEASUREMENTS:
(1) Bismuth; Bi; [7440-69-9] (2) Mercury; Hg; [7439-97-6]	Predel, B.; Rothacker, D. *J. Less-Common Met.* 1966, *10*. 392-401.
VARIABLES: Temperature: (-16)-264°C	PREPARED BY: C. Guminski; Z. Galus

EXPERIMENTAL VALUES:

The liquidus data were presented graphically as a phase diagram; the solubilities were read for each temperature from the curve by the compilers.

$t/°C$	Soly/at %	$t/°C$	Soly/at %
-16	0.9	125	40.6
- 6	1.3	132	44.5
17	1.4	162	55.2
35	3.0	178	63.7
38	4.0	194	72.1
42	5.0	208	77.4
57	8.5	225	84.2
75	16.6	238	88.1
90	22.	245	90.6
97	26.3	252	93.7
105	31.8	257	95.6
121	36.0	264	97.9

AUXILIARY INFORMATION

METHOD/APPARATUS/PROCEDURE:	SOURCE AND PURITY OF MATERIALS:
The amalgams were prepared from the pure metals in evacuated tubes. Temperatures on the liquidus curve were determined by differential thermal analysis.	Both mercury and bismuth were 99.9995% pure.
	ESTIMATED ERROR: Soly: nothing specified. Temp: nothing specified.
	REFERENCES:

COMPONENTS:	ORIGINAL MEASUREMENTS:
(1) Bismuth; Bi; [7440-69-9] (2) Mercury; Hg; [7439-97-6]	1. Petot-Ervas, G.; Allibert, M.; Petot, C.; Desrè, P.; Bonnier, E. *Bull. Soc. Chim. Fr.* 1969, 1477-81. 2. Desrè, P.; Bonnier, E. *C.R. Acad. Sci., Ser. 2* 1965, *261*, 3406-9.
VARIABLES: Temperature: (-39)-240°C	PREPARED BY: C. Guminski; Z. Galus

EXPERIMENTAL VALUES:

Solubility of bismuth in mercury:

Electrochemical Measurements				Thermal Analysis	
$t/°C$	Soly/at %	$t/°C$	Soly/at %	$t/°C$	Soly/at %
-35.4	0.1	37	2	120	30
-30.3	0.15	47	3	135	40
-22.1	0.22	54	4	155	50
-9.85	0.36	62	5	170	60
-2.6	0.46	71	8	200	70
17.6	0.97	79	11	240	90
22.5	1.12	81	13		
32.4	1.75	86	15		
42.2	2.75	90	17		
50.85	4.0	96	20		
61.6	5.8	108	25		
69.5	7.7	118	30		

Eutectic point was determined at 0.072 ± 0.004 at % Bi and -39.10 ± 0.04°C. It was reported that the equilibrium solid-phase consisted of pure Bi.

AUXILIARY INFORMATION

METHOD/APPARATUS/PROCEDURE:	SOURCE AND PURITY OF MATERIALS:
Solubilities were determined by EMF measurements and by thermal analysis. EMF were determined with the concentration cell, $Bi|Bi(III)|Bi(Hg)$. Various electrolytes were used, including: BiI_3-KI, $BiCl_3$-$ZnCl_2$ in glycerine or H_2O, and H_2O-LiCl eutectic mixture. The liquidus temperatures above 393 K were determined by thermal analysis.	Nothing specified.
	ESTIMATED ERROR: Soly: nothing specified; precision no better than few percent (compilers). Temp: precision ± 0.02 K.
	REFERENCES:

COMPONENTS:	ORIGINAL MEASUREMENTS:
(1) Bismuth; Bi; [7440-69-9] (2) Mercury; Hg; [7439-97-6]	Dergacheva, M.B.; Kozin, L.F. *Zh. Fiz. Khim.* 1977, *51*, 417-20.
VARIABLES: Temperature: 25-75°C	PREPARED BY: C. Guminski; Z. Galus

EXPERIMENTAL VALUES:

Solubility of bismuth in mercury:

t/°C	Soly/x(Bi)
25	0.0150
40	0.0244
65	0.0646
75	0.0860

AUXILIARY INFORMATION

METHOD/APPARATUS/PROCEDURE:	SOURCE AND PURITY OF MATERIALS:
The amalgam was prepared electrolytically and was used to construct the cell Bi(Hg)|Bi(III)|xBi(Hg) The concentration of Bi in the left-hand half-cell was kept constant, while that in the right-hand side was varied. At concentrations of the amalgam exceeding the saturation point, the EMF of the cell was independent of the amalgam concentration.	Mercury was chemically purified and distilled twice. Bismuth was 99.999% pure. All other chemicals were specified as very pure.
	ESTIMATED ERROR: Soly: nothing specified. Precision of EMF measurement was $\pm\ 10^{-4}$ V. Temp: nothing specified.
	REFERENCES:

COMPONENTS:	ORIGINAL MEASUREMENTS:
(1) Bismuth; Bi; [7440-69-9] (2) Mercury; Hg; [7439-97-6]	Filippova, L.M.; Zhumakanov, V.Z.; Zebreva, A.I. *Izv. Vyssh. Ucheb. Zaved., Khim. Khim. Tekhnol.* 1978, *21*, 1450-3; 1980, *23*, 204-7.
VARIABLES: One temperature: 25°C	PREPARED BY: C. Guminski; Z. Galus

EXPERIMENTAL VALUES:

The solubility of bismuth in mercury at 25°C was reported to be 1.55 ± 0.05 at %.

AUXILIARY INFORMATION

METHOD/APPARATUS/PROCEDURE:	SOURCE AND PURITY OF MATERIALS:
Heterogeneous amalgam was obtained by addition of bismuth to mercury. The amalgams were titrated with Hg and employing calorimetric end-point detection. The solubility was determined from the change in slope of the plot of the enthalpy of dilution as a function of bismuth content in the amalgams.	Nothing specified.
	ESTIMATED ERROR: Soly: nothing specified; precision no better than several percent (compilers). Temp: nothing specified.
	REFERENCES:

COMPONENTS:	EVALUATOR:
(1) Tellurium; Te; [13494-80-9] (2) Mercury; Hg; [7439-97-6]	C. Guminski; Z. Galus Department of Chemistry University of Warsaw Warsaw, Poland July, 1985

CRITICAL EVALUATION:

The solubility of tellurium in mercury is very low at room temperature. Kozin (1) first predicted a solubility of 2.3×10^{-4} at % at 298 K; he later (2) corrected this estimate to 5×10^{-3} at %. Gladyshev and Kovaleva (3), without giving details of their polarographic method, reported that the solubility is of the order of 10^{-4} at % at room temperature; these authors subsequently reported a solubility of 1.4×10^{-3} at % (34), but this value appears too high. Pajaczkowska and Dziuba (4) determined the solubility of tellurium in the temperature range of 487-943 K, and these authors showed that their data were in good agreement with equations based on ideal solution theory. Part of the results from (4) were subsequently confirmed by Herning (35). Extrapolation of the data of (4) and (35) leads to a solubility of 2×20^{-4} at % at 298 K.

The first report of a phase diagram for the Hg-Te system was that of Pellini and Aureggi (5) who determined the liquidus line in the Te-rich region. These authors found an eutectic at 87.8 at %. Strauss and coworkers (6-9) determined the complete liquidus line by thermoanalysis and found the eutectic at 83.5 at % Te. Levitskaya and coworkers (10), reported the eutectic at an appreciably higher concentration of 91.2 at % while Williams found it at 83.3 at % Te (11). The calculated eutectic points in the Hg and Te-rich regions are 2×10^{-5} at % and 85.4 at % of Te at 234.3 K and 686.7 K, respectively (12). The partial phase diagrams of refs. (4) and (5) are in general agreement with that of Strauss and coworkers. The phase diagram shows only a single congruently melting compound, HgTe. But, the melting point of HgTe has been reported at various values between 873 and 960 K (6-10, 13-31); the wide range of melting points is due to errors arising from the high volatility of HgTe. The most reliable melting point appears to be 943 K (6,7,19,26-28). Other melting points ranging from 929 to 960 K have been reported (29-31), but the experimental conditions were not defined. The low value of 873 K (14) is rejected. It has been demonstrated (32,33) that the melting point of HgTe has a significant dependence on the vapor pressure over the compound; it was observed that the melting point was 888 K at 12.2 kbar, and, as shown in Fig. 2, there was a linear dependence of the melting point on the pressure (33). The measurements of Steininger (25) and of Brebrick and Strauss (7) show that the melting point is at 941 and 943 K at 13.6 and 12.6 kbar, respectively. Slightly different pressure dependence of the phase relations is presented by Omelchenko and Soshnikov (27).

Delves and Lewis (19,21) showed that the Hg-Te system consists of a two-liquid region on the Te-rich side, and Levitskaya and coworkers (10) confirmed this observation at 52.5-55.7 at % Te. The monotectic temperature was found to be 937 ± 2 K by the former authors. The parameters of the immiscibility region and the solubility at low Te contents need further investigations.

The tentative values of the Te solubility in Hg:

T/K	Soly/at %	Reference
500	0.16	[4,35]
600	1.5[a]	[4,35]
684	2.4[a]	[4,8,35]
700	5.5[a]	[4,8,35]
800	15[a]	[4,8]
900	32	[4,8]

Completely miscible above 943 K.

[a] Interpolated from data of cited references.

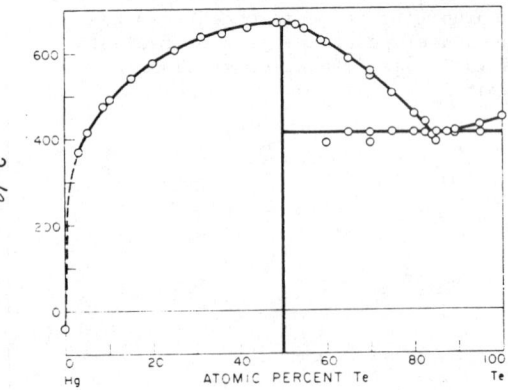

Fig. 1. Phase diagram of the Te-Hg system (8).

COMPONENTS:	EVALUATOR:
(1) Tellurium; Te; [13494-80-9] (2) Mercury; Hg; [7439-97-6]	C. Guminski; Z. Galus Department of Chemistry University of Warsaw Warsaw, Poland July, 1985

CRITICAL EVALUATION:

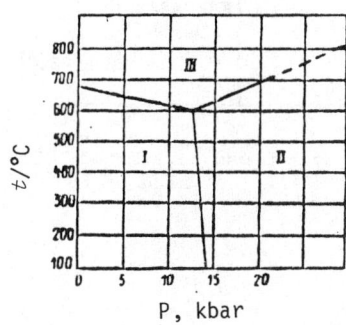

Fig. 2. Dependence of melting point of HgTe on pressure (33).

REFERENCES

1. Kozin, L.F. *Tr. Inst. Khim. Nauk Akad. Nauk Kaz. SSR*, 1962, 9, 101.
2. Kozin, L.F. *Fiziko-Khimicheskie Osnovy Amalgamnoi Metallurgii*, Nauka, Alma-Ata 1964.
3. Gladyshev, V.P.; Kovaleva, S.V. *Dvoinyi Sloi i Adsorbtsia na Tverdykh Elektrodakh, VI Simpozium*, Tartu 1981, 78.
4. Pajaczkowska, A.; Dziuba, E.Z. *J. Cryst. Growth*, 1971, 11, 21.
5. Pellini, G.; Aureggi, C. *Gazz. Chim. Ital.*, 1910, 40 (2),42; *Atti Accad. Naz. Lincei*, 1909, 18 (2), 211.
6. Strauss, A.J.; cited by D. R. Masson, D. F. O'Kane, *International Conference on Semiconductor Physics*, Prague, Academic Press, New York, 1961, 1026.
7. Brebrick, R.F.; Strauss, A.J. *J. Phys. Chem. Solids*, 1965, 26, 989.
8. Strauss, A. J.; cited by T.C. Harman, *Physics and Chemistry of II-VI Compounds*, M. Aven and J.S. Prener, Eds., North-Holland, Amsterdam, 1967, 774.
9. Tung, T.; Golonka, L.; Brebrick, R.F. *J. Electrochem. Soc.*, 1981, 128, 1601.
10. Levitskaya, T.D.; Vanyukov, A.V.; Krestovnikov, A.N.; Bykhanov, I.M. *Izv. Akad. Nauk SSSR, Neorg. Mater.*, 1970, 6, 849.
11. Williams, D.J. *J. Cryst. Growth*, 1982, 58, 657.
12. Tung, T.; Su, C.-H.; Liao, P.-K; Brebrick, R.F. *J. Vac. Sci. Technol.*, 1982, 21, 117.
13. Carlson, R.O. *Phys. Rev., Ser. II*, 1958, 111, 476.
14. Lawson, W.P.; Nielsen, S.; Putley, E.H.; Young, A.S. *Phys. Chem. Solids*, 1959, 9, 325.
15. Rodot, M.; Rodot, H. *C.R. Acad. Sci., Ser 2* 1960, 250, 1447.
16. Woolley, J.C.; Ray, B. *J. Phys. Chem. Solids*, 1960, 13, 151.
17. Usachev, P.V.; Golubkov, A.V.; Volosatova, N.S. *Zh. Prikl. Khim.*, 1960, 33, 2771.
18. Blair, J.; Newnham, R. *Met. Soc. Conf.*, 1961, 12, 393.
19. Delves, R.T.; Lewis, B. *J. Phys. Chem. Solids*, 1963, 24, 549.
20. Gromakov, S.D.; Zoroatskaya, I.V.; Latypov, Z.M.; Khvala, M.A.; Edelman, E.A.; Badygina, L.I.; Zaripova, L.G. *Zh. Neorg. Khim.*, 1964, 9, 2485.
21. Delves, R.T. *Brit. J. Appl. Phys.*, 1965, 16, 343.
22. Ray, B.; Spencer, P.M. *Phys. Stat. Sol.*, 1967, 22, 371.
23. Spencer, P.M.; Ray, B. *Brit. J. Appl. Phys., Ser. 2*, 1968, 1, 299.
24. Vanyarko, V.G.; Zlomov, V.P.; Novoselova, A.V. *Vestn. Moskov. Univer., Khim.*, 1968, 6, 108.
25. Steininger, J. *J. Electron. Mat.*, 1976, 5, 299.
26. Szofran, F.R.; Lehoczky, S.L. *J. Electron. Mater.*, 1981, 10, 1131.
27. Omelchenko, A.V.; Soshnikov, V.I. *Izv. Akad. Nauk SSSR, Neorg. Mater.*, 1982, 18, 685.
28. Vengel, P.F.; Tomashik, V.N.; Mizetskaya, I.P. *Ukr. Khim. Zh.*, 1983, 49, 1247.
29. Kuzmina, G.A. *Izv. Akad. Nauk SSSR, Neorg. Mater.*, 1976, 12, 1121.
30. Gavalenko, N.P.; Gorley, N.P.; Paranchich, S.Yu.; Frasunyak, V.M.; Khomyak, V.V. *Izv. Akad. Nauk SSSR, Neorg. Mater.*, 1983, 19, 327.
31. Babanly, M.B.; Kurbanov, A.A.; Kuliev, A.A. *Zh. Neorg. Khim.*, 1979, 24, 2293.
32. Jayaraman, A.; Klement, W.; Kennedy, G.C. *Phys. Rev.*, 1963, 130, 2277.
33. Rotner, Yu.M.; Suranov, A.V.; Vorona, Yu.V. *Vliyanie Vysokikh Davlenii na Veshchestvo*, Kiev 1978, 72.
34. Gladyshev, V.P.: Kovaleva, S.V.; Sarieva, L.S. *Zh. Anal. Khim.*, 1982, 37, 1762.
35. Herning, P.E. *J. Electron. Mater.*, 1984, 13, 1.

COMPONENTS:	ORIGINAL MEASUREMENTS:
(1) Tellurium; Te; [13494-80-9] (2) Mercury; Hg; [7439-97-6]	Ray, B.; Spencer, P.M.S. *Phys. Stat. Sol.* 1967, 22, 371-372.
VARIABLES: Temperature	PREPARED BY: C. Guminski; Z. Galus

EXPERIMENTAL VALUES:

The liquidus line for the CdTe-HgTe system was determined. The melting point of HgTe, read from the liquidus, was 666°C. A value of 665 ± 2°C was subsequently reported by the same authors (1).

AUXILIARY INFORMATION

METHOD/APPARATUS/PROCEDURE:	SOURCE AND PURITY OF MATERIALS:
The powdered samples were sealed in quartz ampules filled with inert gas under pressure of several atmospheres. The melting point was determined by differential thermal analysis.	High purity HgTe was synthesized from Te (99.9995% pure) from Canadian Copper Refiners, Ltd. and triply distilled Hg.
	ESTIMATED ERROR: Soly: nothing specified. Temp: precision ± 3 K.
	REFERENCES: 1. Spencer, P.M.; Ray, B. *Brit. J. Appl. Phys.*, Ser. 2 1968, 1, 299.

COMPONENTS:	ORIGINAL MEASUREMENTS:
(1) Tellurium; Te; [13494-80-9] (2) Mercury; Hg; [7439-97-6]	Omelchenko, A.V.; Soshnikov, V.I. *Izv. Akad. Nauk SSR, Neorg. Mater.* 1982, 18, 685-6. English translation: *Inorg. Mater.* 1982, 18, 582-84.
VARIABLES:	PREPARED BY:
Pressure	C. Guminski; Z. Galus

EXPERIMENTAL VALUES:

The pressure dependence of the melting point of HgTe is shown in the figure. The melting line shows two linear segments which correspond to the melting points of the I and II phases. The inflection point of the line is at 634°C and 11.11×10^8 Pa.

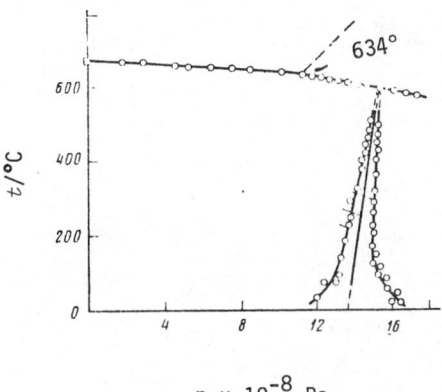

AUXILIARY INFORMATION

METHOD/APPARATUS/PROCEDURE:	SOURCE AND PURITY OF MATERIALS:
Milled crystals of HgTe were pressed into specimens. The pressure was applied to the specimens in an apparatus of piston-cylinder type; argon was used to transmit pressure at temperatures above 550°C and benzene was used at lower temperatures. Phase transition of HgTe in the solid state was determined by dilatometric method, and melting temperatures were determined by thermal analysis with the use of a Chromel-Alumel thermocouple. Pressure was determined by a manganin resistance manometer.	n-type HgTe with $n_e = 3 \times 10^{17}$ cm^{-3}.
	ESTIMATED ERROR: Pressure: accuracy $\pm 1.5 \times 10^7$ Pa. Temp: accuracy ± 2 K.
	REFERENCES:

COMPONENTS:	ORIGINAL MEASUREMENTS:

(1) Tellurium; Te; [13494-80-9]
(2) Mercury; Hg; [7439-97-6]
	1. Delves, R.T.; Lewis, B.
	   J. Phys. Chem. Solids 1963, *24*, 549-556.
	2. Delves, R.T.
	   Brit. J. Appl. Phys. 1965, *16*, 343-351.

VARIABLES: Temperature	PREPARED BY: C. Guminski; Z. Galus

EXPERIMENTAL VALUES:

The melting point of HgTe was determined to be 670 ± 1°C and the monotectic was at 664 ± 2°C. The maximum on the liquidus was observed to be approximately 2.5 to 4 at % on the Te-rich side of HgTe; this may have been caused by a deficiency of 2 at % Hg in the actual composition near HgTe. The eutectic on the Te-rich side was found at 409 ± 2°C. The two-liquids region was found to be between $HgTe_{1.12}$ and $HgTe_{1.25}$.

The HgTe-MnTe system also was investigated.

AUXILIARY INFORMATION

METHOD/APPARATUS/PROCEDURE:	SOURCE AND PURITY OF MATERIALS:
HgTe was prepared by melting mercury and tellurium in an evacuated silica tube. The tube was initially heated to 450°C then slowly heated to 700°C while the tube was continuously rocked to ensure complete mixing of the elements. The tube was then quenched to 550°C, and slowly cooled. Differential thermal analysis of the powdered samples of HgTe was performed. For the determination of the two-liquids region, appropriate amounts of the elements were melted in a silica tube then the melts were directionally frozen in a gradient furnace.	Tellurium was melted in an atmosphere of hydrogen and zone refined. This resulted in a purity of at least 99.99%; Bi, Sb, and Se were the major impurities. Mercury was purified by triple distillation and resulted in a purity of 99.999%.
	ESTIMATED ERROR: Soly: nothing specified. Temp: precision ± 2 K.
	REFERENCES:

COMPONENTS:	ORIGINAL MEASUREMENTS:
(1) Tellurium; Te; [13494-80-9] (2) Mercury; Hg; [7439-97-6]	Pellini, G.; Aureggi, C. *Gazz. Chim. Ital.* 1910, *40* (2), 42-9.
VARIABLES:	PREPARED BY:
Temperature: 408-548°C	C. Guminski; Z. Galus

EXPERIMENTAL VALUES:

The data were presented graphically; the following data were read off the liquidus curve by the compilers:

t/°C	Soly/at %
431	95
422	90
408.5	87.8
464	80
493	75
535	70
548	66.6

An eutectic point for mercury-rich amalgams was also observed but the information is not quantitatively exact. The same results are also reported in (1).

AUXILIARY INFORMATION

METHOD/APPARATUS/PROCEDURE:	SOURCE AND PURITY OF MATERIALS:
The amalgams were prepared by heating the elements in hydrogen atmosphere saturated with mercury vapors. Thermal analyses were made with the use of a Pt-PtRh thermocouple.	Nothing specified.
	ESTIMATED ERROR: Nothing specified.
	REFERENCES: 1. Pellini, G; Aureggi, C. *Atti Accad. Nazl. Lincei* 1909, *18* (2), 211.

COMPONENTS:	ORIGINAL MEASUREMENTS:
(1) Tellurium; Te; [13494-80-9] (2) Mercury; Hg; [7439-97-6]	Brebrick, R.F.; Strauss, A.J. J. Phys. Chem. Solids 1965, 26, 989-1002.
VARIABLES: Temperature; Pressure	PREPARED BY: C. Guminski; Z. Galus

EXPERIMENTAL VALUES:

Liquidus temperatures of tellurium amalgams determined in this work[a] and values abstracted by the compilers from the graphical data in refs. (1) and (2). P is equilibrium pressure of mercury:

T/K	P/atm	Soly/at %	Ref.	T/K	P/atm	Soly/at %	Ref.
643	–	3.1	1	927	–	55.0	1,2
689	–	5.4	1	901	4.4	58.7	a,1,2
748	–	8.8	1	898	–	60.0	1,2
764	–	10.3	1	859	–	65.0	1,2
815	–	15.3	1	828	–	70.0	1,2
849	–	20.3	1	816	–	70.0	1
881	–	25.0	1	776	–	75.0	1,2
911	19	31.1	a,1	729	–	80.3	1,2
918	19	36.1	a,1	708	–	82.7	1,2
932	19	41.8	a,1	687	–	85.3	1
943	16	48.5	a,1	688	–	87.5	1,2
943	–	50.2	1,2	690	–	89.2	1,2
939	9	52.8	a,1,2	706	–	95.0	1,2

AUXILIARY INFORMATION

METHOD/APPARATUS/PROCEDURE:	SOURCE AND PURITY OF MATERIALS:
Thermal analysis was used in refs. (1) and (2) to determine the liquidus. In this work the optical densities of the vapor in equilibrium with liquid and solid amalgams were measured between 2000 and 6000 Å. Samples were sealed in evacuated thick-wall silica optical cells with parallel flat windows and a sidearm. The latter served as the cold spot reservoir for the amalgam. Zero optical density was found with the condensed phases at room temperature. For each run, the optical cell was heated to its measurement temperature and maintained there for a minimum of one hour before the spectral measurements were made.	Zone-refined tellurium of 99.999% purity from Ohio Semiconductors, Inc., and spectrographic grade mercury from Johnson Matthey Co.
	ESTIMATED ERROR: Soly: nothing specified. Temp: precision ± 2 K.
	REFERENCES: 1. Strauss, A.J.; as quoted by T. C. Harman, Physics and Chemistry of II-VI Compounds. M. Aven, J.S. Prenner, eds. North-Holland. Amsterdam. 1967, 774. 2. Tung, T.; Golonka, L.; Brebrick, R.F. J. Electrochem. Soc. 1981, 128, 1601.

COMPONENTS:	ORIGINAL MEASUREMENTS:
(1) Tellurium; Te; [13494-80-9] (2) Mercury; Hg; [7439-97-6]	Pajaczkowska, A.; Dziuba, E.Z. *J. Cryst. Growth* 1971, *11*, 21-4.
VARIABLES:	PREPARED BY:
Temperature: 487-940 K	C. Guminski; Z. Galus

EXPERIMENTAL VALUES:

The solubilities of tellurium in mercury were presented in graphical form. The data points were read from the curve by the compilers.

T/K	Soly/at %
487	0.11
571	0.75
578	1.0
602	1.7
621	2.0
658	3.1
667	3.3
680	4.8
694	5.5
775	11.7
940	50.0

AUXILIARY INFORMATION

METHOD/APPARATUS/PROCEDURE:	SOURCE AND PURITY OF MATERIALS:
Appropriate amounts of tellurium and mercury were placed in quartz tubes which were then evacuated and sealed. The thermal analysis of the samples was performed by measuring the temperature with a constantan-chrome nickel thermocouple. The dissolution and crystallization processes were repeated several times for every concentration, and melting temperatures were taken as the experimental points for the liquidus line.	The metals were of spectroscopic purity.
	ESTIMATED ERROR: Soly: nothing specified. Temp: precision ± 5%, with reference to measured temperature in °C.
	REFERENCES:

COMPONENTS:	ORIGINAL MEASUREMENTS:
(1) Tellurium; Te; [13494-80-9] (2) Mercury; Hg; [7439-97-6]	Steininger, J. *J. Electron. Mater.* 1976, 5, 299-320.
VARIABLES: Temperature: 929-941 K	PREPARED BY: C. Guminski; Z. Galus

EXPERIMENTAL VALUES:

The melting temperatures of HgTe and $Hg_{0.6}Te_{0.4}$ were determined to be 941 and 929 K at Hg vapor pressures of 13.6 and 19.2 atm, respectively.

CdTe-HgTe system was the main purpose of this investigation.

AUXILIARY INFORMATION

METHOD/APPARATUS/PROCEDURE:	SOURCE AND PURITY OF MATERIALS:
Appropriate proportions of the elements, with slight excess of mercury, were placed in a quartz ampule reflux tube. The ampule was placed in a high pressure furnace with a negative temperature gradient along the reflux tube. After evacuation and flushing, the furnace was pressurized with argon and rapidly heated to above the liquidus temperature. Cooling curves under different pressures were recorded with the use of a Chromel-Alumel thermocouple.	99.9999% pure elements were used. Argon purity was 99.999%.
	ESTIMATED ERROR: Soly: nothing specified. Temp: precision $\pm$ 1 K.
	REFERENCES:

COMPONENTS:	ORIGINAL MEASUREMENTS:
(1) Tellurium; Te; [13494-80-9] (2) Mercury; Hg; [7439-97-6]	Szofran, F.R.; Lehoczky, S.L. *J. Electron. Mater.* 1981, *10*, 1131-50.
VARIABLES:	PREPARED BY:
One temperature: 669°C	C. Guminski; Z. Galus

EXPERIMENTAL VALUES:

The melting temperature of HgTe is 699.5°C.

The pseudobinary CdTe-HgTe phase diagram was also investigated.

AUXILIARY INFORMATION

METHOD/APPARATUS/PROCEDURE:	SOURCE AND PURITY OF MATERIALS:
Fused silica ampules were etched in HF solution and annealed at 1423 K in vacuum. The Te bars were etched in Br_2 and rinsed repeatedly in methanol. The ampules were loaded with Hg and Te. They were evacuated and backfilled with He several times before the final evacuation and sealing. The differential thermal analysis curves were recorded, with the use of a calibrated Chromel-Alumel thermocouple, at various rates of cooling and heating.	99.9999% pure Te and 99.99999% pure Hg were used.
	ESTIMATED ERROR:
	Composition: precision better than ± 0.1%. Temp: precision ± 1.7 K.
	REFERENCES:

COMPONENTS:	ORIGINAL MEASUREMENTS:
(1) Tellurium; Te; [13494-80-9] (2) Mercury; Hg; [7439-97-6]	Gladyshev, V.P.; Kovaleva, S.V.; Sarieva, L.S. Zh. Anal. Khim. 1982, 37, 1762-6.
VARIABLES: One temperature: 293 K	PREPARED BY: C. Guminski; Z. Galus

EXPERIMENTAL VALUES:

The solubility of tellurium in mercury at 293 K was reported as 9×10^{-4} mass %. The solubility in atomic % calculated by the compilers is 1.4×10^{-3} at %.

It seems that the result is overstated due to short drop times of the electrode, and the equilibrium between the saturated amalgam and the solid phase is not reached.

AUXILIARY INFORMATION

METHOD/APPARATUS/PROCEDURE:	SOURCE AND PURITY OF MATERIALS:
Te(IV) was reduced on the dropping mercury electrode by direct and alternating current polarography. The background electrolyte contained 1 mol dm^{-3} of NaOH. The results were analyzed on a plot of peak current vs. logarithm of concentration of Te(IV). A bend on the curve corresponds to the saturation concentration of Te in Hg since crystallization of HgTe causes an inflection in the recorded curve.	Very pure TeO_2 and Hg of purity "R-O" were used.
	ESTIMATED ERROR: Soly: ± 10%. Temp.: nothing specified.
	REFERENCES:

COMPONENTS:	ORIGINAL MEASUREMENTS:
(1) Tellurium; Te; [13494-80-9] (2) Mercury; Hg; [7439-97-6]	Herning, P.E. *J. Electron Mater.* 1984, *13*, 1-14.

VARIABLES:
   Temperature: 189-431°C
	PREPARED BY:
	  C. Guminski; Z. Galus

EXPERIMENTAL VALUES:

Solubility of Te in Hg at several temperatures were read off a curve by the compilers.

t/°C	Soly/at %
189	0.1
250	0.4
298	1.0
323	1.5
343	2.0
376	3.0
431	6.0

AUXILIARY INFORMATION

METHOD/APPARATUS/PROCEDURE:	SOURCE AND PURITY OF MATERIALS:
Assuming that the technique of the solubility determination was the same as for CdTe in Hg, described in this paper, the procedure was as follows: A carefully weighed piece of Te was lowered into Hg on a graphite paddle assembly. The melt was stirred for more than 4 hours. After saturating the Hg in this manner, the Te was weighed again; the difference was recorded. Completeness of the saturation was checked by observing the melt surface while slowly lowering the temperature. The amalgams supersaturate only very slightly, probably less than 0.1 K.	Hg was 99.99999% pure; Te purity not specified.
	ESTIMATED ERROR: Temp: accuracy $\pm$ 5 K. Soly: nothing specified, error may be as high as $\pm$ 10% (compilers).
	REFERENCES:

COMPONENTS:	EVALUATOR:
(1) Scandium; Sc; [7440-20-2] (2) Mercury; Hg; [7439-97-6]	C. Guminski; Z. Galus Department of Chemistry University of Warsaw Warsaw, Poland July, 1985

CRITICAL EVALUATION:

There are no experimental data on the solubility of scandium in mercury. Kozin used his semiempirical equations to calculate solubilities of 9.3×10^{-6} (1) and 7.7×10^{-5} (2) at % at 298 K. Further work is needed on this system.

The existence of $ScHg_3$ and $ScHg$ solid phases have been established (3); the liquid amalgam may be in equilibrium with these phases.

References

1. Kozin, L.F. *Tr. Inst. Khim. Nauk Akad. Nauk Kaz. SSR*, 1962, *9*, 101.
2. Kozin, L.F. *Fiziko-Khimicheskie Osnovy Amalgamnoi Metallurgii*, Nauka, Alma-Ata, 1964.
3. Laube, E.; Nowotny, H. *Monatsh. Chem.* 1963, *94*, 851.

COMPONENTS:	EVALUATOR:
(1) Yttrium; Y; [7440-65-5] (2) Mercury; Hg; [7439-97-6]	C. Guminski; Z. Galus Department of Chemistry University of Warsaw Warsaw, Poland July, 1985

CRITICAL EVALUATION:

There is no experimental determination of the solubility of yttrium in mercury. Kozin calculated solubilities of 1×10^{-6} (1) and 1.6×10^{-5} (2) at % at 298 K. Kirchmayr and Lugscheider (3) reported a general schematic phase diagram for the lanthanide-mercury and Y-Hg systems; the phase diagram shows that the saturated amalgams are in equilibrium with Y-Hg intermetallic compounds. YHg_5 was also identified, but no decomposition temperature was reported (4). The estimated solubilities are about 0.2 at % at 423 K (3), 1 at % at 548 K (5), and 2 at % at 723 K (3). These estimated solubilities clearly need experimental confirmation.

References

1. Kozin, L.F. *Tr. Inst. Khim. Nauk Akad. Nauk Kaz. SSR* 1962, *9*, 101.
2. Kozin, L.F. *Fiziko-Khimicheskie Osnovy Amalgamnoi Metallurgii*, Nauka, Alma-Ata, 1964.
3. Kirchmayr, H.R.; Lugscheider, W. *Z. Metallk.* 1966, *57*, 725.
4. Laube, E.; Kusma, I.B. *Monatsh. Chem.* 1964, *95*, 1504.
5. Kirchmayr, H.R.; Jangg, G. *Monatsh. Chem.* 1965, *96*, 1147.

COMPONENTS:	EVALUATOR:
(1) Lanthanum; La; [7439-91-0] (2) Mercury; Hg; [7439-97-6]	C. Guminski; Z. Galus Department of Chemistry University of Warsaw Warsaw, Poland July, 1985

CRITICAL EVALUATION:

Parks and Campanella (1) were the first to analytically determine the solubility of lanthanum in mercury; these authors reported that the solubilities increased from 8.0×10^{-3} to 2.64×10^{-2} at % in the temperature range of 273 to 323 K. Shvedov et al. (2) reported a solubility at 293 K which was sixfold higher than that reported by (1); the amalgam in (2) probably was not in equilibrium and the graphical procedure of the solubility determination from polarographic experiment is questionable. The result of (2) is rejected. More recent works of Zebreva et al. (3,7,12,13), from chronoamperometric oxidation of the amalgams, confirm the results of (1). Zebreva et al. found that the solubility increased from 1.8×10^{-2} to 4.0×10^{-2} at % at 298 to 333 K. Bowersox and Leary (14) determined the solubility at 293, 423 and 523 K by chemical analysis, and although the value at 293 K appears too high, the values at the higher temperatures agree well with those obtained from the extrapolation of the results of (1), (7) and (12). In the high temperature range of 531 to 1351 K the solubility of lanthanum may be obtained from the liquidus curve of the La-Hg phase diagram which was determined by thermal analysis by Bruzzone and Merlo (6). However, the solubilities obtained from the liquidus are approximately one order of magnitude higher than those expected on the basis of solubilities determined at lower temperatures. Kozin's calculated solubility of 3.8×10^{-2} (4) and 5.4×10^{-2} at % (5,8) at 298 K are too high.

The saturated amalgams are in equilibrium with various La-Hg solid phases (6,9-11). Partial phase diagrams have been reported by (6) and (9), but these diagrams are not directly comparable because they were determined at different mercury vapor pressures; Fig. 1 shows that of (6).

The tentative values of the solubility of La in Hg:

T/K	Soly/at %	Reference
273	8×10^{-3}	(1)
298	1.4×10^{-2}	(1)
323	2.6×10^{-2}	(1)
423	0.25	(14)
523	0.4	(14)

Lanthanum

COMPONENTS:	EVALUATOR:
(1) Lanthanum; La; [7439-91-0] (2) Mercury; Hg; [7439-97-6]	C. Guminski; Z. Galus Department of Chemistry University of Warsaw Warsaw, Poland July, 1985

CRITICAL EVALUATION:

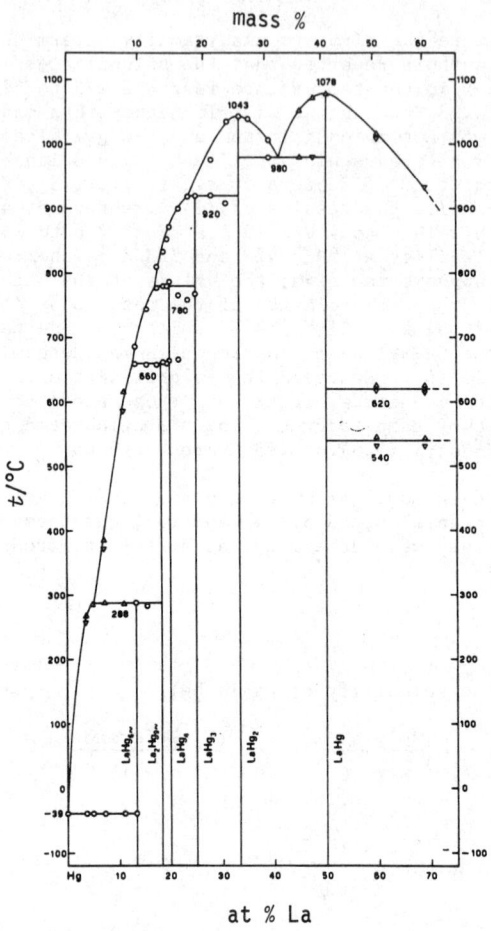

Fig. 1. The La-Hg system (6).

References

1. Parks, W.G.; Campanella, J.L. *J. Phys. Chem.* 1936, 40, 333.
2. Shvedov, V.P.; Frolkov, A.Z.; Nikishin, G.D. *Radiokhimia* 1971, 13, 252.
3. Bulina, V.A.; Zebreva, A.I.; Enikeev, R.Sh. *Izv. Vyssh. Ucheb. Zaved., Khim. Khim. Tekhnol.* 1977, 20, 959.
4. Kozin, L.F. *Tr. Inst. Khim. Nauk Akad. Nauk Kaz. SSR.* 1962, 9, 101.
5. Kozin, L.F. *Fiziko-Khimicheskie Osnovy Amalgamnoi Metallurgii*, Nauka, Alma-Ata, 1964.
6. Bruzzone, G.; Merlo, F. *J. Less-Common Metals* 1976, 44, 259.
7. Sagadieva, K.Zh.; Zebreva, A.I.; Zheldybaeva, B. *Izv. Vyssh. Ucheb. Zaved., Khim. Khim. Tekhnol.* 1973, 16, 47.
8. Kozin, L.F. *Vestn. Akad. Nauk Kaz. SSR*, 1972, No. 3, 34.
9. Kirchmayr, H.R.; Lugscheider, W. *Z. Metallk.* 1966, 57, 725.
10. Flad, D.; Matthes, F. *Z. Chem.* 1964, 4, 466.
11. Merlo, F.; Fornasini, M.L. *J. Less-Common Metals* 1979, 64, 221.
12. Sagadieva, K.Zh.; Dzholdasova, R.M.; Zebreva, A.I. *Uspekhi Polarografii s Nakopleniem*, Tomsk, 1973, p. 104.
13. Sagadieva, K.Zh.; Badavamova, G.L.; Zebreva, A.I. *Izv. Akad. Nauk Kaz. SSR, Ser. Khim.* 1982, No. 2, 59.
14. Bowersox, D.F.; Leary, J.A. *U.S. At. Ener. Comm. Rep.*, LAMS-2518, 1961.

COMPONENTS:	ORIGINAL MEASUREMENTS:
(1) Lanthanum; La; [7439-91-0] (2) Mercury; Hg; [7439-97-6]	Parks, W.G.; Campanella, J.L. *J. Phys. Chem.* 1936, *40*, 333-41.
VARIABLES: Temperature: 0-50°C	PREPARED BY: C. Guminski; Z. Galus

EXPERIMENTAL VALUES:

Solubility of lanthanum in mercury.

$t/°C$	Soly/mass %[a]	Soly/at %[b]
0	$(5.52 \pm 0.08) \times 10^{-3}$	7.97×10^{-3}
12.5	$(9.07 \pm 0.06) \times 10^{-3}$	1.30×10^{-2}
25	$(9.60 \pm 0.06) \times 10^{-3}$	1.38×10^{-2}
37.5	$(1.34 \pm 0.04) \times 10^{-2}$	1.92×10^{-2}
50	$(1.84 \pm 0.05) \times 10^{-2}$	2.64×10^{-2}

[a] original data.
[b] corrected at % by compilers.

The authors state that at % was calculated from mass % by the graphical method described by Ölander (1) and checked by an analytical computation, but the compilers found that there is a mistake in the at % reported in this paper. The empirical formula of the solid phase in equilibrium with the saturated amalgam at 25°C was reported to be La_2Hg_{11}.

AUXILIARY INFORMATION

METHOD/APPARATUS/PROCEDURE:	SOURCE AND PURITY OF MATERIALS:
Amalgams were prepared by electrolysis of concentrated solutions of $LaBr_3 \cdot H_2O$ in absolute ethanol or by the dissolution of an appropriate amount of lanthanum in mercury. The heterogeneous amalgams in quartz flasks were placed in a water thermostat at desired temperatures and shaken at intervals for several days. Amalgams were filtered into a special filter pipette, which was also thermostated, by means of a vacuum pump. After weighing of the samples they were set aside in contact with air for 2 weeks. La(III) hydroxide, with some basic carbonate over the mercury phase, was treated with known amount of 0.1 mol dm^{-3} HCl. The excess of acid was back titrated with NaOH.	Mercury was purified by stirring for 3 days with solution of $HNO_3-Hg_2(NO_3)_2$ then redistilled 4 times, with last distillation under high vacuum. La, $LaCl_3$, HBr and oxalic acid were chemically pure; oxalic acid recrystallized 3 times. Commercial, 95% ethanol distilled several times after treatment with lime and sodium.
	ESTIMATED ERROR: Soly: precision better than ± 2%. Temp: precision ± 0.1 K.
	REFERENCES: 1. Ölander, A. *Ind. Eng. Chem., Anal. Ed.* 1932, *4*, 438.

COMPONENTS:	ORIGINAL MEASUREMENTS:
(1) Lanthanum; La; [7439-91-0] (2) Mercury; Hg; [7439-97-6]	Bowersox, D.F.; Leary, J.A. *U.S. At. Ener. Comm. Rep.*, LAMS-2518, <u>1961</u>.
VARIABLES:	PREPARED BY:
Temperature: 20-250°C	C. Guminski; Z. Galus

EXPERIMENTAL VALUES:

The solubilities of lanthanum in mercury.

$t/°C$	g La/dm^3 Hg	at %^{a}
20	2.87^b	3.1×10^{-2}
150	22.9	0.25
250	37.0	0.41

^{a}by compilers.
^{b}The result at 20° is too high.

AUXILIARY INFORMATION

METHOD/APPARATUS/PROCEDURE:	SOURCE AND PURITY OF MATERIALS:
Hg was outgassed in a reaction vessel at 250°C then cooled to room temperature. A weighed La coupon was added and the vessel was backfilled with He. The evacuation and backfilling of the vessel with He were repeated several times. The mixture of the metals was equilibrated for 24 hr at 350°C, then the vessel was adjusted to the selected temperature. The samples were drawn through a coarse Pyrex frit at intervals of 5 to 90 hr. Each sample was cooled, weighed and analyzed for La content. The procedure gives good results when the filtration is carried out at least 20 hr after adjusting the selected equilibration temperature.	Triple distilled Hg was used. Lanthanum purity not specified.
	ESTIMATED ERROR: Soly: precision better than ± 2%. Temp: not specified.
	REFERENCES:

COMPONENTS:	ORIGINAL MEASUREMENTS:
(1) Lanthanum; La; [7439-91-0] (2) Mercury; Hg; [7439-97-6]	Sagadieva, K.Zh.; Zebreva, A.I.; Zheldybaeva, B. *Izv. Vyssh. Ucheb. Zaved., Khim. Khim. Tekhnol.* 1973, *16*, 47-50.
VARIABLES:	PREPARED BY:
Temperature: 25-60°C	C. Guminski; Z. Galus

EXPERIMENTAL VALUES:

Solubility of lanthanum in mercury at various temperatures is reported.

$t/°C$	Soly/mol dm^{-3}	Soly/at %^{a}
25	$(1.2 \pm 0.1) \times 10^{-2}$	1.8×10^{-2}
30	1.5×10^{-2}	2.2×10^{-2}
40	1.8×10^{-2}	2.7×10^{-2}
50	2.1×10^{-2}	3.1×10^{-2}
60	2.7×10^{-2}	4.0×10^{-2}

^{a}by compilers

The same results were reported in (1).

AUXILIARY INFORMATION

METHOD/APPARATUS/PROCEDURE:	SOURCE AND PURITY OF MATERIALS:
The amalgams were obtained by reduction of La(III) solution with sodium amalgam. Composition of amalgam was established by analysis of solution before and after reduction. The constantly mixed amalgams were then oxidized at -0.10 V vs. SCE and current-time dependences were recorded. The point of transition from a homogeneous to heterogeneous amalgam was determined from the breakpoint in the current-time curve. Concentration of the saturated amalgam was calculated from the charge corresponding to the oxidation of the homogeneous amalgam. Measurements were performed under a hydrogen atmosphere.	Nothing specified.
	ESTIMATED ERROR:
	Soly: precision approximately $\pm$ 10%. Temp: precision $\pm$ 0.5 K.
	REFERENCES:
	1. Sagadieva, K.Zh.; Dzholdasova, R.M.; Zebreva, A.I. *Uspekhi Polarografii s Nakopleniem*, Tomsk, 1973, pp. 104-5.

COMPONENTS:	ORIGINAL MEASUREMENTS:
(1) Lanthanum; La; [7439-91-0] (2) Mercury; Hg; [7439-97-6]	Bruzzone, G.; Merlo, F. J. Less-Common Metals 1976, 44, 259-65.
VARIABLES:	PREPARED BY:
Temperature: 258-1078°C	C. Guminski; Z. Galus

EXPERIMENTAL VALUES:

Data were reported as a phase diagram. The solubility of lanthanum was read from the liquidus data points by the compilers.

t/°C	Soly/at %	t/°C	Soly/at %
258-268	3.5	976	28.0
288	5.0	1033	31.0
371-383	7.0	1043	33.3
587-615	11.0	1038	35.2
686	13.2	1004	39.1
744	15.8	980	41.0
810	17.7	1052	45.0
833	18.7	1071	47.5
852	19.5	1078	50.0
872	20.0	1008-1016	60.0
901	21.8	932	69.1
920	23.5		

AUXILIARY INFORMATION

METHOD/APPARATUS/PROCEDURE:	SOURCE AND PURITY OF MATERIALS:
Appropriate quantities of the two metals were sealed in iron crucibles, for alloys of 0-15% La, and in tantalum crucibles enclosed in iron containers, for alloys of greater than 15% La. Thermal analysis by heating and cooling curves were made with Chromel-Alumel thermocouples. For alloys with less than 15 at % La, thermal analysis was also made at temperatures below 0°C with iron-constantan thermocouples. X-ray analysis was also made on the solid phases.	Lanthanum was 99.6% pure from Koch-Light Labs. Mercury was a commercial product of 99.99% purity.
	ESTIMATED ERROR: Soly: nothing specified. Temp: accuracy ± 5 K.
	REFERENCES:

COMPONENTS:	ORIGINAL MEASUREMENTS:
(1) Lanthanum; La; [7439-91-0] (2) Mercury; Hg; [7439-97-6]	1. Bulina, V.A.; Zebreva, A.I.; Enikeev, R.Sh. *Izv. Vyssh. Ucheb. Zaved., Khim. Khim. Tekhnol.* 1977, *20*, 959-61. 2. Sagadieva, K.Zh.; Badamova, G.L.; Zebreva, A.I. *Izv. Akad. Nauk Kaz. SSR, Ser. Khim.* 1982, No. 2, 59-61.
VARIABLES: One temperature: 25°C	PREPARED BY: C. Guminski; Z. Galus

EXPERIMENTAL VALUES:

The solubility of lanthanum in mercury at 25°C was found to be $(1.1 \pm 0.2) \times 10^{-2}$ and 1.17×10^{-2} mol dm^{-3}, respectively, in (1) and (2). The respective atomic % solubility calculated by the compilers are 1.6×10^{-2} and 1.73×10^{-2} at %.

AUXILIARY INFORMATION

METHOD/APPARATUS/PROCEDURE:	SOURCE AND PURITY OF MATERIALS:
Heterogeneous amalgam in (1) was potentiostatically oxidized at -0.3 V vs. SCE in acetate buffer of pH 3.0. The current-time curve attained a plateau at saturation, and the solubility was calculated from the charge consumed for the oxidation to the breakpoint in the i-t curve. In (2) the amalgam was obtained by reduction of La(III) with Na amalgam; La concentration in amalgam determined by analysis of solution before and after reduction. Amalgam then oxidized chronoamperometrically, at various periods after amalgam preparation, at -0.10 V vs. SCE in an unmixed, buffered acetate solution of 1 mol dm^{-3}. Limiting current, i_d, obtained from current-time curves. Solubility determined from breakpoint in plots of i_d vs. La concentration in amalgam. i_d was constant for amalgams equilibrated over 90 minutes at fixed amalgam concentration.	Nothing specified. ESTIMATED ERROR: Soly: precision 10-20% (compilers). Temp: precision $\pm$ 0.5 K. REFERENCES:

COMPONENTS:	EVALUATOR:
(1) Cerium; Ce; [7440-45-1] (2) Mercury; Hg; [7439-97-6]	C. Guminski; Z. Galus Department of Chemistry University of Warsaw Warsaw, Poland July, 1985

CRITICAL EVALUATION:

There have been a number of reports on the solubility of cerium in mercury near room temperature; nearly all of the determinations were made by electrochemical methods, such as polarography, stripping voltammetry, and chronoamperometry. The 293 K solubilities reported by Shvedov et al., 8.1×10^{-2} at %, (1) and by Sagadieva, 2.3×10^{-2} at %, (19) are too high and are rejected. The report that the amalgam containing 5 at % Ce in liquid Hg at room temperature (24) is clearly in error. Also, Kozin's calculated solubilities of 4.5×10^{-2} (5,11) and 0.42 at % (12) at 298 K are too high. On the other hand the 298 K solubilities reported by several workers are in general agreement: 1.00×10^{-2} at % (2), 9.3×10^{-3} at % (3), 8.7×10^{-3} at % (4,21), 9.0×10^{-3} at % (8-10), 1.0×10^{-2} at % (18), and 8.3×10^{-3} at % (27).

There have been a number of reports on the determination of the solubility of cerium in mercury at various temperatures. Sagadieva et al. (13,22) observed that the solubility increased from 9×10^{-3} to 1.8×10^{-2} at % at 298 to 343 K. The latter group of workers (14) also reported solubilities ranging from 1.3×10^{-2} to 3.6×10^{-2} at % at the same temperatures, but these values are rejected by consensus of the original authors. Bowersox and Leary (23) determined the solubility of cerium at 293, 423 and 523 K by chemical analysis of the equilibrated amalgams. The results of the latter authors are in rough agreement with those of (13), (14) and (22); the temperature dependence of the solubility in (23) is steeper than in (13) and (22), and the solubility at 293 K appears to be slightly high.

Usenova et al. investigated the solubility of cerium in lead and cadmium (15) and other lanthanide amalgams (16).

The liquid amalgam is in equilibrium with various Ce-Hg solid phases (6,20); this is also suggested by the partial phase diagram (17). The compounds $CeHg_5$ (7) and $CeHg_{6.5}$ (25) have been reported, but their temperature stability limits have not been specified. The compound $CeHg_4$ has been shown to be actually Ce_5Hg_{21} (26), similar to other isostructural $LnHg_4$ compounds.

Recommended (r) and tentative values of the solubility of Ce in Hg.

T/K	Soly/at %	Reference
298	9×10^{-3} (r)[a]	[2-4, 8-10, 13,14,18,21-23,27]
323	2×10^{-2} [b]	[13,14,23]
423	0.2	[23]
523	0.8	[23]

[a] mean value from data of cited references.

[b] interpolated value from data of cited references.

References

1. Shvedov, V.P.; Frolkov, A.Z.; Nikishin, G.D. *Radiokhimia* 1971, *13*, 252.
2. Usenova, K.A.; Osipova, G.V.; Krebaeva, Sh.D.; Enikeev, R.Sh. *Radiokhimia* 1974, *16*, 99.
3. Bulina, V.A.; Zebreva, A.I.; Enikeev, R.Sh. *Izv. Vyssh. Ucheb. Zaved., Khim. Khim. Tekhnol.* 1977, *20*, 959.
4. Sagadieva, K.Zh.; Zebreva, A.I.; Badavamova, G.L. *Elektrokhimia* 1979, *15*, 210.
5. Kozin, L.F. *Fiziko-Khimicheskie Osnovy Amalgamnoi Metallurgii*, Nauka, Alma-Ata, 1964.
6. Flad, D.; Matthes, F. *Z. Chem.* 1964, *4*, 466.
7. Olcese, G.L. *Atti Accad. Naz. Lincei, Rend. Cl. Sci. Fiz. Mat. Nat.* 1963, *35*, 48.
8. Zebreva, A.I.; Sagadieva, K.Zh.; Dzholdasova, R.M. *Issl. v Obl. Khim. Redkozeml. Elementov* 1975, 77.
9. Dzholdasova, R.M.; Sagadieva, K.Zh.; Zebreva, A.I. *Izv. Akad. Nauk Kaz. SSR, Ser. Khim.* 1976, No. 3, 63.

(Continued next page)

COMPONENTS:	EVALUATOR:
(1) Cerium; Ce; [7440-45-1] (2) Mercury; Hg; [7439-97-6]	C. Guminski; Z. Galus Department of Chemistry University of Warsaw Warsaw, Poland July, 1985

CRITICAL EVALUATION: (Continued)

10. Sagadieva, K.Zh.; Zebreva, A.I. *Izv. Akad. Nauk. Kaz. SSR, Ser. Khim.* 1977, No. 5, 28.
11. Kozin, L.F. *Vestn. Akad. Nauk. Kaz. SSR* 1972, No. 3, 34.
12. Kozin, L.F. *Tr. Inst. Khim. Nauk Akad. Nauk Kaz. SSR* 1962, *9*, 101.
13. Sagadieva, K.Zh.; Zebreva, A.I.; Dzholdasova, R.M.; Mukhamedieva, Sh.M. *Sbor. Rab. Khim.*, Alma-Ata, 1973, No. 3, 341.
14. Sagadieva, K.Zh.; Zebreva, A.I.; Dzholdasova, R.M.; Mamutova, Z.A. *Prikl. Teoret. Khim.*, Alma-Ata 1973, 152.
15. Usenova, K.A.; Aygaraeva, M.M.; Osipova, G.V.; Enikeev, R.Sh. *Radiokhimia* 1973, *15*, 826.
16. Usenova, K.A.; Krebaeva, Sh.D.; Osipova, G.V.; Enikeev, R.Sh. *Radiokhimia* 1974, *16*, 104.
17. Kirchmayr, H.R.; Lugscheider, W. *Z. Metallk.* 1966, *57*, 725.
18. Usenova, K.A.; Osipova, G.V. *Sbor. Rabot Khim.*, Alma-Ata, 1973, No. 3, 364.
19. Sagadieva, K.Zh. cited by M.T. Kozlovskii, A.I. Zebreva, V.P. Gladyshev, *Amalgamy i ikh Primienienie*, Nauka, Alma-Ata, 1971, p. 19.
20. Merlo, F.; Fornasini, M.L. *J. Less-Common Metals* 1979, *64*, 221.
21. Sagadieva, K.Zh.; Badavamova, G.L. *VII Vsesoyuznoe Soveshchanie po Polarografii*, Nauka, Moskva, 1978, p. 122.
22. Sagadieva, K.Zh.; Dzholdasova, R.M.; Zebreva, A.I. *Uspekhi Polarografii s Nakopleniem*, Tomsk, 1973, p. 104.
23. Bowersox, D.F.; Leary, J.A. *U.S. At. Ener. Comm. Rep.*, LAMS-2518 1961.
24. Muthman, W.; Beck, H. *Ann. Chem.* 1904, *331*, 46.
25. Iandelli, A.; Palenzona, A. *Handbook on the Physics and Chemistry of Rare Earths*, K. A. Gschneider, L. Eyring, Eds., North-Holland, Amsterdam 1978, Ch. I.
26. Berndt, A.F. *J. Less-Common Metals* 1967, *13*, 366.
27. Sagadieva, K.Zh.; Badavamova, G.L.; Zebreva, A.I. *Izv. Vyssh. Ucheb. Zaved., Khim. Khim. Tekhnol.* 1984, *27*, 329.

COMPONENTS:	ORIGINAL MEASUREMENTS:
(1) Cerium; Ce; [7440-45-1] (2) Mercury; Hg; [7439-97-6]	Bowersox, D.F.; Leary, J.A. *U.S. At. Ener. Comm. Rep.*, *LAMS-2518*, <u>1961</u>.
VARIABLES:	PREPARED BY:
Temperature: 20-250°C	C. Guminski; Z. Galus

EXPERIMENTAL VALUES:

The solubilities of Ce in Hg.

$t/°C$	g Ce/dm^3 Hg	Soly/at %^{a}
20	1.31	1.38×10^{-2}
150	19.8	0.214
250	74.5	0.817

^{a}by compilers.

AUXILIARY INFORMATION

METHOD/APPARATUS/PROCEDURE:	SOURCE AND PURITY OF MATERIALS:
Hg was outgassed in a reaction vessel at 523 K then cooled to room temperature. The vessel was filled with He and a weighed quantity of Ce coupons was added. The evacuation and refilling of the vessel with He were repeated several times. The mixture of the metals was equilibrated for 24 h at 623 K. Then temperature of the vessel was adjusted to a selected level and samples of the amalgam were drawn through a coarse Pyrex frit at intervals of 5 to 90 h. Each sample was cooled, weighed and analyzed for Ce content. The procedure gives good results when filtration is carried out 20 h after adjustment of equilibration temperature.	Triply distilled Hg was used. Cerium purity not specified.
	ESTIMATED ERROR: Soly: precision better than ± 2%. Temp: nothing specified.
	REFERENCES:

COMPONENTS:	ORIGINAL MEASUREMENTS:
(1) Cerium; Ce; [7440-45-1] (2) Mercury; Hg; [7439-97-6]	Sagadieva, K.Zh.; Zebreva, A.I.; Dzholdasova, R.M.; Mukhamedieva, Sh.M. *Sbor. Rabot Khim.*, Alma-Ata, 1973, No. 3, 341-6.
VARIABLES: Temperature: 25-70°C	PREPARED BY: C. Guminski; Z. Galus

EXPERIMENTAL VALUES:

Solubility of cerium at 25-70°C.

$t/°C$	Soly/mol dm^{-3} x 10^3	Soly/at % x 10^{3a}
25	6	9
40	8	12
55	10	15
70	12	18

^{a}by compilers.

The same results are also reported in (1).

AUXILIARY INFORMATION

METHOD/APPARATUS/PROCEDURE:	SOURCE AND PURITY OF MATERIALS:
The amalgams were obtained by reduction of Ce(III) with sodium amalgam. Cerium amalgams were oxidized potentiostatically at -0.1 V vs. SCE, and current-time dependencies were recorded. Current was initially constant, then it decreased exponentially with time. The concentration of the saturated amalgam was calculated from the charge corresponding to the oxidation in the region of exponential dependence of the current.	Nothing specified.
	ESTIMATED ERROR: Soly: precision $\pm$ 10% (compilers). Temp: nothing specified.
	REFERENCES: 1. Sagadieva, K.Zh.; Dzholdasova, R.M.; Zebreva, A.I. *Uspekhi Polarografii s Nakopleniem*, Tomsk, 1973, pp. 104-5.

COMPONENTS:	ORIGINAL MEASUREMENTS:
(1) Cerium; Ce; [7440-45-1] (2) Mercury; Hg; [7439-97-6]	1. Usenova, K.A.; Osipova, G.V.; Krebaeva, Sh.D.; Enikeev, R.Sh. *Radiokhimia* 1974, *16*, 99-103. 2. Dzholdasova, R.M.; Sagadieva, K.Zh.; Zebreva, A.I. *Izv. Akad. Nauk. Kaz. SSR, Ser. Khim.* 1976, No. 3, 63-6.
VARIABLES: One temperature: 25°C	PREPARED BY: C. Guminski; Z. Galus

EXPERIMENTAL VALUES:

The solubility of cerium in mercury at 25°C was found to be $(6.8 \pm 0.7) \times 10^{-3}$ and $(6.0 \pm 1.0) \times 10^{-3}$ mol dm^{-3} in (1) and (2), respectively.

The respective atomic % solubility calculated by the compilers are 1.00×10^{-2} and 9.0×10^{-3} at %.

Refs. (3)-(5) reported identical solubility as in (2), while (6) reported a slightly higher value of 7×10^{-3} mol dm^{-3} (read from a curve by the compilers).

AUXILIARY INFORMATION

METHOD/APPARATUS/PROCEDURE:	SOURCE AND PURITY OF MATERIALS:
The amalgams were prepared by reduction of Ce(III) with Na amalgam; the Ce content in the amalgams was determined by analysis of solution before and after reduction. Constantly stirred acetate-buffered solutions were oxidized potentiostatically: at -0.5 V vs. SCE at pH 3 in (1) and at -0.1 V vs. SCE in (2). Kinetic decomposition curves were recorded in (1) and the inflection on the plot of half-decomposition time vs. Ce concentration corresponded to the saturation point; concentration change was traced with Ce^{144} radioisotope. In (2) the solubility was determined from the exponential part of the current-time curve which corresponds to the charge consumed for the oxidation of the homogeneous amalgam. All measurements were made in hydrogen atmosphere.	Nothing specified.
	ESTIMATED ERROR: Soly: precision approximately $\pm$ 10% in (1) and $\pm$ 15% in (2). Temp: precision $\pm$ 0.5 K.

REFERENCES:

3. Sagadieva, K.Zh.; Zebreva, A.I. *Izv. Akad. Nauk Kaz. SSR, Ser. Khim.* 1977, No. 5, 28.
4. Zebreva, A.I.; Sagadieva, K.Zh.; Dzholdasova, R.M. *Issl. v Obl. Khim. Redkozeml. Elementov* 1975, 77.
5. Sagadieva, K.Zh.; Zebreva, A.I.; Dzholdasova, R.M.; Mukhamedieva, Sh.M. *Sbor. Rabot Khim.*, Alma-Ata, 1973, No. 3, 341.
6. Dzholdasova, R.M.; Sagadieva, K.Zh.; Zebreva, A.I. *Prikl. Teoret. Khim.*, Alma-Ata, 1974, 206.

COMPONENTS:	ORIGINAL MEASUREMENTS:
(1) Cerium; Ce; [7440-45-1] (2) Mercury; Hg; [7439-97-6]	1. Bulina, V.A.; Zebreva, A.I.; Enikeev, R.Sh. *Izv. Vyssh. Ucheb. Zaved., Khim. Khim. Tekhnol.* 1977, *20*, 959-61. 2. Sagadieva, K.Zh.; Zebreva, A.I.; Badavamova, G.L. *Elektrokhimia* 1979, *15*, 210-3.
VARIABLES: One temperature: 25°C	PREPARED BY: C. Guminski; Z. Galus

EXPERIMENTAL VALUES:

The solubility of cerium in mercury at 25°C was found to be $(6.3 \pm 0.5) \times 10^{-3}$ and 5.9×10^{-3} mol dm^{-3}, respectively, in (1) and (2). The respective atomic % solubility calculated by the compilers are 9.3×10^{-3} and 8.7×10^{-3} at %. The result of (2) was previously reported in (3).

AUXILIARY INFORMATION

METHOD/APPARATUS/PROCEDURE:	SOURCE AND PURITY OF MATERIALS:
Amalgams in (2) were prepared by reduction of Ce(III) with Na amalgam. Amalgams of various concentrations were oxidized potentiostatically at limiting-current potential of -0.1 V vs. SCE, and current recorded as function of time. Measurements were made in static buffered solution of 0.2 mol dm^{-3} NaOAc + 0.04 mol dm^{-3} HCl at pH 5. Limiting current, i_d, was obtained from semilog plot of current vs. time; i_d increased linearly with Ce concentration, and was nearly constant above saturation. Solubility determined from intercept of i_d vs. concentration plot. Heterogeneous amalgam in (1) also oxidized potentiostatically, and transition from oxidation of heterogeneous to homogeneous amalgam was indicated by inflection in current-time curve. Solubility determined from charge used for oxidation of homogeneous amalgam.	Nothing specified.
	ESTIMATED ERROR: Soly: precision 10-20% (compilers). Temp: precision probably $\pm$ 0.5 K (compilers).
	REFERENCES: 3. Sagadieva, K.Zh.; Badavamova, G.L. *VII Vsesoyuznoe Soveshchanie po Polarografii*, Nauka, Moskva, 1978, p. 122.

COMPONENTS:	ORIGINAL MEASUREMENTS:
(1) Cerium; Ce; [7440-45-1] (2) Mercury; Hg; [7439-97-6]	Sagadieva, K.Zh.; Badavamova, G.L.; Zebreva, A.I. *Izv. Vyssh. Ucheb. Zaved., Khim. Khim. Tekhnol.* 1984, *27*, 329-33.
VARIABLES:	PREPARED BY:
One temperature: 25°C	C. Guminski; Z. Galus

EXPERIMENTAL VALUES:

Solubility of Ce in Hg at 25°C was reported to be 5.6×10^{-3} mol dm^{-3}.
The atomic % solubility calculated by the compilers is 8.3×10^{-3} at %.

AUXILIARY INFORMATION

METHOD/APPARATUS/PROCEDURE:	SOURCE AND PURITY OF MATERIALS:
The amalgam was prepared by electro-reduction of Ce(III) in citrate buffer (pH 5-6) on the hanging-mercury-drop electrode with Pt base. The electrolysis was carried out in potentiostatic conditions at potentials changing between -1.7 and -2.5 V vs. SCE in atmosphere of an inert gas. The Ce amalgam was then oxidized by stripping voltammetry. The total amount of Ce in Hg was found by integration of area under the voltammetric peak. A break on the plot relating the anodic peak current against Ce concentration corresponds to the saturation of the amalgam.	Purity of Hg drop electrode was tested by stripping analysis without the depolarizer in the solution. Cerium purity not specified.
	ESTIMATED ERROR: Soly: precision no better than ± 10% (compilers). Temp: nothing specified.
	REFERENCES:

COMPONENTS:	EVALUATOR:
(1) Praseodymium; Pr; [7440-10-0] (2) Mercury, Hg; [7439-97-6]	C. Guminski; Z. Galus Department of Chemistry University of Warsaw Warsaw, Poland July, 1985

CRITICAL EVALUATION:

The most reliable solubility of praseodymium in mercury at 298 K has been determined by electroanalytical measurements by Zebreva and coworkers: 8.3×10^{-3} (6), 1.15×10^{-2} (7,10,16), 1.12×10^{-2} (8,14), and 1.41×10^{-2} at % (8,14). At 343 K, Zebreva et al. reported solubilities of 2.35×10^{-2} (8,14), 2.36×10^{-2} (10) and 2.41×10^{-2} at % (8,14). Other experimental determinations at 298 K are rejected because the solubilities were too high (3.1×10^{-2} at % (3,13) and 5.6×10^{-2} at % (4)), or too low (5.9×10^{-3} at % (5)). Kozin's calculated solubilities of 0.13 (1) and 0.15 (2) at % at 298 K are clearly too high.

The evaluators' plot of the logarithm of solubility vs. reciprocal temperature for the Pr-Hg system shows a significantly lower slope as compared to the same type of plots for the other lanthanide-Hg systems. The lower slope also suggests a lower enthalpy of solution for the Pr-Hg system as compared to the other Ln-Hg systems.

Griffin and Gschneider (9) determined the liquidus in the Pr-rich region, while Kirchmayr and Lugscheider (12) presented a schematic phase diagram at 760 mm Hg vapor pressure. The compound, $PrHg_{6.5}$, has been established (17), but its decomposition temperature is not known. Existence of other compounds also has been reported (9,11,12,15,17).

Tentative values of the solubility of Pr in Hg:

T/K	Soly/at %	Reference
298	1.1×10^{-2}	[7,8,10,16]
323	1.7×10^{-2}	[8,10]
973	84	[9]
1073	91	[9]
1173	97.5	[9]

References

1. Kozin, L.F. *Tr. Inst. Khim. Nauk Akad. Nauk Kaz. SSR* 1962, *9*, 101.
2. Kozin, L.F. *Fiziko-Khimicheskie Osnovy Amalgamnoi Metalurgii*, Nauka, Alma-Ata, 1964.
3. Usenova, K.A.; Osipova, G.V.; Krebaeva, Sh.D.; Enikeev, R.Sh. *Radiokhimia* 1974, *16*, 99.
4. Shvedov, V.P.; Frolkov, A.Z.; Nikishin, G.D. *Radiokhimia* 1971, *13*, 252.
5. Bulina, V.A.; Zebreva, A.I.; Enikeev, R.Sh. *Izv. Vyssh. Ucheb. Zaved., Khim. Khim. Tekhnol.* 1977, *20*, 959.
6. Bulina, V.A.; Guminichenko, L.V.; Zebreva, A.I.; Enikeev, P.Sh. *Radiokhimia* 1977, *19*, 89.
7. Sagadieva, K.Zh.; Zebreva, A.I.; Badavamova, G.L. *Elektrokhimia* 1979, *15*, 210.
8. Sagadieva, K.Zh.; Zebreva, A.I.; Oteeva, G.Z. *Izv. Akad. Nauk Kaz. SSR, Ser. Khim.* 1977, No. 6, 22.
9. Griffin, R.B.; Gschneider, K.A. *Met. Trans.* 1971, *2*, 2517.
10. Sagadieva, K.Zh.; Zebreva, A.I.; Badavamova, G.L. *Izv. Akad. Nauk Kaz. SSR, Ser. Khim.* 1978, No. 3, 74.
11. Flad, D.; Matthes. F. *Z. Chem.* 1964, *4*, 466.
12. Kirchmayr, H.R.; Lugscheider, W. *Z. Metallk.* 1966, *57*, 725.
13. Usenova, K.A.; Osipova, G.V. *Sbor. Rab. Khim.*, Alma-Ata, 1973, No. 3, 364.
14. Sagadieva, K.Zh.; Zebreva, A.I. *Dep. VINITI*, 1355-77, 1977; abstracted in *Izv. Vyssh. Ucheb. Zaved., Khim. Khim. Tekhnol.* 1978, *21*, 157.
15. Merlo, F.; Fornasini, M.L. *J. Less-Common Metals* 1979, *64*, 221.
16. Sagadieva, K.Zh.; Badavamova, G.L. *VII Vsesoyuznoe Soveshchanie po Polarografii*, Nauka, Moskva, 1978, p. 122.
17. Iandelli, A.; Palenzona, A. *Handbook on Physics and Chemistry of Rare Earths*, Gschneider, K.A., Eyring, L., Eds., North-Holland, Amsterdam 1978, Ch. I.

COMPONENTS:	ORIGINAL MEASUREMENTS:
(1) Praseodymium; Pr; [7440-10-0] (2) Mercury; Hg; [7439-97-6]	Griffin, R.B.; Gschneider, K.A. *Met. Trans.* 1971, *2*, 2517-24.
VARIABLES: Temperature: 650-932°C	PREPARED BY: C. Guminski; Z. Galus

EXPERIMENTAL VALUES:

Data were presented as a phase diagram in the Pr-rich region. The solubility of Hg in Pr was taken from the liquidus data points by compilers.

t/°C	Soly/at %
932	0.15
930	0.3
926	0.6
920	1.1
916	1.3
907	2.0
900	2.6
890	3.3
866	5.1
839	6.8
807	9.1
792	10.2
770	12.4
713	15.2
675	17.3
650	18.9

AUXILIARY INFORMATION

METHOD/APPARATUS/PROCEDURE:	SOURCE AND PURITY OF MATERIALS:
The alloys were prepared by adding weighed amounts of mercury to a tantalum crucible containing praseodymium or previously prepared amalgam with lower content of mercury. The crucibles were sealed by arc-welding in a helium atmosphere. The samples were then melted at 200 to 250°C above the liquidus temperature for 1 hour. Differential thermal analysis cooling curves were recorded with the use of a calibrated Chromel-Alumel thermocouple.	Praseodymium was 99.6% pure, with 0.02% calcium and 0.3% other lanthanides. Mercury was triply distilled.
	ESTIMATED ERROR: Soly: nothing specified. Temp: precision ± 2 K.
	REFERENCES:

COMPONENTS:	ORIGINAL MEASUREMENTS:
(1) Praseodymium; Pr; [7440-10-0] (2) Mercury; Hg; [7439-97-6]	1. Bulina, V.A.; Guminichenko, L.V.; Zebreva, A.I.; Enikeev, R.Sh. *Radiokhimia* 1977, *19*, 89-93. 2. Sagadieva, K.Zh.; Zebreva, A.I.; Badavamova, G.L. *Elektrokhimia* 1979, *15*, 210-13.
VARIABLES: One temperature: 25°C	PREPARED BY: C. Guminski; Z. Galus

EXPERIMENTAL VALUES:

The solubility of praseodymium in mercury at 25°C was reported as follows:

Soly/mol dm^{-3} x 10^3	Soly/at % x 10^{3a}	Reference
5.6 $\pm$ 1.0	8.3	(1)
7.8	11.5	(2)

^{a}by compilers.

Result of (2) was also reported in (3) and (4).

AUXILIARY INFORMATION

METHOD/APPARATUS/PROCEDURE:	SOURCE AND PURITY OF MATERIALS:
Heterogeneous amalgams were prepared by reduction of Pr(III) with Na-amalgam from buffered solutions of pH 4. The amalgams of various concentrations were oxidized potentiostatically: the oxidation in (2) was made at limiting-current potential of -0.1 V vs. SCE at pH 5 in soln. of 0.2 mol dm^{-3} NaOAc + 0.04 mol dm^{-3} HCl. Limiting current, i_d, was obtained from semilog plot of current vs. time; i_d increased linearly with Pr concentration, and became nearly constant above saturation. Solubility was determined from the inflection point of curve of i_d vs. Pr concentration. In (1) the solubility was obtained from the charge used to oxidize the homogeneous amalgam up to the saturation point.	Nothing specified.
	ESTIMATED ERROR: Soly: nothing specified; precision no better than 10-20% (compilers). Temp: nothing specified.
	REFERENCES: 3. Sagadieva, K.Zh.; Badavamova, G.L. *VII Vsesoyuznoe Soveshchanie po Polarografii*, Nauka, Moskva, 1978, p. 122-3. 4. Sagadieva, K.Zh.; Zebreva, A.I.; Badavamova, G.L. *Izv. Akad. Nauk Kaz. SSR, Ser. Khim.* 1978, No. 3, 74-6.

COMPONENTS:	ORIGINAL MEASUREMENTS:
(1) Praseodymium; Pr; [7440-10-0] (2) Mercury; Hg; [7439-97-6]	1. Sagadieva, K.Zh.; Zebreva, A.I.; Oteeva, G.Z. *Izv. Akad. Nauk Kaz. SSR, Ser. Khim.* 1977, No. 6, 22-4. 2. Sagadieva, K.Zh.; Zebreva, A.I.; Badavamova, G.L. *Izv. Akad. Nauk Kaz. SSR, Ser. Khim.* 1978, No. 3, 74-6.
VARIABLES:	PREPARED BY:
Temperature: 25-70°C	C. Guminski; Z. Galus

EXPERIMENTAL VALUES:

Solubility of praseodymium in mercury was determined by chronoamperometry and coulometry.

	By Chronoamperometry			By Coulometry[b]	
	Soly			Soly	
$t/°C$	mol dm^{-3}	at %[a]	Ref.	mol dm^{-3}	at %[a]
25	0.0095	0.0141	(1)	0.0076	0.0112
	0.0078	0.0115	(2)		
40	0.0125	0.0125	(1)	0.0098	0.0145
	0.0097	0.0144	(2)		
55	0.0145	0.0215	(1)	0.0115	0.0170
	0.0112	0.0166	(2)		
70	0.0163	0.0241	(1)	0.0159	0.0235
	0.0160	0.0236	(2)		

[a] by compilers.

[b] ref. (1).

Results of ref. (1) also reported in (3); authors in (1) state that coulometric results are more accurate.

AUXILIARY INFORMATION

METHOD/APPARATUS/PROCEDURE:	SOURCE AND PURITY OF MATERIALS:
Amalgams were prepared by reduction of Pr(III) with Na-amalgam. Chronoamperometric curves were recorded in (1); magnitude of limiting current, i_d, is proportional to Pr concentration, N_{Pr}, in homogeneous amalgam. Inflection point on i_d vs. N_{Pr} curve corresponds to solubility of Pr. After all crystals in amalgam are dissolved by chronoamperometric oxidation, the solubility is calculated by integrating the charge corresponding to exponentially decreasing current (coulometry). In (2), N_{Pr} varied from 1.8-75.0 x 10^{-3} mol dm^{-3}. Amalgams allowed to stand 6 hr before chronoamperometric oxidation at -0.1 V vs. SCE; solubility obtained as in (1).	Nothing specified.
	ESTIMATED ERROR:
	Soly: precision no better than ± 10% (compilers). Temp: nothing specified.
	REFERENCES:
	3. Sagadieva, K.Zh.; Zebreva, A.I. *Dep. VINITI*, 1355-77, 1977; abstracted in *Izv. Vyssh. Ucheb. Zaved., Khim. Khim. Tekhnol.* 1978, 21, 157.

COMPONENTS:	EVALUATOR:
(1) Neodymium; Nd; [7440-00-8] (2) Mercury; Hg; [7439-97-6]	C. Guminski; Z. Galus Department of Chemistry University of Warsaw Warsaw, Poland July, 1985

CRITICAL EVALUATION:

Kozin calculated neodymium solubilities in mercury at 298 K of 0.044 (1) and 0.058 (2,8) at %. Experimental determinations show that the latter values are too high. Bulina et al. reported a solubility of 8.1×10^{-3} at % (4) and 4.6×10^{-3} at % (7) at 298 K, while Usenova et al. (5,10) reported 1.6×10^{-2} at %. Sagadieva et al. (14), from electroanalytical methods, reported that the neodymium solubility increased from 4.9×10^{-3} to 1.1×10^{-2} at % at 298 to 343 K; the results at 298 and 313 K are in good agreement with those reported by (3), (4) and (7), but the solubilities at higher temperatures were too low as compared to other measurements. Shvedov et al. (6) reported a solubility of 0.055 at % at 293 K, but this value is rejected because it is too high. Messing and Dean (3) determined the solubilities at 368 to 573 K and they reported that the solubility increased from 0.0496 to 0.680 at % over this temperature range.

The saturated neodymium amalgams are in equilibrium with Nd-Hg intermetallic compounds (9,11,12); a schematic phase diagram has been presented by Kirchmayr and Lugscheider (11). $NdHg_{6.5}$ solid phase was also determined, but no temperature range of its stability was reported (13).

The tentative values of neodymium solubility in mercury.

T/K	Soly/at %	Reference
298	6×10^{-3} [a]	[4,7,14]
323	1.2×10^{-2} [b]	[3]
373	4×10^{-2}	[3]
473	0.2	[3]
573	0.6	[3]

[a] mean value from data of cited references.
[b] extrapolated from data of (3).

References

1. Kozin, L.F. *Tr. Inst. Khim. Nauk Akad. Nauk Kaz. SSR* 1962, 9, 101.
2. Kozin, L.F. *Fiziko-Khimicheskie Osnovy Amalgamnoi Metalurgii*, Nauka, Alma-Ata, 1964.
3. Messing, A.F.; Dean, O.C. *U.S. At. Ener. Comm. Rep.*, ORNL-2871, 1960.
4. Bulina, V.A.; Guminichenko, L.V.; Zebreva, A.I.; Enikeev, R.Sh. *Radiokhimia*, 1977, 19, 89.
5. Usenova, K.A.; Osipova, G.V.; Krebaeva, Sh.D.; Enikeev, R.Sh. *Radiokhimia* 1974, 16, 99.
6. Shvedov, V.P.; Frolkov, A.Z.; Nikishin, G.D. *Radiokhimia* 1971, 13, 252.
7. Bulina, V.A.; Zebreva, A.I.; Enikeev, R.Sh. *Izv. Vyssh. Ucheb. Zaved., Khim. Khim. Tekhnol.* 1977, 20, 959.
8. Kozin, L.F. *Vestn. Akad. Nauk Kaz. SSR* 1972, No. 3, 34.
9. Flad, D.; Matthes, F. *Z. Chem.* 1964, 4, 466.
10. Usenova, K.A.; Osipova, G.V. *Sborn. Rab. Khim. Kaz. Univer.*, Alma-Ata, 1973, No. 3, 364.
11. Kirchmayr, H.S.; Lugscheider, W. *Z. Metallk.* 1966, 57, 725.
12. Merlo, F.; Fornasini, M.L. *J. Less-Common Metals* 1979, 64, 221.
13. Iandelli, A.; Palenzona, A. *Handbook on the Physics and Chemistry of Rare Earths*, K. A. Gschneider, L. Eyring, Eds., North-Holland, 1978, Amsterdam, Ch. I.
14. Sagadieva, K.Zh.; Badavamova, G.L.; Zebreva, A.I. *Izv. Akad. Nauk Kaz. SSR, Ser. Khim.*, 1983, No. 6, 29.

COMPONENTS:	ORIGINAL MEASUREMENTS:
(1) Neodymium; Nd; [7440-00-8] (2) Mercury; Hg; [7439-97-6]	Messing, A.F.; Dean, O.C. *U.S. At. Ener. Comm. Rep.*, ORNL-2871, 1960.
VARIABLES: Temperature: 313-523 K	PREPARED BY: C. Guminski; Z. Galus

EXPERIMENTAL VALUES:

The solubility of neodymium in mercury at various temperatures.

T/K	Soly/mass %	Soly/at %
313	0.00898[a]	0.0125
368-373	0.0357	0.0496
378-383	0.0296	0.0411
433-438	0.0746	0.104
473-477	0.188	0.261
483-488	0.193	0.268
518-523	0.196	0.272
521-523	0.262	0.364
568-573	0.418	0.680
629	0.633[a]	0.877

[a] calculated from least squares equation by the authors.

AUXILIARY INFORMATION

METHOD/APPARATUS/PROCEDURE:	SOURCE AND PURITY OF MATERIALS:
Mercury and neodymium, after drying and outgassing in the stainless steel dissolver, were kept for several days at the desired temperature. A similar equilibration period was allowed after each temperature change. After equilibration, a sample of liquid amalgam was forced through the filter, and the collected sample was dissolved in nitric acid and submitted for analysis for neodymium and mercury.	Not given.
	ESTIMATED ERROR: Soly: standard deviation of least squares fit 0.1063. Temp: precision ± 5 K.
	REFERENCES:

COMPONENTS:	ORIGINAL MEASUREMENTS:
(1) Neodymium; Nd; [7440-00-8] (2) Mercury; Hg; [7439-97-6]	Usenova, K.A.; Osipova, G.V.; Krebaeva, Sh.D.; Enikeev, R.Sh. *Radiokhimia* 1974, *16*, 99-103.
VARIABLES: One temperature: 25°C	PREPARED BY: C. Guminski; Z. Galus

EXPERIMENTAL VALUES:

The solubility of neodymium in mercury at 25°C was found to be $(1.1 \pm 0.2) \times 10^{-2}$ mol dm^{-3}. The atomic % solubility calculated by the compilers is 1.6×10^{-2} at %.

The same result is also reported in (1).

AUXILIARY INFORMATION

METHOD/APPARATUS/PROCEDURE:	SOURCE AND PURITY OF MATERIALS:
The amalgam was prepared by reduction of Nd(III) with sodium amalgam. The amalgam was then oxidized potentiostatically, at -0.5 V vs. SCE, in a stirred system containing the amalgam and an acetate buffer of pH = 3. Based on analysis of the kinetic decomposition curves, the oxidation of homogeneous and heterogeneous amalgams could be distinguished. An inflection on the plot of half-decomposition time versus concentration of Nd corresponds to the saturation point. Changes of concentration were determined with the use of Nd^{147} radioisotope. The experiments were performed in H$_2$ atmosphere.	Not specified.
	ESTIMATED ERROR: Soly: precision approximately $\pm$ 20%. Temp: precision $\pm$ 0.5 K.
	REFERENCES: 1. Usenova, K.A.; Osipova, G.V. *Sborn. Rab. Khim.* Alma-Ata, 1973, No. 3, 364.

COMPONENTS:	ORIGINAL MEASUREMENTS:
(1) Neodymium; Nd; [7440-00-8] (2) Mercury; Hg; [7439-97-6]	1. Bulina, V.A.; Zebreva, A.I.; Enikeev, R.Sh. *Izv. Vyssh. Ucheb. Zaved., Khim. Khim. Tekhnol.* 1977, *20*, 959-61. 2. Bulina, V.A.; Guminichenko, L.V.; Zebreva, A.I.; Enikeev, R.Sh. *Radiokhimia* 1977, *19*, 89-93.
VARIABLES: One temperature: 25°C	PREPARED BY: C. Guminski; Z. Galus

EXPERIMENTAL VALUES:

The solubility of neodymium in mercury at 25°C was found to be $(3.1 \pm 0.5) \times 10^{-3}$ and $(5.5 \pm 0.8) \times 10^{-3}$ mol dm^{-3}, respectively, in (1) and (2). The corresponding atomic % solubility calculated by the compilers were 4.6×10^{-3} and 8.1×10^{-3} at %.

AUXILIARY INFORMATION

METHOD/APPARATUS/PROCEDURE:	SOURCE AND PURITY OF MATERIALS:
The method of preparation of amalgam in (1) was not specified. In (2) the amalgam was prepared by reduction of Nd(III) with Na-amalgam from chloride-acetate buffered solution of pH = 4; the Nd content in the amalgam was determined from the Nd(III) in the solution before and after reduction. The amalgam in (1) was oxidized potentiostatically at -0.3 V vs. SCE, and the solubility was obtained from the i-t curve and the charge for oxidation of the saturated amalgam. The solubility in (2) was obtained by a similar method.	Nothing specified.
	ESTIMATED ERROR: Soly: precision approximately $\pm$ 20%. Temp: nothing specified.
	REFERENCES:

COMPONENTS:	ORIGINAL MEASUREMENTS:
(1) Neodymium; Nd; [7440-00-8] (2) Mercury; Hg; [7439-97-6]	Sagadieva, K.Zh.; Badavamova, G.L.; Zebreva, A.I. *Izv. Akad. Nauk Kaz. SSR, Ser. Khim.* 1983, No. 6, 29-32.
VARIABLES: Temperature	PREPARED BY: C. Guminski; Z. Galus

EXPERIMENTAL VALUES:

The solubilities of Nd in Hg at various temperatures were determined by coulometry and chronoamperometry.

$t/°C$	25	40	55^b	70^b	
Soly/10^3 mol dm^{-3}	3.3	5.4	6.5	6.9	
Soly/10^3 at %^{a}	4.9	8.0	9.6	10	by coulometry
Soly/10^3 mol dm^{-3}	3.4	5.6	6.0	7.5	
Soly/10^3 at %^{a}	5.0	8.3	8.9	11	by chronoamperometry

^{a} by compilers.
^{b} results at 55 and 70°C are understated.

AUXILIARY INFORMATION

METHOD/APPARATUS/PROCEDURE:	SOURCE AND PURITY OF MATERIALS:
The amalgams were prepared by reduction of Nd(III) by sodium amalgam in chloride-acetate solution at pH = 4. The amount of Nd introduced into Hg was determined from analysis of the Nd(III) solution before and after reduction. Chronoamperometric oxidation of the amalgam was then carried out at 0.1 V vs. SCE. The saturation point was estimated from a bend on the current vs. Nd concentration curve (chrono-amperometry), as well as by the integration of the diffusion current as a function of time (couloumetry).	Nothing specified.
	ESTIMATED ERROR: Soly: nothing specified; precision ± 10% (compilers). Temp: nothing specified.
	REFERENCES:

COMPONENTS:	EVALUATOR:
(1) Samarium; Sm; [7440-19-9] (2) Mercury; Hg; [7439-97-6]	C. Guminski; Z. Galus Department of Chemistry University of Warsaw Warsaw, Poland July, 1985

CRITICAL EVALUATION:

The first experimental determination of the solubility of samarium in mercury was reported by Messing and Dean (3); these authors found that the solubility increased from 0.0501 to 0.652 at % at 358 to 578 K, and the extrapolated solubility at 298 K, from their fitted solubility equation, was 1.2×10^{-2} at %. Because the solubility determinations by these authors for other lanthanide metals in mercury are generally acceptable, the evaluators tentatively accept the data for the Sm-Hg system reported in (3). Kozin reported predicted solubilities of 3.2×10^{-2} (1) and 4.5×10^{-2} at % (2) at 298 K. The solubility determination of Zebreva and coworkers (6-8, 11), 4.6×10^{-2} at % at 298 K, is in rough agreement with the calculated results of Kozin and with the extrapolated value of Messing and Dean. Other determinations of the solubility near room temperature are: 0.15 at % at 298 K by potentiometry (4,12); 0.164 at % at 293 K by amalgam polarography (5); 0.13 at % at 293 K by anodic oxidation of a thin film of the amalgam (15); and 0.124 at % at 293 K by an unspecified method (9); all of these latter values are too high and are rejected. Dzholdasova (16) reported solubilities of 0.045, 0.060 and 0.094 at % at 286, 298 and 313 K, respectively; however, experimental details were not described.

As shown by the schematic phase diagram for the Ln-Hg system (13,14,17), the saturated amalgam is in equilibrium with intermetallic solid phases. Also, $SmHg_{6.5}$ was established (17) but no decomposition temperature is known.

The tentative values of Sm solubility in Hg:

T/K	Soly/at %	Reference
298	2×10^{-2} [a]	(3)
323	3×10^{-2} [a]	(3)
373	6×10^{-2}	(3)
473	0.2	(3)
473	0.6	(3)

[a] extrapolated values from data of (3).

References

1. Kozin, L.F. *Tr. Inst. Khim. Nauk Akad. Nauk Kaz. SSR* 1962, 9, 101.
2. Kozin, L.F. *Fiziko-Khimicheskie Osnovy Amalgamnoi Metallurgii*, Nauka, Alma-Ata, 1964.
3. Messing, A.F.; Dean, O.C. *U.S. At. Ener. Comm. Rep.*, ORNL-2871, 1960.
4. Usenova, K.A.; Krebaeva, Sh.D.; Osipova, G.V.; Enikeev, R.Sh. *Radiokhimia*, 1974, 16, 104.
5. Shvedov, V.P.; Frolkov, A.Z.; Nikishin, G.D. *Radiokhimia*, 1971, 13, 252.
6. Sagadieva, K.Zh.; Zebreva, A.I.; Badavamova, G.L. *Elektrokhimia*, 1979, 15, 210.
7. Sagadieva, K.Zh.; Zebreva, A.I.; Dzholdasova, R.M.; Mertke, I. *Dep. VINITI*, 2573-75, 1975.
8. Sagadieva, K.Zh.; Zebreva, A.I. *Izv. Akad. Nauk Kaz. SSR, Ser Khim.* 1977, No. 5, 28.
9. Gerasimov, Ya.I.; Krestovnikov, N.N.; Kuznetsov, F.A.; Lavrentev, V.I.; Lomov, A.L. *Khimicheskaya Termodynamika v Tsvetnoi Metallurgii*, Vol. 7, Metallurgia, Moskva, 1975; as cited by Kozin, L.F.; Nigmetova, R.Sh.; Dergacheva, M.B. *Termodinamika Binarnykh Amalgamnykh System*, Nauka, Alma-Ata, 1977, p. 192.
10. Kozin, L.F. *Vestn. Akad. Nauk Kaz. SSR* 1972, No. 3, 94.
11. Zebreva, A.I.; Sagadieva, K.Zh.; Dzholdasova, R.M.*Issl. Obl. Khim. Redkozeml. Elementov*, 1975, 77.
12. Usenova, K.A.; Krebaeva, Sh.D. *Sbor. Rab. Khim.*, Alma-Ata, 1973, No. 3, 369.

(Continued next page)

COMPONENTS:	EVALUATOR:
(1) Samarium; Sm; [7440-19-9] (2) Mercury; Hg; [7439-97-6]	C. Guminski; Z. Galus Department of Chemistry University of Warsaw Warsaw, Poland July, 1985

CRITICAL EVALUATION: (continued)

13. Kirchmayr, H.R.; Lugscheider, W. *Z. Metallk.* 1960, *57*, 725.
14. Merlo, F.; Fornasini, M.L. *J. Less-Common Metals* 1979, *64*, 221.
15. Perov, E.I.; Akimov, V.P.; Serebrennikov, V.V. *Tr. Tomsk. Univ.* 1971, *204*, 90.
16. Dzholdasova, R.M. *Dissertation*, Alma-Ata, 1979; as cited by Korshunov, V.I. *Itogi Nauki i Tekhniki, Elektrokhimia* 1981, *17*, 85.
17. Iandelli, A.; Palenzona, A. *Handbook on the Physics and Chemistry of Rare Earths*, K. A. Gschneider, L. Eyring, Eds., North-Holland, 1978, Amsterdam, Ch. I.

COMPONENTS:	ORIGINAL MEASUREMENTS:
(1) Samarium; Sm; [7440-19-9] (2) Mercury; Hg; [7439-97-6]	Messing, A.F.; Dean, O.C. *U.S. At. Ener. Comm. Rep. ORNL-2871, 1960.*
VARIABLES:	PREPARED BY:
Temperature: 313-578 K	C. Guminski; Z. Galus

EXPERIMENTAL VALUES:

The solubility of samarium in mercury was determined at various temperatures.

T/K	Soly/mass %	Soly/at %
313	[a]0.0131	0.0175
358-363	0.0376	0.0501
373-383	0.0627	0.0834
418-423	0.0834	0.111
433-438	0.142	0.189
443-448	0.104	0.139
468-478	0.168	0.224
498-503	0.213	0.284
523-528	0.202	0.269
573-578	0.467	0.621
573-578	0.490	0.652
629	[a]0.618	0.822

[a] Calculated from least squares-fitted equation by the authors.

AUXILIARY INFORMATION

METHOD/APPARATUS/PROCEDURE:	SOURCE AND PURITY OF MATERIALS:
Mercury and the test metal, after drying and outgassing in the stainless steel dissolver, were kept for several days at the desired temperature. A similar equilibration period was allowed after each temperature change. After equilibration, a sample of liquid amalgam was forced through the filter. The filtrate was collected, dissolved in nitric acid, and submitted for analysis for samarium and mercury.	Nothing specified.
	ESTIMATED ERROR: Soly: least squares fit standard deviation 0.07606. Temp: precision ± 5 K.
	REFERENCES:

COMPONENTS:	ORIGINAL MEASUREMENTS:
(1) Samarium; Sm; [7440-19-9] (2) Mercury; Hg; [7439-97-6]	Sagadieva, K.Zh.; Zebreva, A.I. *Izv. Akad. Nauk Kaz. SSR, Ser. Khim.* <u>1977</u>, No. 5, 28-32.
VARIABLES: One temperature: 25°C	PREPARED BY: C. Guminski; Z. Galus

EXPERIMENTAL VALUES:

The solubility of samarium in mercury at 25°C was found to be $(3.1 \pm 0.4) \times 10^{-2}$ mol dm^{-3}. The atomic % solubility calculated by the compilers is 4.6×10^{-2} at %.

The same result is given also in (1-3).

AUXILIARY INFORMATION

METHOD/APPARATUS/PROCEDURE:	SOURCE AND PURITY OF MATERIALS:
The amalgams were presumably prepared by reduction of Sm(II) with sodium amalgam. Amalgams of various concentrations were oxidized potentiostatically at the limiting current potential of -0.1 V vs. SCE, and the current was recorded as a function of time. The measurements were made in a static, buffered solution of 0.2 mol-dm^{-3} NaOAc + 0.04 mol-dm^{-3} HCl at pH = 5. The limiting current, i_d, was obtained from a semilog plot of current vs. time; i_d increased linearly with concentration of Sm, and was nearly constant above saturation. The solubility was determined from the intercept of the i_d vs. concentration plot.	Nothing specified.
	ESTIMATED ERROR: Soly: nothing specified; precision approximately ± 10% (compilers). Temp: precision ± 0.5 K.
	REFERENCES: 1. Zebreva, A.I.; Sagadieva, K.Zh.; Dzholdasova, K.M. *Issl. Obl. Khim. Redkozeml. Elementov* <u>1975</u>, 77. 2. Sagadieva, K.Zh.; Zebreva, A.I.; Dhzoldasova, R.M.; Mertke, I. *Dep. VINITI* 2573-75, <u>1975</u>. 3. Sagadieva, K.Zh.; Zebreva, A.I.; Badavamova, G.L. *Elektrokhimia* <u>1979</u>, *15*, 210.

COMPONENTS:	EVALUATOR:
(1) Europium; Eu; [7440-53-1] (2) Mercury; Hg; [7439-97-6]	C. Guminski; Z. Galus Department of Chemistry University of Warsaw Warsaw, Poland July, 1985

CRITICAL EVALUATION:

Kozin estimated from his semiempirical treatment that the solubility of europium in mercury at 298 K is 0.12 (1) and 0.14 at % (2,10). Some recent experimental determinations of the solubility appear to confirm Kozin's estimates. Sagadieva and coworkers reported solubility at 298 K of 0.098 (3,7,8,17) and 0.112 at % (7,17). At 293 K, Shvedov et al. (4) reported a solubility of 0.142 at %, while Perov and coworkers (14) reported a solubility of 0.152 at %, and Gerasimov et al. (9) reported a solubility of 0.165 at % without describing the method of determination. There have been other reports (5,6,11,15) of the solubility near room temperature, but these are rejected because they are nearly an order of magnitude too high because of supersaturation of the amalgams in these studies.

The saturated amalgam is in equilibrium with various compounds (6,12,13,16): $EuHg_{10}$, $EuHg_5$, $EuHg_{3.6}$, $EuHg_3$, $EuHg_2$, $EuHg$, and Eu_3Hg_2. No systematic thermal analysis of the Eu-Hg system has been made.

The tentative solubility of europium in mercury at 298 K is 0.1 at % (3,7,17).

References

1. Kozin, L.F. *Tr. Inst. Khim. Nauk Akad. Nauk Kaz. SSR* 1962, *9*, 101.
2. Kozin, L.F. *Fiziko-Khimicheskie Osnovy Amalgamnoi Metalurgii*, Nauka, Alma-Ata, 1964.
3. Sagadieva, K.Zh.; Zebreva, A.I.; Badavamova, G.L. *Elektrokhimia* 1979, *15*, 210.
4. Shvedov, V.P.; Frolkov, A.Z.; Nikishin, G.D. *Radiokhimia* 1971, *13*, 252.
5. Usenova, K.A.; Krebaeva, Sh.D.; Osipova, G.V.; Enikeev, R.Sh. *Radiokhimia* 1974, *16*, 104.
6. McCoy, H.N. *J. Am. Chem. Soc.* 1941, *63*, 1622.
7. Sagadieva, K.Zh.; Zebreva, A.I.; Khanapina, K. *Izv. Vyssh. Ucheb. Zaved., Khim. Khim. Tekhnol.* 1977, *20*, 1263.
8. Sagadieva, K.Zh.; Zebreva, A.I. *Izv. Akad. Nauk Kaz. SSR, Ser. Khim.* 1977, No. 5, 28.
9. Gerasimov, Ya.I.; Krestovnikov, A.N.; Kuznetsov, F.A.; Lavrentev, V.I.; Lomov, A.L. *Khimicheskaya Termodinamika v Tsvetnoi Metallurgii, Vol. 7*, Metallurgia, Moskva, 1975; as cited from Kozin, L.F.; Nigmetova, R.Sh.; Dergacheva, M.B. *Termodinamika Binarnykh Amalgamnykh Sistem*, Nauka, Alma-Ata, 1977, p. 192.
10. Kozin, L.F. *Vestn. Akad. Nauk Kaz. SSR* 1972, No. 3, 34.
11. Usenova, K.A.; Krebaeva, Sh.D. *Sbor. Rabot Khim.*, Alma-Ata, 1973, No. 3, 369.
12. Iandelli, A.; Palenzona, A. *Atti Acad. Naz. Lincei, Cl. Sci. Fiz. Mat. Nat., Rend.* 1964, *37*, 165.
13. Merlo, F.; Fornasini, M.L. *J. Less-Common Metals* 1979, *64*, 221.
14. Perov, E.I.; Akimov, V.P.; Serebrennikov, V.V. *Tr. Tomsk. Univ.* 1971, *204*, 90.
15. Udris, E.Ya.; Korshunov, V.N. *Elektrokhimia* 1982, *18*, 636.
16. Iandelli, A.; Palenzona, A. *Handbook on the Physics and Chemistry of Rare Earths*, Gschneider, K.A.; Eyring, L., Eds., North-Holland, 1978, Amsterdam, Ch. I.
17. Sagadieva, K.Zh.; Zebreva, A.I.; Khanapina, K. *Dep. VINITI*, 2234-76, 1976.

COMPONENTS:	ORIGINAL MEASUREMENTS:
(1) Europium; Eu; [7440-53-1] (2) Mercury; Hg; [7439-97-6]	Shvedov, V.P.; Frolkov, A.Z.; Nikishin, G.D. *Radiokhimia* 1971, *13*, 252-5.
VARIABLES: One temperature: 20°C	PREPARED BY: C. Guminski; Z. Galus

EXPERIMENTAL VALUES:

The solubility of europium in mercury at 20°C was found to be 0.142 at %.

AUXILIARY INFORMATION

METHOD/APPARATUS/PROCEDURE:	SOURCE AND PURITY OF MATERIALS:
The amalgams were prepared by reduction of Eu(III) with sodium amalgam, and then oxidized under polarographic conditions. The wave height increased with the amalgam concentration up to the saturation point; further increase of amalgam concentration caused a decrease of the wave height. The maximum on the wave height-amalgam concentration dependency indicated the concentration of the saturated amalgam. Solutions were deoxygenated by bubbling H_2 or by adding Na_2SO_3 before the experiment. Concentrations of Eu in the amalgam were determined radiochemically or by complexometric titration with Trilon B. Back titration method was used with $Zn(CH_3COO)_2$ in ammonia buffer. All work with amalgams was carried out under a layer of dehydrated ethanol or acetone.	Nothing specified.
	ESTIMATED ERROR: Soly: precision ± 10% (compilers). Temp: not specified.
	REFERENCES:

COMPONENTS:	ORIGINAL MEASUREMENTS:
(1) Europium; Eu; [7440-53-1] (2) Mercury; Hg; [7439-97-6]	Perov, E.I.; Akimov, V.P.; Serebrennikov, V.V. *Tr. Tomsk. Univ.* 1971, *204*, 90-3.
VARIABLES: One temperature: 20°C	PREPARED BY: C. Guminski; Z. Galus

EXPERIMENTAL VALUES:

The solubility of europium in mercury at 20°C was reported to be 0.114 mass %. The atomic % solubility calculated by the compilers is 0.152 at %.

It is not certain if the liquid amalgam and the solid phase reached equilibrium.

AUXILIARY INFORMATION

METHOD/APPARATUS/PROCEDURE:	SOURCE AND PURITY OF MATERIALS:
Thin film mercury electrode on Ag base was polarized in citrate-alkaline solution containing Eu(III). The electrolysis was performed under a hydrogen atmosphere, then the electrode was washed in water and dipped in 0.1 mol dm^{-3} HCl solution. The Eu content in the latter solution was analyzed by photocolorimetry. The solubility of Eu in Hg was then calculated from amount of Eu(III) in the solution and volume of Hg electrode.	Nothing specified.
	ESTIMATED ERROR: Soly: nothing specified; no better than ± 10% (compilers). Temp: nothing specified.
	REFERENCES:

COMPONENTS:	ORIGINAL MEASUREMENTS:
(1) Europium; Eu; [7440-53-1] (2) Mercury; Hg; [7439-97-6]	Sagadieva, K.Zh.; Zebreva, A.I.; Khanapina, K. *Izv. Vyssh. Ucheb. Zaved., Khim. Khim. Tekhnol.* 1977, *20*, 1263-6.
VARIABLES: One temperature: 25°C	PREPARED BY: C. Guminski; Z. Galus

EXPERIMENTAL VALUES:

Solubility of europium in mercury at 25°C was reported to be:

$(7.5 \pm 0.5) \times 10^{-2}$ mol dm^{-3}, or 0.112 at %, from chronoamperometry;

$(6.6 \pm 0.7) \times 10^{-2}$ mol dm^{-3}, or 0.098 at %, from coulometry.

AUXILIARY INFORMATION

METHOD/APPARATUS/PROCEDURE:	SOURCE AND PURITY OF MATERIALS:
The Eu amalgams were presumably prepared by reduction of Eu(III) with sodium amalgam. Amalgams of various concentrations were then oxidized under potentiostatic conditions at -1.0 V vs. SCE. When the metal content in the amalgam exceeded its solubility in mercury, the anodic limiting current, i_s, was independent of the amalgam concentration, N. On the basis of a plot of i_s vs. N, the solubility was estimated from the breakpoint of the curve by chronoamperometry. For the current-time dependence for the oxidation of the heterogeneous amalgam, there was an exponential component attributed to the oxidation of the homogeneous amalgam. The content of Eu in the saturated amalgam by coulometry was obtained by integration of the i_s vs. t curve for the oxidation.	Nothing specified.
	ESTIMATED ERROR: Soly: precision approximately ± 10%. Temp: precision ± 0.5 K.
	REFERENCES:

COMPONENTS:	ORIGINAL MEASUREMENTS:
(1) Europium; Eu; [7440-53-1] (2) Mercury; Hg; [7439-97-6]	Sagadieva, K.Zh.; Zebreva, A.I.; Badavamova, G.L. *Elektrokhimia* 1979, *15*, 210-3.
VARIABLES: One temperature: 25°C	PREPARED BY: C. Guminski; Z. Galus

EXPERIMENTAL VALUES:

The solubility of europium in mercury at 25°C was found to be 6.6×10^{-2} mol dm^{-3} (9.8×10^{-2} at % by compilers).

The same result is reported also in (1,2).

AUXILIARY INFORMATION

METHOD/APPARATUS/PROCEDURE:	SOURCE AND PURITY OF MATERIALS:
The amalgams were presumably prepared by reduction of Eu(III) with sodium amalgam. Amalgams of various concentrations were oxidized potentiostatically at the limiting current potential of -0.1 V vs. SCE, and the current was recorded as a function of time. The measurements were made in a static, buffered solution of 0.2 mol dm^{-3} NaOAc + 0.04 mol dm^{-3} HCl at pH = 5. The limiting current, i_d, was obtained from a semilog plot of current vs. time; i_d increased linearly with concentration of Eu and was nearly constant above saturation. The solubility was determined from the intercept of the i_d vs. concentration plot.	Nothing specified.
	ESTIMATED ERROR: Soly: nothing specified; precision probably about ± 10% (compilers). Temp: not specified.
	REFERENCES: 1. Sagadieva, K.Zh.; Zebreva, A.I.; Khanapina, K. *Izv. Vyssh. Ucheb. Zaved., Khim. Khim. Tekhnol.* 1977, *20*, 1263. 2. Sagadieva, K.Zh.; Zebreva, A.I. *Izv. Akad. Nauk Kaz. SSR, Ser. Khim.* 1977, No. 5, 28.

COMPONENTS:	EVALUATOR:
(1) Gadolinium; Gd; [7440-54-2] (2) Mercury; Hg; [7439-97-6]	C. Guminski; Z. Galus Department of Chemistry University of Warsaw Warsaw, Poland July, 1985

CRITICAL EVALUATION:

Kozin initially calculated (1) a solubility of 3.4×10^{-5} at % at 298 K for gadolinium in mercury, but he subsequently corrected (2) his estimate to 1.96×10^{-4} at %. The experimental determinations of the gadolinium solubility are appreciably higher than the estimates of Kozin. Bulina and coworkers employed electrochemical oxidation and reported solubilities of 9.8×10^{-3} (4) and 5.3×10^{-3} at % (5) at 298 K. Messing and Dean (3) equilibrated the saturated amalgams at 363 to 618 K then analyzed the filtered liquid phase; they found that the solubility of gadolinium increased monotonically from 0.0377 to 0.967 at % in this temperature range. The extrapolated solubility at 298 K, from the least-squares fitted equation of (3), is 8.2×10^{-3} at %. The latter solubility is in good agreement with those reported by (4) and (5). Sayun and Vokhrysheva (6) observed that the gadolinium amalgam of concentration 2.1×10^{-2} at % was a homogeneous liquid at temperatures higher than 293 K. The latter authors subsequently reported solubilities of 0.042 to 0.061 at % in the temperature range of 293 to 353 K, respectively (9,10), but all of their data are rejected because they are clearly too high.

Schematic, partial phase diagrams for the Gd-Hg system have been presented by (7,8).

The tentative values of gadolinium solubility in mercury:

T/K	Soly/at %	Reference
298	7×10^{-3} [a]	[4,5]
323	1.5×10^{-2} [b]	[3]
373	5×10^{-2}	[3]
473	0.24	[3]
573	0.65 [b]	[3]

[a] mean value from data of cited references.
[b] interpolated data of (3).

References

1. Kozin, L.F. *Tr. Inst. Khim. Nauk Akad. Nauk Kaz. SSR* 1962, *9*, 101.
2. Kozin, L.F. *Fiziko-Khimicheskie Osnovy Amalgamnoi Metalurgii*, Nauka, Alma-Ata, 1964.
3. Messing, A.F.; Dean, O.C. *U.S. At. Ener. Comm. Rep.*, ORNL-2871, 1960.
4. Bulina, V.A.; Zebreva, A.I.; Enikeev, R.Sh. *Izv. Vyssh. Ucheb. Zaved., Khim. Khim. Tekhnol.* 1977, *20*, 959.
5. Bulina, V.A.; Guminichenko, L.V.; Zebreva, A.I. *Radiokhimia* 1977, *19*, 89.
6. Sayun, M.G.; Vokhrysheva, L.E. *Elektrokhimia* 1975, *11*, 1679.
7. Kirchmayr, H.R.; Lugscheider, W. *Z. Metallk.* 1966, *57*, 725.
8. Merlo, F.; Fornasini, M.L. *J. Less-Common Metals* 1979, *64*, 221.
9. Vokhrysheva, L.E.; Sayun, M.G. *Izv. Akad. Nauk Kaz. SSR, Ser. Khim.* 1976, No. 5, 64.
10. Vokhrysheva, L.E.; Sayun, M.G. *Dep. VINITI*, 146-77, 1977; *Novosti Polarografii*, Zinatne, Riga, 1975, p. 52.

COMPONENTS:	ORIGINAL MEASUREMENTS:
(1) Gadolinium; Gd; [7440-54-2] (2) Mercury; Hg; [7439-97-6]	Messing, A.F.; Dean. O.C. U.S. At. Ener. Comm. Rep., ORNL-2871, 1960.
VARIABLES:	PREPARED BY:
Temperature: 313-629 K	C. Guminski; Z. Galus

EXPERIMENTAL VALUES:

The solubility of gadolinium in mercury was determined at various temperatures.

T/K	Soly/mass %	Soly/at %
313	[a]8.95×10^{-3}	0.0114
363-368	0.0296	0.0377
403-408	0.0635	0.0810
418-423	0.0948	0.121
478-483	0.212	0.270
483-493	0.215	0.274
553-558	0.443	0.664
553-563	0.419	0.533
608-618	0.760	0.967
629	[a]0.785	1.000

[a] Calculated from least squares equation by the authors.

AUXILIARY INFORMATION

METHOD/APPARATUS/PROCEDURE:	SOURCE AND PURITY OF MATERIALS:
Mercury and gadolinium, after drying and outgassing in the stainless steel dissolver, were kept for several days at the desired temperature. A similar equilibration period was allowed after each temperature change. After equilibration, a sample of liquid amalgam was forced through the filter. The filtrate was collected, dissolved in nitric acid, and submitted for analysis for gadolinium and mercury.	Nothing specified.
	ESTIMATED ERROR: Soly: standard deviation of least squares fit = 0.03539. Temp: precision ± 5 K.
	REFERENCES:

COMPONENTS:	ORIGINAL MEASUREMENTS:
(1) Gadolinium; Gd; [7440-54-2] (2) Mercury; Hg; [7439-97-6]	1. Bulina, V.A.; Guminichenko, L.V.; Zebreva, A.I.; Enikeev, R.Sh. *Radiokhimia* 1977, *19*, 89-93. 2. Bulina, V.A.; Zebreva, A.I.; Enikeev, R.Sh. *Izv. Vyssh. Ucheb. Zaved., Khim. Khim. Tekhnol.* 1977, *20*, 959-61.
VARIABLES:	PREPARED BY:
One temperature: 25°C	C. Guminski; Z. Galus

EXPERIMENTAL VALUES:

The solubility of gadolinium in mercury at 25°C was found to be $(6.6 \pm 1.2) \times 10^{-3}$ and $(3.6 \pm 0.4) \times 10^{-3}$ mol dm^{-3} in (1) and (2), respectively. The atomic % solubility calculated by the compilers are 9.8×10^{-3} and 5.3×10^{-3} at %, respectively.

AUXILIARY INFORMATION

METHOD/APPARATUS/PROCEDURE:	SOURCE AND PURITY OF MATERIALS:
The heterogeneous amalgam in (1) was prepared by reduction of Gd(III) with Na amalgam from buffered solution of pH = 4. The amalgam was potentiostatically oxidized at 0.1 V vs. SCE. The solubility was determined from the charge, from the i-t curve, used to oxidize the homogeneous amalgam; the curve attained a plateau at saturation. In (2) the heterogeneous amalgam with small admixture of Na was kept 2-3 days under purified benzene. The liquid phase was carefully decanted and oxidized chronoamperometrically at 0.1 V vs. SCE. The solubility was calculated from the charge consumed for the oxidation of the homogeneous amalgam.	Nothing specified.
	ESTIMATED ERROR: Soly: precision approximately $\pm$ 20%. Temp: precision $\pm$ 0.5 K.
	REFERENCES:

COMPONENTS:	EVALUATOR:
(1) Terbium; Tb; [7440-27-9] (2) Mercury; Hg; [7439-97-6]	C. Guminski; Z. Galus Department of Chemistry University of Warsaw Warsaw, Poland July, 1985

CRITICAL EVALUATION:

Kozin first predicted a terbium solubility of 7.4×10^{-6} at % at 298 K (1), and subsequently corrected his estimate to 5.2×10^{-5} at %. These estimated solubilities are at least two orders of magnitude lower than the best experimental determinations. Bulina et al. (3) and Sagadieva et al. (6) reported 298 K solubilities of 1.5×10^{-3} and 1.1×10^{-3} at %, respectively. The latter values agree within the experimental errors; also, these solubilities are nearer to the solubilities of the neighboring lanthanides at the same temperature, as compared to the rejected solubility of less than 10^{-4} at % at 293 K (4). Kirchmayr and Lugscheider (5) stated that the solubility of terbium should be similar to those of Nd, Sm and Gd.

The saturated amalgam is in equilibrium with Tb-Hg solid phases, as shown by the schematic phase diagram reported by (5).

The tentative solubility of terbium at 298 K, taking the mean value from (3) and (6), is 1.3×10^{-3} at %.

References

1. Kozin, L.F. *Tr. Inst. Khim. Nauk Akad. Nauk Kaz. SSR* 1962, *9*, 101.
2. Kozin, L.F. *Fiziko-Khimicheskie Osnovy Amalgamnoi Metallurgii*, Nauka, Alma-Ata, 1964.
3. Bulina, V.A.; Zebreva, A.I.; Enikeev, R.Sh. *Izv. Vyssh. Ucheb. Zaved., Khim. Khim. Tekhnol.* 1977, *20*, 959.
4. Shvedov, V.P.; Frolkov, A.Z.; Nikishin, G.D. *Radiokhimia* 1971, *13*, 252.
5. Kirchmayr, H.R.; Lugscheider, W. *Z. Metallk.* 1968, *59*, 296.
6. Sagadieva, K.Zh.; Badavamova, G.L.; Zebreva, A.I. *Izv. Vyssh. Ucheb. Zaved., Khim. Khim. Tekhnol.* 1984, *27*, 329.

COMPONENTS:	ORIGINAL MEASUREMENTS:
(1) Terbium; Tb; [7440-27-9] (2) Mercury; Hg; [7439-97-6]	Sagadieva, K.Zh.; Badavamova, G.L.; Zebreva, A.I. *Izv. Vyssh. Ucheb. Zaved., Khim. Khim. Tekhnol.* <u>1984</u>, *27*, 329-33.
VARIABLES: One temperature: 25°C	PREPARED BY: C. Guminski; Z. Galus

EXPERIMENTAL VALUES:

The solubility of terbium in mercury at 25°C was reported to be 7.3×10^{-4} mol dm^{-3}. The atomic % solubility calculated by the compilers is 1.1×10^{-3} at %.

AUXILIARY INFORMATION

METHOD/APPARATUS/PROCEDURE:	SOURCE AND PURITY OF MATERIALS:
The amalgam was prepared by electro-reduction of Tb(III) in citrate buffer (pH 5-6) on the hanging mercury drop electrode with Pt base. The electrolysis was carried out under potentiostatic conditions at potentials ranging between -1.7 and -2.5 V vs. SCE in an inert gas atmosphere. The amalgam was then oxidized by stripping voltammetry. The total amount of Tb in Hg was found by integration of the area under the voltammetric peak. The breakpoint on the plot relating the anodic peak current against Tb concentration corresponds to the saturation of the amalgam.	Purity of Hg-drop electrode was tested by the stripping analysis without the depolarizer; impurities were below 10^{-6} at % (compilers).
	ESTIMATED ERROR: Soly: nothing specified; precision no better than ± 10% (compilers). Temp: nothing specified.
	REFERENCES:

COMPONENTS:	ORIGINAL MEASUREMENTS:
(1) Terbium; Tb; [7440-27-9] (2) Mercury; Hg; [7439-97-6]	Bulina, V.A.; Zebreva, A.I.; Enikeev, R.Sh. *Izv. Vyssh. Ucheb. Zaved., Khim. Khim. Tekhnol.* 1977, 20, 959-61.
VARIABLES: One temperature: 25°C	PREPARED BY: C. Guminski; Z. Galus

EXPERIMENTAL VALUES:

The solubility of terbium in mercury at 25°C was found to be $(1.0 \pm 0.2) \times 10^{-3}$ g-atom dm^{-3}. The atomic % solubility calculated by the compilers is 1.5×10^{-3} at %.

AUXILIARY INFORMATION

METHOD/APPARATUS/PROCEDURE:	SOURCE AND PURITY OF MATERIALS:
The heterogeneous amalgam of terbium, with small admixtures of sodium, was kept for 2-3 days under purified benzene, then the liquid phase was carefully separated by decantation. The amalgam was oxidized chronoamperometrically at +0.1 V vs. SCE. The solubility was calculated from the charge consumed for the oxidation of the saturated amalgam.	Nothing specified.
	ESTIMATED ERROR: Soly: precision $\pm$ 20%. Temp: nothing specified.
	REFERENCES:

COMPONENTS:	EVALUATOR:
(1) Dysprosium; Dy; [7429-91-6] (2) Mercury; Hg; [7439-97-6]	C. Guminski; Z. Galus Department of Chemistry University of Warsaw Warsaw, Poland July, 1985

CRITICAL EVALUATION:

Bulina et al. determined dysprosium solubilities of 1.9×10^{-2} (4,5) and 1.2×10^{-3} at % (3) at 298 K. In the opinion of the evaluators the second result appears to be more reliable than the first because of its similarity to the solubility of the neighboring rare earths. Kozin's predicted solubilities of 1.0×10^{-6} (1) and 1.6×10^{-5} at % (2) at 298 K are too low.

The schematic phase diagram for the Dy-Hg system shows that the liquid is in equilibrium with Dy-Hg intermetallic phases (6,7).

The tentative solubility of dysprosium at 298 K is 1.2×10^{-3} at % (3).

References

1. Kozin, L.F. *Tr. Inst. Khim. Nauk Akad. Nauk Kaz. SSR* 1962, *9*, 101.
2. Kozin, L.F. *Fiziko-Khimicheskie Osnovy Amalgamnoi Metallurgii*, Nauka, Alma-Ata, 1964.
3. Bulina, V.A.; Zebreva, A.I.; Enikeev, R.Sh. *Izv. Vyssh. Ucheb. Zaved., Khim. Khim. Tekhnol.* 1977, *20*, 959.
4. Bulina, V.A.; Usenova, K.A.; Zebreva, A.I.; Enikeev, R.Sh. *Issl. Obl. Khim. Redkozeml. Elementov* 1975, 78.
5. Bulina, V.A.; Zebreva, A.I.; Enikeev, R.Sh. *Khim. Khim. Tekhnol.*, Alma-Ata, 1974, No. 16, 189.
6. Kirchmayr, H.R.; Lugscheider, W. *Z. Metallk.* 1966, *57*, 725.
7. Merlo, F.; Fornasini, M.L. *J. Less-Common Metals* 1979, *64*, 221.

COMPONENTS:	ORIGINAL MEASUREMENTS:
(1) Dysprosium; Dy; [7429-91-6] (2) Mercury; Hg; [7439-97-6]	1. Bulina, V.A.; Zebreva, A.I.; Enikeev, R.Sh. *Khim. Khim. Tekhnol.*, Alma-Ata <u>1974</u>, No. 16, 189-91. 2. Same authors *Izv. Vyssh. Ucheb. Zaved., Khim. Khim. Tekhnol.* <u>1977</u>, *20*, 959-61.
VARIABLES:	PREPARED BY:
One temperature: 25°C	C. Guminski; Z. Galus

EXPERIMENTAL VALUES:

The solubility of dysprosium in mercury at 25°C was reported to be $(1.3 \pm 0.5) \times 10^{-2}$ mol dm^{-3} and $(0.8 \pm 0.2) \times 10^{-3}$ g-atom dm^{-3}. The respective atomic % solubility calculated by the compilers are 1.9×10^{-2} and 1.2×10^{-3} at %.

AUXILIARY INFORMATION

METHOD/APPARATUS/PROCEDURE:	SOURCE AND PURITY OF MATERIALS:
Dy amalgams in (1) were obtained by reduction of Dy(III) with Li amalgam in acetate buffer, and the Dy contents were determined by decomposition of the amalgam with acetic acid, followed by complexometric titration with trilon. In (2) the heterogeneous amalgam, with small admixture of Na, was kept for 2-3 days under purified benzene; the liquid phase was carefully separated by decantation. The amalgams were oxidized chronoamperometrically: -0.1 V vs. silver chloride electrode in (1); +0.1 V vs. SCE in (2). The solubility in (1) was calculated from the charge corresponding to the exponential part of the i-t curve; the solubility in (2) was calculated from the charge consumed for the oxidation of the saturated amalgam.	Nothing specified.
	ESTIMATED ERROR: Soly: precision approximately $\pm$ 40% in (1) and $\pm$ 25% in (2) (compilers). Temp: nothing specified.
	REFERENCES:

COMPONENTS:	EVALUATOR:
(1) Holmium; Ho; [7440-60-0] (2) Mercury; Hg; [7439-97-6]	C. Guminski; Z. Galus Department of Chemistry University of Warsaw Warsaw, Poland July, 1985

CRITICAL EVALUATION:

There has been only one experimental determination of the solubility of holmium in mercury (3); at 298 K the solubility was reported to be 9×10^{-4} at %. Kozin's calculated values (1,2) were at least an order of magnitude lower than that found by (3).

The schematic phase diagram for this system, which shows that the liquid is in equilibrium with various intermetallic phases, has been presented by (4). However, Bulina and coworkers (5) reported that the saturated amalgam at 298 K is in equilibrium with HoHg rather than with $HoHg_3$ as suggested by (4).

The tentative value of the solubility of Ho in Hg at 298 K is 9×10^{-4} at % (3).

References

1. Kozin, L.F. *Tr. Inst. Khim. Nauk Akad. Nauk Kaz. SSR*, 1962, *9*, 101.
2. Kozin, L.F. *Fiziko-Khimicheskie Osnovy Amalgamnoi Metallurgii*, Nauka, Alma-Ata, 1964.
3. Bulina, V.A.; Zebreva, A.I.; Enikeev, R.Sh. *Izv. Vyssh. Ucheb. Zaved., Khim. Khim. Tekhnol.* 1977, *20*, 959.
4. Kirchmayr, H.R.; Lugscheider, W. *Z. Metallk.* 1966, *57*, 725.
5. Bulina, V.A.; Zebreva, A.I.; Enikeev, R.Sh. *Izv. Vyssh. Ucheb. Zaved., Khim. Khim. Tekhnol.* 1977, *20*, 522.

COMPONENTS:	ORIGINAL MEASUREMENTS:
(1) Holmium; Ho; [7440-60-0] (2) Mercury; Hg; [7439-97-6]	Bulina, V.A.; Zebreva, A.I.; Enikeev, R.Sh. *Izv. Vyssh. Ucheb. Zaved., Khim. Khim. Tekhnol.* 1977, *20*, 959-61.
VARIABLES: One temperature: 25°C	PREPARED BY: C. Guminski; Z. Galus

EXPERIMENTAL VALUES:

The solubility of holmium in mercury at 25°C was reported to be $(0.6 \pm 0.2) \times 10^{-3}$ g-atom dm^{-3}. The atomic % solubility calculated by the compilers is 9×10^{-4} at %.

AUXILIARY INFORMATION

METHOD/APPARATUS/PROCEDURE:	SOURCE AND PURITY OF MATERIALS:
The heterogeneous amalgam of Ho, with small admixture of sodium, was kept for 2-3 days under purified benzene, then the liquid phase was carefully separated by decantation. The amalgam was oxidized chronoamperometrically at +0.1 V vs. SCE. The solubility was calculated from the charge consumed for the oxidation of the saturated amalgam.	Nothing specified.
	ESTIMATED ERROR: Soly: precision ± 35%. Temp: nothing specified.
	REFERENCES:

COMPONENTS:	EVALUATOR:
(1) Erbium; Er; [7440-52-0] (2) Mercury; Hg; [7439-97-6]	C. Guminski; Z. Galus Department of Chemistry University of Warsaw Warsaw, Poland July, 1985

CRITICAL EVALUATION:

There has been only one experimental determination of the solubility of erbium in mercury; the solubility at 298 K was reported to be 6×10^{-4} at % (3). Kozin's calculated solubilities (1,2) were more than an order of magnitude lower than that found by (3).

The schematic phase diagram for this system, which shows that the liquid is in equilibrium with intermetallic phases, has been presented by (5). Flad and Matthes (4) found the compound, Er_3Hg; therefore, the report by Bulina et al. (6), that the saturated amalgam at 298 K is in equilibrium with ErHg, is questionable.

The tentative solubility of Er in Hg at 298 K is 6×10^{-4} at % (3).

References

1. Kozin, L.F. *Tr. Inst. Khim. Nauk Akad. Nauk Kaz. SSR* 1962, *9*, 101.
2. Kozin, L.F. *Fiziko-Khimicheskie Osnovy Amalgamnoi Metallurgii*, Nauka, Alma-Ata, 1964.
3. Bulina, V.A.; Zebreva, A.I.; Enikeev, R.Sh. *Izv. Vyssh. Ucheb. Zaved., Khim. Khim. Tekhnol.* 1977, *20*, 959.
4. Flad, D.; Matthes, F. *Z. Chem.* 1964, *4*, 466.
5. Kirchmayr, H.R.; Lugscheider, W. *Z. Metallk.* 1966, *57*, 725.
6. Bulina, V.A.; Zebreva, A.I.; Enikeev, R.Sh. *Izv. Vyssh. Ucheb. Zaved., Khim. Khim. Tekhnol.* 1977, *20*, 522.

COMPONENTS:	ORIGINAL MEASUREMENTS:
(1) Erbium; Er; [7440-52-0] (2) Mercury; Hg [7439-97-6]	Bulina, V.A.; Zebreva, A.I.; Enikeev, R.Sh. *Izv. Vyssh. Ucheb. Zaved., Khim. Khim. Tekhnol.* 1977, *20*, 959-61.
VARIABLES: One temperature: 25°C	PREPARED BY: C. Guminski; Z. Galus

EXPERIMENTAL VALUES:

The solubility of erbium in mercury at 25°C was reported to be $(0.4 \pm 0.1) \times 10^{-3}$ g-atom dm^{-3}. The atomic % solubility calculated by the compilers is 6×10^{-4} at %.

AUXILIARY INFORMATION

METHOD/APPARATUS/PROCEDURE:	SOURCE AND PURITY OF MATERIALS:
The heterogeneous amalgam of erbium, with small admixture of sodium, was kept for 2-3 days under purified benzene, then the liquid phase was carefully separated by decantation. The amalgam was oxidized chronoamperometrically at +0.1 V vs. SCE. The solubility was calculated from the charge consumed for the oxidation of the saturated amalgam.	Nothing specified.
	ESTIMATED ERROR: Soly: precision approximately $\pm$ 25%. Temp: nothing specified.
	REFERENCES:

COMPONENTS:	EVALUATOR:
(1) Thulium; Tm; [7440-30-4] (2) Mercury; Hg; 7439-97-6]	C. Guminski; Z. Galus Department of Chemistry University of Warsaw Warsaw, Poland July, 1985

CRITICAL EVALUATION:

Bulina et al. (1) reported the only experimental determination of the solubility of thulium in mercury; these authors reported a 298 K solubility of 4×10^{-4} at %. Kozin's (2,3) calculated solubilities were lower than the experimental value by at least two orders of magnitude.

References

1. Bulina, V.A.; Zebreva, A.I.; Enikeev, R.Sh. *Izv. Vyssh. Ucheb. Zaved., Khim. Khim. Tekhnol.* 1977, 20, 959.
2. Kozin, L.F. *Tr. Inst. Khim. Nauk Akad. Nauk Kaz. SSR* 1962, 9, 101.
3. Kozin, L.F. *Fiziko-Khimicheskie Osnovy Amalgamnoi Metallurgii*, Nauka, Alma-Ata, 1964.

COMPONENTS:	ORIGINAL MEASUREMENTS:
(1) Thulium; Tm; [7440-30-4] (2) Mercury; Hg; [7439-97-6]	Bulina, V.A.; Zebreva, A.I.; Enikeev, R.Sh. *Izv. Vyssh. Ucheb. Zaved., Khim. Khim. Tekhnol.* 1977, *20*, 959-61.
VARIABLES: One temperature: 25°C	PREPARED BY: C. Guminski; Z. Galus

EXPERIMENTAL VALUES:

The solubility of thulium in mercury at 25°C was found to be $(0.3 \pm 0.1) \times 10^{-3}$ g-atom dm^{-3}. The atomic % solubility calculated by the compilers is 4×10^{-4} at %.

AUXILIARY INFORMATION

METHOD/APPARATUS/PROCEDURE:	SOURCE AND PURITY OF MATERIALS:
The heterogeneous amalgam of thulium, with small admixture of sodium, was kept for 2-3 days under purified benzene, then the liquid phase was carefully separated by decantation. The amalgam was oxidized at +0.1 V vs. SCE under chronoamperometric conditions. The solubility was calculated from the charge consumed for the oxidation of the saturated amalgam.	Nothing specified.
	ESTIMATED ERROR: Soly: precision $\pm$ 35%. Temp: not specified.
	REFERENCES:

COMPONENTS:	EVALUATOR:
(1) Ytterbium; Yb; [7440-64-4] (2) Mercury; Hg; [7439-97-6]	C. Guminski; Z. Galus Department of Chemistry University of Warsaw Warsaw, Poland July, 1985

CRITICAL EVALUATION:

Experimental determination of the solubility of ytterbium in mercury has been reported to be 0.128 at % at 293 K by Shvedov et al. (3), and 1.32 at % at 298 K by Usenova et al. (4,5). Kozin reported calculated solubilities of 0.40 (1) and 0.42 at % (2) at 298 K. In the opinion of the evaluators, the result of Shvedov et al. appears to be the most accurate; their solubility is more consistent by comparison with the solubilities of other lanthanides in mercury. The solubility of Usenova et al. appears to be too high. Kirchmayr and Lugscheider (6), by analogy to that of Nd, Sm and Gd, estimated that the solubility of Yb is of the order of 10^{-2} at % at 298 K.

The saturated ytterbium amalgams are in equilibrium with Yb-Hg intermetallic phases (6,7); a schematic phase diagram has been reported by (6).

Tentative value of the solubility of Yb in Hg at 298 K is 0.13 at % (3).

References

1. Kozin, L.F. *Tr. Inst. Khim. Nauk Akad. Nauk Kaz. SSR* 1962, *9*, 101.
2. Kozin, L.F. *Fiziko Khimicheskie Osnovy Amalgamnoi Metallurgii*, Nauka, Alma-Ata, 1964.
3. Shvedov, V.P.; Frolkov, A.Z.; Nikishin, G.D. *Radiokhimia* 1971, *13*, 252.
4. Usenova, K.A.; Krebaeva, Sh.D. *Sbor. Rabot. Khim.* Alma-Ata, 1973, No. 3, 369.
5. Usenova, K.A.; Krebaeva, Sh.D.; Osipova, G.V.; Enikeev, R.Sh. *Radiokhimia*, 1974, *16*, 104.
6. Kirchmayr, H.R.; Lugscheider, W. *Z. Metallk.* 1968, *59*, 296.
7. Merlo, F.; Fornasini, M.L. *J. Less-Common Metals* 1979, *64*, 221.

COMPONENTS:	ORIGINAL MEASUREMENTS:
(1) Ytterbium; Yb; [7440-64-4] (2) Mercury; Hg; [7439-97-6]	Svedov, V.P.; Frolkov, A.Z.; Nikishin, G.D. *Radiokhimia* 1971, *13*, 252-5; *Soviet Radiochemistry* 1971, *13*, 251-3.
VARIABLES:	PREPARED BY:
One temperature: 20°C	C. Guminski; Z. Galus; M. Salomon

EXPERIMENTAL VALUES:

The solubility of ytterbium in mercury at 20°C was found to be 1.28×10^{-1} at %.

AUXILIARY INFORMATION

METHOD/APPARATUS/PROCEDURE:	SOURCE AND PURITY OF MATERIALS:
Soly determined polarographically by measuring the anodic limiting current, i_d, of amalgams of varying composition. For unsaturated amalgams, i_d increases linearly with increasing Yb concentration: for saturated amalgams, i_d decreases linearly with increasing Yb concentration apparently due to the formation of solid phases and increase in amalgam viscosity (1). The soly of Yb is obtained from the intercept of the two linear plots of i_d vs. concentration. Experimental details: supporting electrolyte was 0.1 mol dm^{-3} LiCl or KCl; solutions deoxygenated by bubbling H$_2$ or by adding Na$_2$SO$_3$ to solution before the experiment; the capillary constant, $(m^{2/3}t^{1/6})$, was "practically" independent of Yb concentration. Presumably all polarograms were run in alcohol or acetone solutions. Experimental error said not to exceed 10% (compiler assumes this to mean accuracy).	Yb amalgams prepared by reducing Yb(III) acetate with Na amalgam (details not given). All work with amalgams done under a layer of dehydrated alcohol or acetone (source and purity of solvents not specified). Concentration of amalgam determined radiochemically or complexometrically by back-titration of excess Trilon B with Zn acetate in ammonia buffer (indicator eriochrome black).
	ESTIMATED ERROR:
	Soly: accuracy ± 10%. Temp: not specified.
	REFERENCES:
	1. Zebreva, A.I.; Kozlovskii, M.T. *Coll. Czech Chem. Commun.* 1960, *25*, 3188.

COMPONENTS:	ORIGINAL MEASUREMENTS:
(1) Ytterbium; Yb; [7440-64-4] (2) Mercury; Hg; [7439-97-6]	Usenova, K.A.; Krebaeva, Sh.D.; Osipova, G.V.; Enikeev, R.Sh. *Radiokhimia* 1974, *16*, 104-6.
VARIABLES:	PREPARED BY:
One temperature: 25°C	C. Guminski; Z. Galus

EXPERIMENTAL VALUES:

The solubility of ytterbium in mercury at 25°C was found to be 0.88 ± 0.07 mol dm^{-3}. The atomic % solubility calculated by the compilers is 1.3 at %.

AUXILIARY INFORMATION

METHOD/APPARATUS/PROCEDURE:	SOURCE AND PURITY OF MATERIALS:
The amalgam was prepared by reduction of Yb(III) with sodium amalgam. The solubility of Yb in Hg was determined from the variation of the potential of the amalgam cell as a function of contact time of the cell with aqueous solutions of pH = 3. It was observed that the potential of homogeneous amalgams changed rapidly toward positive values, whereas the potential of heterogeneous amalgams remained nearly constant upon contact with the aqueous solution. The solubility of Yb was determined by measuring the potential of amalgams of various Yb content and observing the point of constant potential at saturation.	Nothing specified.
	ESTIMATED ERROR: Soly: precision approximately $\pm$ 10%. Temp: not specified.
	REFERENCES:

COMPONENTS:	EVALUATOR:
(1) Lutetium; Lu; [7439-94-3] (2) Mercury; Hg; [7439-97-6]	C. Guminski; Z. Galus Department of Chemistry University of Warsaw Warsaw, Poland July, 1985

CRITICAL EVALUATION:

There has been only one experimental determination of the solubility of lutetium in mercury; at 298 K Bulina et al. (1) reported a solubility of 3×10^{-4} at %. Kozin's (2,3) calculated solubilities were more than three orders of magnitude lower than that of the experimental value.

Although the Lu-Hg phase diagram is not known, the compounds $LuHg_3$ and $LuHg$ have been reported (4); the liquid amalgam probably is in equilibrium with these compounds in the appropriate temperature range.

Tentative value of the solubility of Lu in Hg at 298 K is 3×10^{-4} at % (1).

References

1. Bulina, V.A.; Zebreva, A.I.; Enikeev, R.Sh. *Izv. Vyssh. Ucheb. Zaved., Khim. Khim. Tekhnol.* 1977, *20*, 959.
2. Kozin, L.F. *Tr. Inst. Khim. Nauk Akad. Nauk Kaz. SSR* 1962, *9*, 101.
3. Kozin, L.F. *Fiziko-Khimicheskie Osnovy Amalgamnoi Metallurgii*, Nauka, Alma-Ata, 1964.
4. Iandelli, A.; Palenzona, A. *Handbook on the Physics and Chemistry of Rare Earths*, Gschneider, K. A.; Eyring, L., Eds., North-Holland, 1978, Amsterdam, Ch. I.

COMPONENTS:	ORIGINAL MEASUREMENTS:
(1) Lutetium; Lu; [7439-94-3] (2) Mercury; Hg; [7439-97-6]	Bulina, V.A.; Zebreva, A.I.; Enikeev, R.Sh. *Izv. Vyssh. Ucheb. Zaved., Khim. Khim. Teknol.* 1977, *20*, 959-61.
VARIABLES: One temperature: 25°C	PREPARED BY: C. Guminski; Z. Galus

EXPERIMENTAL VALUES:

The solubility of lutetium in mercury at 25°C was reported to be $(0.2 \pm 0.1) \times 10^{-3}$ g-atom dm^{-3}. The atomic % solubility calculated by the compilers is 3×10^{-4} at %.

AUXILIARY INFORMATION

METHOD/APPARATUS/PROCEDURE:	SOURCE AND PURITY OF MATERIALS:
The heterogeneous amalgam of lutetium, with small admixture of sodium, was kept for 2-3 days under purified benzene, then the liquid phase was separated carefully by decantation. The amalgam was oxidized chronoamperometrically at +0.1 V vs. SCE. The solubility was calculated from the charge consumed for the oxidation of the saturated amalgam.	Nothing specified.
	ESTIMATED ERROR: Soly: precision approximately $\pm$ 50%. Temp: nothing specified.
	REFERENCES:

COMPONENTS:	EVALUATOR:
(1) Titanium; Ti; [7440-32-6] (2) Mercury; Hg; [7439-97-6]	C. Guminski; Z. Galus Department of Chemistry University of Warsaw Warsaw, Poland July, 1985

CRITICAL EVALUATION:

The first reports on the solubility of titanium in mercury stated only that the solubility at room temperature is less than 4×10^{-5} (1) and 4×10^{-3} at % (2).

Weeks and Fink (3-6) determined the solubility of titanium in the temperature range of 773-968 K, while Jangg and Palman (7) determined the solubility in the range of 293-839 K. All of the data of (7) were reported graphically, and all except one point, 9.0×10^{-2} at % at 923 K (4), were also reported graphically by (3-6). The data of (7) are preferred because the measurements are more precise than those of (3-6); the scatter in the data of (3-6) is as high as an order of magnitude from the curve fitted to the data. The temperature dependence of the solubility reported by (3-6) has a steeper slope than that of (7). Wang's (8) result of 5.5×10^{-3} at % for the titanium solubility at 644 K is slightly lower than that of (7).

Nejedlik (9) reported titanium solubilities of 8×10^{-4} to 1.7×10^{-3} at % at four temperatures between 755 and 912 K, respectively, but these results are an order of magnitude lower than those of (3-7), and are rejected. The report of increasing solubilities ranging from 8.4×10^{-4} to 2.1×10^{-3} at % at 763 to 889 K also are too low and are rejected; no experimental details were presented for these determinations (10). Reid's (11) solubility limit of 2.5×10^{-4} at % at 623 K is much too low. Kozin's (12) estimate of 9.3×10^{-5} at % for the titanium solubility at 298 K is higher than the experimental value (7).

The following solid phases have been reported (13) to be in equilibrium with the saturated amalgams: $TiHg_3$, $TiHg$, $Ti_{1.73}Hg$ and Ti_3Hg. At a pressure of one atmosphere these compounds decompose at 401, 638, 647 and 709 K, respectively.

Tentative values of the solubility of titanium in mercury:

T/K	Soly/at %	Reference
293	1.7×10^{-5}	[7]
298	2×10^{-5} [a]	[7]
323	4×10^{-5}	[7]
373	1.6×10^{-4}	[7]
473	1.1×10^{-3}	[7]
573	3.5×10^{-3}	[7]
673	8×10^{-3}	[7]
773	1.5×10^{-2}	[7]
873	2.5×10^{-2}	[3]

[a] Interpolated value from data of (7).

(continued next page)

COMPONENTS:	EVALUATOR:
(1) Titanium; Ti; [7440-32-6] (2) Mercury; Hg; [7439-97-6]	C. Guminski; Z. Galus Department of Chemistry University of Warsaw Warsaw, Poland July, 1985

CRITICAL EVALUATION: (continued)

References

1. Irvin, N.M.; Russell, A.S.   *J. Chem. Soc.* <u>1932</u>, 891.
2. Strachan, J.F.; Harris, N.L.   *J. Inst. Metals* <u>1956-57</u>, *85*, 17.
3. Weeks, J.R.   *Corrosion* <u>1967</u>, *23*, 98.
4. Weeks, J.R.; Fink, S.   *U.S. At. Ener. Comm. Rep.*, *BNL-900*, <u>1964</u>, p. 136.
5. Weeks, J.R.; Fink, S.   *U.S. At. Ener. Comm. Rep.*, *BNL-782*, <u>1962</u>, p. 73.
6. Weeks, J.R.; Fink, S.   *U.S. At. Ener. Comm. Rep.*, *BNL-799*, <u>1963</u>, p. 85.
7. Jangg, G.; Palman, H.   *Z. Metallk.* <u>1963</u>, *54*, 364.
8. Wang, J.Y.N.   *Nucl. Sci. Eng.* <u>1964</u>, *18*, 18.
9. Nejedlik, J.F.   *U.S. At. Ener. Comm. Rep.*, *NAA-SR-6306*, <u>1961</u>.
10. General Electric Co.; cited by (5).
11. Reid, R.C.   *Techn. Paper 51-S-13, presented at the 1951 Meeting of the ASME;* cited by A. H. Fleitman, A. J. Romano, C. J. Klamut, *J. Electrochem. Soc.* <u>1963</u>, *110*, 964.
12. Kozin, L.F.   *Fiziko-Khimicheskie Osnovy Amalgamnoi Metallurgii*, Nauka, Alma Ata, <u>1964</u>.
13. Lugscheider, E.; Jangg, G.   *Z. Metallk.* <u>1973</u>, *64*, 711.

COMPONENTS:	ORIGINAL MEASUREMENTS:
(1) Titanium; Ti; [7440-32-6] (2) Mercury; Hg; [7439-97-6]	Jangg, G.; Palman, H. Z. Metallk. 1963, 54, 364-69.
VARIABLES:	PREPARED BY:
Temperature: 20-566°C	C. Guminski; Z. Galus

EXPERIMENTAL VALUES:

The mass % solubility was presented graphically as a function of temperature; the compilers read the data points from the curve and recalculated the solubility to atomic %.

$t/°C$	Soly/mass %	Soly/at %	$t/°C$	Soly/mass %	Soly/at %
20	4.2×10^{-6}	1.8×10^{-5}	300	9.2×10^{-4}	3.8×10^{-3}
50	9.6×10^{-6}	4.0×10^{-5}	332	1.12×10^{-3}	4.7×10^{-3}
100	3.9×10^{-5}	1.6×10^{-4}	350	1.3×10^{-3}	5.4×10^{-3}
127	7.8×10^{-5}	3.3×10^{-4}	369	1.6×10^{-3}	6.7×10^{-3}
150	1.14×10^{-4}	4.8×10^{-4}	403	2.0×10^{-3}	8.4×10^{-3}
164	1.62×10^{-4}	6.8×10^{-4}	432	2.4×10^{-3}	1.0×10^{-2}
200	2.7×10^{-4}	1.1×10^{-3}	454	2.6×10^{-3}	1.1×10^{-2}
242	4.5×10^{-4}	1.9×10^{-3}	505	3.8×10^{-3}	1.6×10^{-2}
250	5.1×10^{-4}	2.1×10^{-3}	566	4.9×10^{-3}	2.0×10^{-2}
291	7.8×10^{-4}	3.3×10^{-3}			

AUXILIARY INFORMATION

METHOD/APPARATUS/PROCEDURE:	SOURCE AND PURITY OF MATERIALS:
The heterogeneous amalgam was introduced into a specially constructed apparatus made of refractory chromium steel. Such an apparatus could be used because the Cr_2O_3 film inhibits the wetting of the steel by Hg. After 12 hrs of equilibration at the experimental temperature, the amalgam was filtered through a sintered iron frit under a pressure of purified nitrogen. Usually 3- to 4-fold filtration was necessary. The Ti content in the saturated, filtered amalgam was analytically determined by an unspecified method. For experiments below 320°C the amalgam was equilibrated in a glass vessel.	Nothing specified.
	ESTIMATED ERROR:
	Soly: precision ± 5%. Temp: precision ± 2 K.
	REFERENCES:

COMPONENTS:	ORIGINAL MEASUREMENTS:
(1) Titanium; Ti; [7440-32-6] (2) Mercury; Hg; [7439-97-6]	Wang, J.Y.N. *Nucl. Sci. Eng.* 1964, *18*, 18-30.
VARIABLES: Temperature: 644 K	PREPARED BY: C. Guminski; Z. Galus

EXPERIMENTAL VALUES:

The solubility of titanium in mercury at 644 K was reported to be 13.2 mg/kg solution. The corresponding atomic % solubility calculated by the compilers is 5.53×10^{-3} at %.

Examination of surface films of Ti immersed in Hg showed the presence of Ti_3Hg.

AUXILIARY INFORMATION

METHOD/APPARATUS/PROCEDURE:	SOURCE AND PURITY OF MATERIALS:
The amalgam was prepared by dissolution of cleaned, degreased and vacuum-dried titanium in mercury. An acid extraction analysis was carried out after the equilibration. No further details described.	Mercury was triple-vacuum-distilled. Titanium was of high purity.
	ESTIMATED ERROR: Soly: nothing specified; precision no better than ± 10% (compilers). Temp: not specified.
	REFERENCES:

COMPONENTS:	ORIGINAL MEASUREMENTS:
(1) Titanium; Ti; [7440-32-6] (2) Mercury; Hg; [7439-97-6]	Weeks, J.R. *Corrosion* 1967, *23*, 98-106.
VARIABLES: Temperature: 500-695°C	PREPARED BY: C. Guminski; Z. Galus

EXPERIMENTAL VALUES:

The mass % solubility was reported graphically as a function of temperature. The data points were read from the curve and recalculated to atomic % by the compilers. The same data are also reported in refs. (1-3).

t/°C	Soly/mass %	Soly/at %
695	1.75×10^{-1}	7.3×10^{-1}
^{a}650	2.15×10^{-2}	9.2×10^{-2}
625	7.6×10^{-3}	3.2×10^{-2}
600	5.7×10^{-3}	2.4×10^{-2}
600	2.2×10^{-3}	9.2×10^{-3}
575	2.3×10^{-3}	9.6×10^{-3}
545	2.4×10^{-3}	1.0×10^{-2}
550	7.6×10^{-4}	3.2×10^{-3}
525	6.7×10^{-4}	2.8×10^{-3}
500	4.1×10^{-4}	1.7×10^{-3}

^{a}Numerical value reported in ref. (1).

AUXILIARY INFORMATION

METHOD/APPARATUS/PROCEDURE:	SOURCE AND PURITY OF MATERIALS:
Mercury and titanium were sealed in the upper chamber of an evacuated quartz tube. The two chambers were separated by a coarse quartz filter. The filled tubes were equilibrated for 72 hrs at each temperature, then centrifuged at temperature to force the liquid through the filter. The mercury was distilled from the known quantity of amalgam, and the residue was dissolved in HF-HNO$_3$ or aqua regia and analyzed spectrographically.	Titanium purity not specified. Mercury was triple-distilled, reagent grade.
	ESTIMATED ERROR: Soly: nothing specified. Temp: precision $\pm$ 2 K.
	REFERENCES: 1. Weeks, J.R.; Fink, S. *U.S. At. Ener. Comm. Rep.*, *BNL-900*, 1964, p. 136. 2. Same authors. *U.S. At. Ener. Comm. Rep.*, *BNL-787*, 1962, p. 73. 3. Same authors. *U.S. At. Ener. Comm. Rep.*, *BNL-799*, 1963, p. 84.

COMPONENTS:	EVALUATOR:
(1) Zirconium; Zr; [7440-67-7] (2) Mercury; Hg; [7439-97-6]	C. Guminski; Z. Galus Department of Chemistry University of Warsaw Warsaw, Poland July, 1985

CRITICAL EVALUATION:

The solubility of zirconium in mercury at room temperature is very low, but no experimental value may be suggested, even tentatively. The solubility reported by Strachan and Harris (1), 7×10^{-3} at %, is much too high by comparison with other measurements at higher temperatures. At 298 K, Kozin predicted a solubility of 1.4×10^{-5} at % (2) and an improbably low value of 2×10^{-19} at % (3).

Weeks and Fink (4-6) determined the solubility of zirconium between 773 and 973 K; the only numerical value reported by these workers was a solubility of 0.043 at % at 923 K. The data of Weeks and Fink are tentatively acceptable. Nerad (7), without giving the experimental details, reported a zirconium solubility of 1.1×10^{-3} without 3.5×10^{-3} at % at 623 and 823 K, respectively. The latter solubilities are in agreement with those of Weeks (4-6). Leary (8,9) reported a solubility of 1.6×10^{-3} at % at 623 K; this value is higher than that of Nerad (7). In a later report, Bowersox and Leary (10) reported that the upper limit of the solubility at 623 K is 1.7×10^{-5} at %; this solubility is much too low as compared to other determinations. The low solubility found by the latter authors is attributed to the method of solubility determination whereby the solubility was determined from weight loss of a zirconium coupon which had been equilibrated with the mercury; the formation of Zr-Hg compounds on the surface of the coupon leads to decreased solubility of the zirconium. In other determinations at high temperatures, Nejedlik (11) reported a solubility of 2.2×10^{-3} at % at 755 K, while Fleitman and Brandon (12) reported a solubility of 0.4 at % at 1033 K. No experimental details were presented in the latter two reports, but the results are in good agreement with those of Weeks (3).

Wang (13) reported a solubility of 5.3×10^{-2} at % at 644 K, but this value is too high and is rejected.

Messing and Dean (14) investigated the solubility of zirconium in mercury which had been saturated with uranium at 361 to 629 K, but these authors could not detect any dissolution of zirconium at their detection limit of 4×10^{-4} at %. The solubility of Nb-Zr alloys in mercury also have been investigated (15).

The solid phases in equilibrium with the amalgam consist of the intermetallic compounds (16): $ZrHg_3$, $ZrHg$, and Zr_3Hg stable up to 678, 696 and 833, respectively.

Tentative values of Zr solubility in Hg:

T/K	Soly/at %	Reference
623	1×10^{-3}	[7]
773	2.5×10^{-3}	[4,11]

References

1. Strachan, J.F.; Harris, N.L. *J. Inst. Metals* 1956-57, *85*, 17.
2. Kozin, L.F. *Fiziko-Khimicheskie Osnovy Amalgamnoi Metallurgii*, Nauka, Alma-Ata, 1964.
3. Kozin, L.F. *Tr. Inst. Khim. Nauk Akad. Nauk Kaz. SSR* 1962, *9*, 101.
4. Weeks, J.R. *Corrosion* 1967, *23*, 98.
5. Weeks, J.R.; Fink, S. *U.S. At. Ener. Comm. Rep.*, BNL-900, 1964, p. 136.
6. Weeks, J.R.; Fink, S. *U.S. At. Ener. Comm. Rep.*, BNL-782, 1962, p. 73.
7. Nerad, A.J.; as cited by Kelman, L.R.; Wilkinson, W.D.; Yagee, F.L. *U.S. At. Ener. Comm. Rep.*, ANL-4417, 1950.

(Continued next page)

COMPONENTS:	EVALUATOR:
(1) Zirconium; Zr; [7440-67-7] (2) Mercury; Hg; [7439-97-6]	C. Guminski, Z. Galus Department of Chemistry University of Warsaw Warsaw, Poland July, 1985

CRITICAL EVALUATION: (continued)

8. Leary, J.A.; Benz, R.; Bowersox, D.F.; Bjorklund, C.W.; Johnson, K.W.R.; Maraman, W.J.; Mullins, L.J.; Reavis, J.G. *Proc. of the Second U.N. Intern. Conf. on the Peaceful Uses of At. Ener.*, Geneva, 1958, 17, p. 376; P/529.
9. Leary, J.A. *U.S. At. Ener. Comm. Rep.*, LA-2218, 1958, as cited by (3).
10. Bowersox, D.F.; Leary, J.A. *U.S. At. Ener. Comm. Rep.*, LAMS-2518, 1961.
11. Nejedlik, J.F. *U.S. At. Ener. Comm. Rep.*, NAA-SR-6306, 1961.
12. Fleitman, A.H.; Brandon, J. *U.S. At. Ener. Comm. Rep.*, BNL-799, 1963, p. 76.
13. Wang, J.Y.N. *Nucl. Sci. Techn.* 1964, 18, 18.
14. Messing, A.F.; Dean, O.C. *U.S. At. Ener. Comm. Rep.*, ORNL-2871, 1960.
15. Fleitman, A.H.; Romano, A.J.; Klamut, C.J. *Trans. Am. Nucl. Soc.* 1965, 8, 15.
16. Lugscheider, E.; Jangg, G. *Z. Metallk.* 1973, 64, 711.

COMPONENTS:	ORIGINAL MEASUREMENTS:
(1) Zirconium; Zr; [7440-67-7] (2) Mercury; Hg; [9439-97-6]	Leary, J.A.; Benz, R.; Bowersox, D.F.; Bjorklund, C.W.; Johnson, K.W.R.; Maraman, W.J.; Mullins, L.J.; Reavis, J.G. *Proc. 2nd U.N.Intern. Conf. Peaceful Uses of At. Ener.*, Geneva, 1958, *16*, p. 376; P/529.
VARIABLES: One temperature: 350°C	PREPARED BY: C. Guminski; Z. Galus

EXPERIMENTAL VALUES:

The solubility of Zr in Hg at 350°C was reported to be 0.093 g Zr/dm^3 Hg, or 7×10^{-4} mass %. The corresponding atomic % solubility calculated by the compilers is 1.6×10^{-3} at %. The same result was also reported in (1). In the subsequent work by Bowersox and Leary (2) the upper limit of the solubility was reported to be 0.001 g Zr/dm^3 Hg. This difference of nearly two orders of magnitude is probably due to the different methods for the determination of dissolved Zr; the latter authors determined the solubility from the loss in weight of the Zr coupon. The authors did not comment on the difference in solubility determinations. In the compilers' opinion, the analysis of the filtrate is correct since the surface layer of Zr coupon may be transformed into ZrHg$_3$ or ZrHg, and the dissolution of Zr may be compensated by bonding to Hg.

AUXILIARY INFORMATION

METHOD/APPARATUS/PROCEDURE:	SOURCE AND PURITY OF MATERIALS:
A sample of Zr was equilibrated with Hg at 623 K and the mixture was filtered at this temperature. The filtrate was then cooled, weighed and analyzed for Zr content.	Triple distilled Hg was used. Zr purity not specified.
	ESTIMATED ERROR: Soly: nothing specified; precision probably no better than $\pm$ 10% (compilers). Temp: nothing specified.

	REFERENCES:
	1. Leary, J.A.
	   U.S. At. Ener. Comm. Rep., LA-2218, 1958.
	2. Bowersox, D.F.; Leary, J.A.
	   U.S. At. Ener. Comm. Rep., LAMS-2518, 1961.

COMPONENTS:	ORIGINAL MEASUREMENTS:
(1) Zirconium; Zr; [7440-67-7] (2) Mercury; Hg; [7439-97-6]	Weeks, J.R. *Corrosion* 1967, *23*, 98-106.
VARIABLES:	PREPARED BY:
Temperature: 500-700°C	C. Guminski; Z. Galus

EXPERIMENTAL VALUES:

The dependence of the mass % solubility of zirconium in mercury was reported graphically as a function of temperature. The data points were read off the curve and the conversion to atomic % was made by the compilers.

$t/°C$	Soly/mass % x 10^3	Soly/at % x 10^3
500	1.2	2.5
525	2.5	5.5
545	1.6	3.4
572	1.7	3.7
600	4.5	9.9
600	5.4	12
625	15	33
650[a]	19.5	43
700	66	145

[a] Numerical value reported in (1); above data also reported in (1) and (2).

AUXILIARY INFORMATION

METHOD/APPARATUS/PROCEDURE:	SOURCE AND PURITY OF MATERIALS:
Mercury and zirconium were placed in a quartz capsule which was sealed under vacuum and placed in the stainless steel capsule. The amalgam mixture was equilibrated for 72 hours at the desired temperature. A centrifuge was used at the end of the equilibration period to force a sample of the liquid alloy through the filter. The mercury was distilled from the homogeneous amalgam and the residue was dissolved in HF-HNO$_3$ and analyzed spectrographically.	Mercury was triple distilled, reagent grade. Zirconium purity not specified.
	ESTIMATED ERROR: Soly: nothing specified. Temp: precision $\pm$ 2 K.
	REFERENCES: 1. Weeks, J.R.; Fink, S. *U.S. At. Ener. Comm. Rep.*, BNL-900, 1964, pp. 136-8. 2. Same authors *U.S. At. Ener. Comm. Rep.*, BNL-782, 1962, pp. 73-5.

COMPONENTS:	EVALUATOR:
(1) Hafnium; Hf; [7440-58-6] (2) Mercury; Hg; [7439-97-6]	C. Guminski; Z. Galus Department of Chemistry University of Warsaw Warsaw, Poland July, 1985

CRITICAL EVALUATION:

There are no experimental data on the solubility of hafnium in mercury. Kozin estimated 298 K solubilities of 1.0×10^{-11} (1) and 3.0×10^{-8} at % (2); the second value appears to be the more reliable.

The saturated amalgam is in equilibrium with Hf_2Hg (3). Further work is needed in this system.

References

1. Kozin, L.F. *Tr. Inst. Khim. Nauk Akad. Nauk Kaz. SSR* 1962, *9*, 101.
2. Kozin, L.F. *Fiziko Khimicheskie Osnovy Amalgamnoi Metallurgii*, Nauka, Alma-Ata, 1964.
3. Lugscheider, E.; Jangg, G. *Z. Metallk.* 1973, *64*, 711.

COMPONENTS:	EVALUATOR:
(1) Vanadium; V; [7440-62-2] (2) Mercury; Hg; [7439-97-6]	C. Guminski; Z. Galus Department of Chemistry University of Warsaw Warsaw, Poland July, 1985

CRITICAL EVALUATION:

Tammann and Hinnüber (1) reported that the solubility of vanadium in mercury is very low at 291 K, and Irvin and Russell (2) could only demonstrate that the solubility was lower than their experimentally detectable limit of 2×10^{-4} at % at 293 K. Kozin (3) predicted a solubility of 4.8×10^{-6} at % at 298 K. Strachan and Harris (4) reported an erroneously high solubility of 0.161 at % at room temperature. No reliable experimental data are available for the solubility of vanadium in mercury near room temperature.

At high temperatures, Weeks (5) found that the solubility of vanadium increased from 2.4×10^{-5} to 5.2×10^{-4} at % as the temperature increased from 778 to 955 K. As compared to (5), Parkman reported lower solubilities of 3.2×10^{-5} to 1.2×10^{-4} at % between 811 and 911 K (6). In subsequent reports by Parkman and Whaley (7,8), even lower values of 1.6×10^{-5} to 6×10^{-5} at % were presented for the same temperature range as in (6); the same experimental method was used in (7,8), and the authors did not give any explanation for the lower results in the later measurements. Weeks (5) reported greater confidence in the higher solubilities because of reaction between the silica capsules and the vanadium amalgams.

Extrapolation of the high temperature data of refs. (5) to (8) to 298 K yields a solubility of the order of 10^{-10} at %. However, it should be noted that the solubilities of (5) and (7) differed by nearly an order of magnitude at 900 K.

Because no intermetallic compounds were found in this system, the saturated amalgam should be in equilibrium with solid vanadium (9).

The solubility of vanadium-based alloys in mercury also have been reported (6,8).

The tentative solubility of vanadium in mercury at 773 K is 2×10^{-5} at % (5,6).

References

1. Tammann, G.; Hinnüber, J. *Z. Anorg. Chem.* 1927, *160*, 249.
2. Irvin, N.M.; Russell, A.S. *J. Chem. Soc.* 1932, 891.
3. Kozin, L.F. *Fiziko-Khimicheskie Osnovy Amalgamnoi Metallurgii*, Nauka, Alma-Ata, 1964.
4. Strachan, J.F.; Harris, N.L. *J. Inst. Metals* 1956-57, *85*, 17.
5. Weeks, J.R. *Corrosion* 1967, *23*, 98.
6. Parkman, M.F. *U.S. At. Ener. Comm. Rep.*, TID-7626, 1962, Pt. I, p. 35.
7. Parkman, M.F.; Whaley, D.K. *Aerojet-General Nucleonics*, Rep. AN-957, 1963; cited by (5).
8. Parkman, M.F. *Ext. Abst., Electrothermics and Metallurgy Div.*, Vol. 2, No. 2, The Electrochemical Soc. 1964, pp. 16-21.
9. Jangg, G. *Metall* 1978, *32*, 798.

COMPONENTS:	ORIGINAL MEASUREMENTS:
(1) Vanadium; V; [7440-62-2] (2) Mercury; Hg; [7439-97-6]	1. Parkman, M.F. *Extended Abst., Electrothermics and Metallurgy Div.*, Vol. 2, No. 2, The Electrochemical Soc. 1964, pp. 16-21. 2. Same author *U.S. At. Ener. Comm. Rep.*, TID-7626 1962, Pt. I, pp. 35-41.
VARIABLES: Temperature: 811-911 K	PREPARED BY: C. Guminski; Z. Galus

EXPERIMENTAL VALUES:

The mass % solubility of vanadium in mercury was reported graphically as a semi-logarithmic plot against the reciprocal temperature. The solubility data points were read off the curve and the conversion made to atomic % by the compilers.

T/K	Soly/mass %	Soly/at %
811	4×10^{-6}	1.6×10^{-5}
873	1.1×10^{-5}	4×10^{-5}
911	1.5×10^{-5}	6×10^{-5}

The original data also were reported in ref. (3).

Numerical solubility values reported in (2) were 8×10^{-6} and 3×10^{-5} mass % at 811 and 911 K, respectively. The corresponding atomic % solubilities calculated by the compilers are 3.2×10^{-5} and 1.2×10^{-4} at %, respectively.

AUXILIARY INFORMATION

METHOD/APPARATUS/PROCEDURE:	SOURCE AND PURITY OF MATERIALS:
Specimen of vanadium was placed in contact with Hg in a glass capsule. The capsule was sealed under a vacuum after at least 16 hours of outgassing of the mercury. The capsules were heated to the desired temperature and equilibrated for 16 hours. A sample of the solution was collected and cooled. Mercury was separated from the sample by molecular distillation and the residue was taken into acid solution, dried, and analyzed by emission spectroscopy. It appears that the same method was used in all three reports.	Vanadium was chemically pure. Triple-distilled Hg free of detectable impurities was used.
	ESTIMATED ERROR: Soly: precision $\pm$ 10% in (2). Temp: precision $\pm$ 3 K in (1).
	REFERENCES: 3. Parkman, M.F.; Whaley, D.K. *Aerojet-General Nucleonics*, Rep. AN-957 1963.

COMPONENTS:	ORIGINAL MEASUREMENTS:
(1) Vanadium; V; [7440-62-2] (2) Mercury; Hg; [7439-97-6]	Weeks, J.R. *Corrosion* 1967, *23*, 98-106.
VARIABLES:	PREPARED BY:
Temperature: 505-682°C	C. Guminski; Z. Galus

EXPERIMENTAL VALUES:

The mass % solubility was presented graphically as a function of temperature. The data points were read off the curve and the conversion to atomic % made by the compilers.

$t/°C$	Soly/mass %	Soly/at %
682	1.3×10^{-4}	5.2×10^{-4}
620	1.0×10^{-4}	4.0×10^{-4}
555	5.0×10^{-5}	2.0×10^{-4}
505	6.0×10^{-6}	2.4×10^{-5}

Four other data points were presented, but these were one order of magnitude lower than those reported above.

The authors state that the higher values of the solubilities are most dependable. There was a reaction between the vanadium amalgam and the silica capsules, and the capsule walls were coated with a deposit which could not be removed by heating in vacuum.

AUXILIARY INFORMATION

METHOD/APPARATUS/PROCEDURE:	SOURCE AND PURITY OF MATERIALS:
Mercury and vanadium were placed in Vycor capsule which was sealed under vacuum and put in the stainless steel capsule. The amalgam mixtures were equilibrated for 72 hours at the desired temperature. A centrifuge was used at the end of the equilibration period to force a sample of the liquid alloy through the filter. The Hg was distilled from the sample of the homogeneous amalgam, and the residue was dissolved in HNO_3-HF or aqua regia and analyzed spectrographically.	Mercury was triple-distilled reagent grade. Vanadium purity not specified.
	ESTIMATED ERROR: Soly: nothing specified. Temp: precision $\pm$ 2 K.
	REFERENCES:

COMPONENTS:	EVALUATOR:
(1) Niobium; Nb; [7440-03-1] (2) Mercury; Hg; [7439-97-6]	C. Guminski; Z. Galus Department of Chemistry University of Warsaw Warsaw, Poland July, 1985

CRITICAL EVALUATION:

Limited measurements show that the solubility of niobium in mercury is very low. Strachan and Harris (1) could not detect any niobium in mercury at the detection limit of 2×10^{-3} at % at room temperature. Weeks (2) studied the solubility at temperatures up to 1023 K and concluded that the solubility is lower than the detection limit of 6×10^{-6} at %. However, Weeks and Fink (3) earlier reported a solubility of 8×10^{-6} at % at 923 K. Bowersox and Leary (4) showed that the solubility is lower than 1.6×10^{-5} at % at 623 K. Fleitman and coworkers (5,6) reported that Nb is not affected by Hg at 866 and 976 K; similar observation was noticed by Nejedlik and Vargo (7) at 719 K. Kozin predicted very low solubilities; e.g., 1.6×10^{-18} (8) and 1.3×10^{-12} at % (9) at 298 K. The solubilities of Nb-Zr alloys in Hg also were investigated (2,5) and were found to be below the detection limits.

References

1. Strachan, J.F.; Harris, N.L. *J. Inst. Metals* 1956-57, *85*, 17.
2. Weeks, J.R. *Corrosion* 1967, *23*, 98.
3. Weeks, J.R.; Fink, S. *U.S. At. Ener. Comm. Rep.*, BNL-900, 1964, p. 136.
4. Bowersox, D.F.; Leary, J.A. *U.S. At. Ener. Comm. Rep.*, LAMS-2518, 1961.
5. Fleitman, A.H.; Romano, A.J.; Klamut, C.J. *Trans. Am. Nucl. Soc.* 1965, *8*, 15.
6. Fleitman, A.H.; Brandon, J. *U.S. At. Ener. Comm. Rep.*, BNL-799, 1963, p. 75.
7. Nejedlik, J.F.; Vargo, E.J. *Electrochem. Technol.* 1965, *3*, 250.
8. Kozin, L.F. *Tr. Inst. Khim. Nauk Akad. Nauk Kaz. SSR* 1962, *9*, 101.
9. Kozin, L.F. *Fiziko-Khimicheskie Osnovy Amalgamnoi Metallurgii*, Nauka, Alma-Ata, 1964.

COMPONENTS:	ORIGINAL MEASUREMENTS:
(1) Niobium; Nb; [7440-03-1] (2) Mercury; Hg; [7439-97-6]	Bowersox, D.F.; Leary, J.A. *U.S. At. Ener. Comm. Rep.*, LAMS-2518, <u>1961</u>.
VARIABLES:	PREPARED BY:
One temperature: 350°C	C. Guminski; Z. Galus

EXPERIMENTAL VALUES:

The solubility of Nb in Hg at 350°C is lower than 0.001 g of Nb in 1 dm^3 of Hg. The corresponding atomic percent detection limit calculated by the compilers is 2×10^{-5} at %.

AUXILIARY INFORMATION

METHOD/APPARATUS/PROCEDURE:	SOURCE AND PURITY OF MATERIALS:
The solubility was determined by immersing a weighed coupon of niobium into definite amount of boiling mercury and periodically reweighing the coupon. The weight loss corresponds to the part of the niobium which dissolved.	Triple-distilled Hg was used. Niobium purity not specified.
	ESTIMATED ERROR:
	Soly: detection limit was 1 mg of Nb. Temp: nothing specified.
	REFERENCES:

Niobium

COMPONENTS:	ORIGINAL MEASUREMENTS:
(1) Niobium; Nb; [7440-03-1] (2) Mercury; Hg; [7439-97-6]	Weeks, J.R. *Corrosion* 1967, *23*, 98-106.
VARIABLES:	PREPARED BY:
Temperature: 500-750°C	C. Guminski; Z. Galus

EXPERIMENTAL VALUES:

The solubility of niobium in mercury was presented graphically as a function of temperature. All of the solubility data between 500 and 750°C fall close to the detection limit of 3×10^{-6} mass %. The compilers calculated the atomic percent detection limit as 6×10^{-6} at %.

AUXILIARY INFORMATION

METHOD/APPARATUS/PROCEDURE:	SOURCE AND PURITY OF MATERIALS:
Mercury and niobium were placed in a quartz capsule which was sealed under vacuum and put in a stainless steel capsule. The mixture was equilibrated for 72 hours at the desired temperature. The centrifuge was used at the end of the equilibration period to force a sample of the liquid alloy through the filter. The mercury was distilled from the homogeneous amalgam and the residue was dissolved in HF-HNO$_3$ or aqua regia, and analyzed spectrographically.	Mercury was triple-distilled, reagent grade. Niobium purity not specified.
	ESTIMATED ERROR: Soly: nothing specified. Temp: precision $\pm$ 2 K.
	REFERENCES:

COMPONENTS:	EVALUATOR:
(1) Tantalum; Ta; [7440-25-7] (2) Mercury; Hg; [7439-97-6]	C. Guminski; Z. Galus Department of Chemistry University of Warsaw Warsaw, Poland July, 1985

CRITICAL EVALUATION:

The solubility of tantalum in mercury is very low. Kozin predicted solubilities of 4.3×10^{-26} (1) and 1.7×10^{-16} at % (2) at 298 K. The solubility of 0.011 at % reported by Strachan and Harris (3) is much too high and is rejected. Bowersox and Leary (4) found that the solubility is lower than their detection limit of 8×10^{-6} at % at 623 K. At 873 to 973 K, Weeks (5,6) showed that the solubility of tantalum is less than 2×10^{-7} at %. While investigating the Pr-Hg system, Griffin and Gschneider (7) observed no dissolution of tantalum in mercury at 873 to 1403 K. Similar observations of the inertness of tantalum towards mercury were reported earlier by Bolton (8). Fleitman and coworkers (9), Nejedlik and Vargo (10) and Kirchmayr (11) also could not detect any dissolution of Ta in Hg at 630-1300 K.

References

1. Kozin, L.F. *Tr. Inst. Khim. Nauk Akad. Nauk Kaz. SSR* 1962, *9*, 101.
2. Kozin, L.F. *Fiziko-Khimicheskie Osnovy Amalgamnoi Metallurgii*, Nauka, Alma-Ata, 1964.
3. Strachan, J.F.; Harris, N.L. *J. Inst. Metals* 1956-57, *85*, 17.
4. Bowersox, D.F.; Leary, J.A. *U.S. At. Ener. Comm. Rep.*, LAMS-2518, 1961.
5. Weeks, J.R. *Corrosion* 1967, *23*, 98.
6. Weeks, J.R.; Fink, S. *U.S. At. Ener. Comm. Rep.*, BNL-900, 1964, p. 136.
7. Griffin, R.B.; Gschneider, K.A. *Met. Trans.* 1971, *2*, 2517.
8. Bolton, W. *Z. Elektrochem.* 1905, *11*, 51.
9. Fleitman, A.H.; Romano, A.J.; Klamut, C.J. *Trans. Am. Nucl. Soc.* 1965, *8*, 15.
10. Nejedlik, J.F.; Vargo, E.J. *Electrochem. Technol.* 1965, *3*, 250.
11. Kirchmayr, H.R. *Z. Metallk.* 1965, *56*, 767.

COMPONENTS:	ORIGINAL MEASUREMENTS:
(1) Tantalum; Ta; [7440-25-7] (2) Mercury; Hg; [7439-97-6]	Bowersox, D.F.; Leary, J.A. *U.S. At. Ener. Comm. Rep.*, LAMS-2518, 1961.
VARIABLES: One temperature: 350°C	PREPARED BY: C. Guminski; Z. Galus

EXPERIMENTAL VALUES:

The solubility of Ta in Hg at 350°C was reported to be lower than 0.001 g of Ta in 1 dm^3 of Hg. The corresponding atomic % solubility limit calculated by the compilers is 8×10^{-6} at %.

AUXILIARY INFORMATION

METHOD/APPARATUS/PROCEDURE:	SOURCE AND PURITY OF MATERIALS:
The solubility was determined by immersing a weighed coupon of tantalum into a definite amount of boiling Hg and periodically measuring the coupon weight. The weight loss corresponds to the part of the Ta which dissolved.	Triple distilled Hg was used. Tantalum purity not specified.
	ESTIMATED ERROR: Soly: detection limit was 1 mg. Temp: nothing specified.
	REFERENCES:

COMPONENTS:	ORIGINAL MEASUREMENTS:
(1) Tantalum; Ta; [7440-25-7] (2) Mercury; Hg; [7439-97-6]	Weeks, J.R. *Corrosion* 1967, *23*, 98-106.
VARIABLES: Temperature: 600-700°C	PREPARED BY: C. Guminski; Z. Galus

EXPERIMENTAL VALUES:

The solubility of tantalum in mercury in the temperature range of 600-700°C was found to be below the detection limit of 2×10^{-7} mass %. The same result was reported in (1).

AUXILIARY INFORMATION

METHOD/APPARATUS/PROCEDURE:	SOURCE AND PURITY OF MATERIALS:
Mercury and irradiated tantalum were placed in a quartz capsule which was sealed under vacuum and put in a stainless steel capsule. Hg and the solute metal were equilibrated for 72 hours at the desired temperature. The centrifuge was used at the end of the equilibration period to force a sample of the liquid alloy through the filter. The filtrate was analyzed for radioactivity of tantalum.	Triple-distilled, reagent grade mercury was used. Tantalum specimens were irradiated in the Brookhaven Graphite Research Reactor.
	ESTIMATED ERROR: Soly: nothing specified. Temp: precision $\pm$ 2 K.
	REFERENCES: 1. Weeks, J.R.; Fink, S. *U.S. At. Ener. Comm. Rep.*, *BNL-900*, 1964, p. 136.

COMPONENTS:	EVALUATOR:
(1) Chromium; Cr; [7440-47-3] (2) Mercury; Hg; [7439-97-6]	C. Guminski; Z. Galus Department of Chemistry University of Warsaw Warsaw, Poland July, 1985

CRITICAL EVALUATION:

The solubility of chromium in mercury near room temperature is very low. Irvin and Russell (1) reported that it is below their analytically detectable limit of 2×10^{-4} at % at 293 K, while DeWet and Haul (2) reported that it is below 1.5×10^{-6} at % at 303 K. Jangg and Palman (3) extrapolated their high temperature measurements and estimated a solubility of 7.7×10^{-7} at % at 293 K. The solubility of 1.2×10^{-10} at % at 291 K, reported by Tammann and Hinnüber (4) from EMF measurements, is too low, while the solubility of 7×10^{-3} at % at room temperature, reported by Strachan and Harris (5), is much too high. Kozin (6) predicted a solubility of 5.2×10^{-4} at % at 298 K.

High temperature solubility measurements have been reported by several authors. Jangg and Palman (3) determined solubilities of 2.6×10^{-4} and 3.1×10^{-4} at % at 773 and 823 K, respectively, and they estimated solubilities at lower temperatures by extrapolating the two high temperature values; the solubilities obtained from this extrapolation appear to be dubious. Weeks (7), without presenting experimental details, reported a solubility of 2.6×10^{-3} at % at 873 K. Weeks and Fink (8-10) extended the solubility measurements over the temperature range of 778 to 923 K and these authors reported their results in graphical form; the only numerical value reported was 1.1×10^{-2} at % at 923 K (10). Parkman (11) reported a solubility of 2.1×10^{-4} at % at 866 K; the latter result is in better agreement with that of (3) than with (8).

As seen from the high temperature measurements, the solubility of chromium in mercury is very low over a wide temperature range. There is some disagreement in the high temperature measurements reported by refs. (3), (7), (8-10), and (11), and a plot of the logarithm of solubility against reciprocal temperature shows different slopes for these sets of data.

The liquid amalgam is in equilibrium with pure chromium, and no intermetallic compounds are formed with mercury (2, 12).

Solubilities of Cr-containing alloys in mercury were reported in (11, 13, 14).

The tentative value for the solubility of chromium in mercury at 773 K is 2×10^{-4} at %; this value is the mean from refs. (3) and (8).

References

1. Irvin, N.M.; Russell, A.S. *J. Chem. Soc.* 1932, 891.
2. de Wet, J.F.; Haul, R.A.W. *Z. Anorg. Chem.* 1954, 277, 96.
3. Jangg, G.; Palman, H. *Z. Metallk.* 1963, 54, 364.
4. Tammann, G.; Hinnüber, J. *Z. Anorg. Chem.* 1927, 160, 249.
5. Strachan, J.F.; Harris, N.L. *J. Inst. Metals* 1956-57, 85, 17.
6. Kozin, L.F. *Fiziko-Khimicheskie Osnovy Amalgamnoi Metallurgii*, Nauka, Alma-Ata, 1964.
7. Weeks, J.R. *U.S. At. Ener. Comm. Rep.*, NASA-SP-41, 1963, p. 21; *U.S. At. Ener. Comm. Rep.*, BNL-7553, 1963.
8. Weeks, J.R. *Corrosion* 1967, 23, 98.
9. Weeks, J.R.; Fink, S. *U.S. At. Ener. Comm. Rep.*, BNL-782 1962, p. 73.
10. Weeks, J.R.; Fink, S. *U.S. At. Ener. Comm. Rep.*, BNL-900 1964, p. 136.
11. Parkman, M.F. *Ext. Abst., Electrothermics and Metallurgy Div.*, Vol. 2, No. 2, The Electrochemical Soc. 1964, pp. 16-21.
12. Jangg, G.; Burger, E. *Electrochim. Acta* 1972, 17, 1883.
13. Parkman, M.F.; Whaley, D.K. *Aerojet-General Nucleonics*, Rep. AN-957, 1963; as cited by Schulze, R.C.; Vargo, E.J. *U.S. At. Ener. Comm. Rep.*, TRW-690-33, 1968.
14. Parkman, M.F. *U.S. At. Ener. Comm. Rep.*, TID-7626, 1962, Pt. I, p. 35.

COMPONENTS:	ORIGINAL MEASUREMENTS:
(1) Chromium; Cr; [7440-47-3] (2) Mercury; Hg; [7439-97-6]	de Wet, J.F.; Haul, R.A.W. Z. Anorg. Chem. 1954, 277, 96-112.

VARIABLES:	PREPARED BY:
One temperature: 303 K	C. Guminski; Z. Galus

EXPERIMENTAL VALUES:

The solubility of chromium in mercury at 303.2 K was reported to be below 4×10^{-7} mass %. The corresponding atomic % solubility limit calculated by the compilers is 1.5×10^{-6} at %.

Pure chromium was found to be in equilibrium with mercury.

AUXILIARY INFORMATION

METHOD/APPARATUS/PROCEDURE:	SOURCE AND PURITY OF MATERIALS:
Heterogeneous amalgam was obtained by electrolysis of Cr(III)-sulfate at the mercury cathode. The amalgam was washed, dried and sealed off under vacuum in glass ampules. Later, a portion of the amalgam was centrifuged and 10-12 g of the homogeneous amalgam was taken for analysis. The mercury was carefully distilled off and the residue was dissolved in HCl. The resulting solution was analyzed spectrochemically for chromium.	$Cr_2(SO_4)_3 \cdot 18H_2O$ was BDH Lab. Reag. Purified Hg was distilled under vacuum and was found to be spectrochemically free of chromium. All other chemicals and apparatus were carefully cleaned.
	ESTIMATED ERROR: Detection limit for Cr was 4×10^{-7} mass %. Temp: nothing specified.
	REFERENCES:

COMPONENTS:	ORIGINAL MEASUREMENTS:
(1) Chromium; Cr; [7440-47-3] (2) Mercury; Hg; [7439-97-6]	Jangg, G.; Palman, H. Z. Metallk. 1963, 54, 364-69.
VARIABLES: Temperature: 20-550°C	PREPARED BY: C. Guminski; Z. Galus

EXPERIMENTAL VALUES:

The solubility of chromium in mercury:

$t/°C$	Soly/mass %	Soly/at %[a]
550	8.0×10^{-5}	3.1×10^{-4}
500	6.8×10^{-5}	2.6×10^{-4}
400[b]	3.3×10^{-5}	1.3×10^{-4}
300[b]	2×10^{-5}	7.7×10^{-5}
200[b]	7.2×10^{-6}	2.8×10^{-5}
100[b]	1.5×10^{-6}	5.8×10^{-6}
20[b]	2×10^{-7}	7.7×10^{-7}

---

[a] by compilers.

[b] Only the 500 and 550°C solubilities were experimentally determined; the values at lower temperatures were estimated by extrapolation of the two experimental values.

AUXILIARY INFORMATION

METHOD/APPARATUS/PROCEDURE:	SOURCE AND PURITY OF MATERIALS:
The heterogeneous amalgam was introduced into specially constructed apparatus made of refractory chromium steel. Such steel apparatus could be used because the solubility of iron in mercury is very low and the Cr(III)-oxide film inhibits the wetting of the steel by mercury. After 12 hr of equilibration at the temperature of the experiment, the amalgam was filtered through a sintered iron frit in an atmosphere of purified nitrogen. The chromium content in the filtered, saturated amalgam was determined by an unspecified method.	Nothing specified.
	ESTIMATED ERROR: Soly: accuracy ± 5%. Temp: precision ± 2 K.
	REFERENCES:

COMPONENTS:	ORIGINAL MEASUREMENTS:
(1) Chromium; Cr; [7440-47-3] (2) Mercury; Hg; [7439-97-6]	Parkman, M.F. *Ext. Abst., Electrothermics and Metallurgy Div.*, The Electrochemical Soc., *Vol. 2, No. 2*, 1964, pp. 16-21.
VARIABLES: One temperature: 866 K	PREPARED BY: C. Guminski; Z. Galus

EXPERIMENTAL VALUES:

The mass % solubility of chromium in mercury was reported graphically; a value of 5.5×10^{-5} mass % at 866 K was read off the curve by the compilers. The corresponding atomic % solubility calculated by the compilers is 2.1×10^{-4} at %.

AUXILIARY INFORMATION

METHOD/APPARATUS/PROCEDURE:	SOURCE AND PURITY OF MATERIALS:
Specimen of Cr was placed in contact with Hg in a glass capsule. The capsule was sealed under vacuum after at least 16 hr outgassing of the Hg. The temperature of the capsule was raised to the desired level and held there for 16 hr. A sample of the solution was then collected and cooled. Hg was separated from the sample by molecular distillation. The residue was taken into acid solution, dried, and analyzed by emission spectroscopy.	Pure chromium from AGN. Mercury purity not specified, but probably triple-distilled.
	ESTIMATED ERROR: Soly: nothing specified. Temp: precision $\pm$ 3 K.
	REFERENCES:

COMPONENTS:	ORIGINAL MEASUREMENTS:
(1) Chromium; Cr; [7440-47-3] (2) Mercury; Hg; [7439-97-6]	Weeks, J.R. *Corrosion* 1967, *23*, 98-106.
VARIABLES:	PREPARED BY:
Temperature: 505-650°C	C. Guminski; Z. Galus

EXPERIMENTAL VALUES:

The mass % solubility of chromium in mercury was presented graphically as a function of temperature. The data points were read from the curve and the solubility converted to atomic % by the compilers.

$t/°C$	Soly/at %
650[a]	1.1×10^{-2}
605	4.6×10^{-3}
575	9.6×10^{-4}
555	5.0×10^{-4}
550	4.2×10^{-4}
530	3.6×10^{-4}
510	2.5×10^{-4}
505	1.2×10^{-4}

[a] This value also reported in (1); the other data also reported in (2).

AUXILIARY INFORMATION

METHOD/APPARATUS/PROCEDURE:	SOURCE AND PURITY OF MATERIALS:
Mercury and chromium were sealed in the upper chamber of an evacuated quartz tube. The two chambers were separated by a coarse quartz filter. The filled tubes were equilibrated for 72 hr at each temperature, then centrifuged at temperature to force the liquid through the filter. The mercury was distilled from the known quantity of amalgam, and the residue was dissolved in HF-HNO$_3$ or aqua regia and analyzed spectrographically.	Mercury was triple-distilled. Chromium purity not specified.
	ESTIMATED ERROR: Soly: nothing specified. Temp: precision ± 2 K.
	REFERENCES: 1. Weeks, J.R.; Fink, S. *U.S. At. Ener. Comm. Rep.*, BNL-900, 1964, pp. 136-9. 2. Same authors *U.S. At. Ener. Comm. Rep.*, BNL-782, 1962, pp. 73-5.

COMPONENTS:	EVALUATOR:
(1) Molybdenum; Mo; [7439-98-7] (2) Mercury; Hg; [7439-97-6]	C. Guminski; Z. Galus Department of Chemistry University of Warsaw Warsaw, Poland July, 1985

CRITICAL EVALUATION:

Published experimental values show low solubility of molybdenum in mercury. Tammann and Hinnüber (1) could not detect any dissolution of molybdenum, and Irvin and Russell (2) concluded that the solubility at 293 K should be lower than 4×10^{-5} at %. Strachan and Harris (3) could not detect any molybdenum in mercury by their analytical procedure which had a detection limit of 2×10^{-3} at %. Kozin first predicted a solubility of 7.5×10^{-20} at % at 298 K (4), but this was later revised to an estimate of 2.5×10^{-13} at % (5). Based on regular solution theory, Brewer and Lamoreaux (6) derived the equation,

$$\ln N = 3 - 20000/(T/K)$$

where N is the solubility of molybdenum in at %; the solubility calculated at 298 K from this equation is 10^{-28} at %.

At high temperatures, Bowersox and Leary (7,8) reported that the solubility is lower than 1.5×10^{-5} at % at 623 K, while Messing and Dean (9) found that the solubility of molybdenum in saturated uranium amalgam is also below the detection limit of 1.1×10^{-4} at % at 629 K.

No corrosion of Mo by Hg was observed after their contact at 719 K for more than 30 days (10). This means that the solubility of Mo in Hg should be very low, probably of similar order of magnitude as for Nb.

Férée (11) reported the intermetallic compounds, $MoHg_9$, Mo_2Hg_3 and $MoHg_2$, but these results are questionable and further experimental work is needed on this system.

References

1. Tammann, G.; Hinnüber, J. Z. Anorg. Chem. 1927, 160, 249.
2. Irvin, N.M.; Russell, A.S. J. Chem. Soc. 1932, 891.
3. Strachan, J.F.; Harris, N.L. J. Inst. Metals 1956-57, 85, 17.
4. Kozin, L.F. Tr. Inst. Khim. Nauk Akad. Nauk Kaz. SSR 1962, 9, 101.
5. Kozin, L.F. Fiziko Khimicheskie Osnovy Amalgamnoi Metallurgii, Nauka, Alma-Ata, 1964.
6. Brewer, L.; Lamoreaux, R.H. At. Ener. Rev., Spec. Issue, 1980, 7, 119, 203, 259.
7. Leary, J.A. U.S. At. Ener. Comm. Rep., LA-2218, 1958; as cited by (9).
8. Bowersox, D.F.; Leary, J.A. U.S. At. Ener. Comm. Rep., LAMS-2518, 1961.
9. Messing, A.F.; Dean, O.C. U.S. At. Ener. Comm. Rep., ORNL-2871, 1960.
10. Nejedlik, J.F.; Vargo, E.J. Electrochem. Technol. 1965, 3, 250.
11. Férée, J. C.R. Acad. Sci., Ser. 2 1896, 122, 733.

COMPONENTS:	ORIGINAL MEASUREMENTS:
(1) Molybdenum; Mo; [7439-98-7] (2) Mercury; Hg; [7439-97-6]	Bowersox, D.F.; Leary, J.A. *U.S. At. Ener. Comm. Rep.*, LAMS-2518, 1961.
VARIABLES: One temperature: 350°C	PREPARED BY: C. Guminski; Z. Galus

EXPERIMENTAL VALUES:

The solubility of Mo in Hg at 350°C is lower than 0.001 g of Mo in 1 dm^3 of Hg. The corresponding solubility limit calculated by the compilers is 1.5×10^{-5} at %.

AUXILIARY INFORMATION

METHOD/APPARATUS/PROCEDURE:	SOURCE AND PURITY OF MATERIALS:
The solubility was determined by immersing a weighed coupon of molybdenum into a definite amount of boiling mercury and periodically measuring the coupon weight. Thus, the weight loss corresponds to the part of the molybdenum that dissolved.	Triple-distilled mercury was used. Molybdenum purity not specified.
	ESTIMATED ERROR: Soly: detection limit was 1 mg of Mo. Temp: nothing specified.
	REFERENCES:

COMPONENTS:	EVALUATOR:
(1) Tungsten; W; [7440-33-7] (2) Mercury; Hg; [7439-97-6]	C. Guminski; Z. Galus Department of Chemistry University of Warsaw Warsaw, Poland July, 1985

CRITICAL EVALUATION:

The solubility of tungsten in mercury is very low, and no accurate measurements have been reported. Irvin and Russell (1) have shown that the solubility at 293 K is lower than 1×10^{-5} at % and Strachan and Harris (2) could not detect tungsten in mercury at their detection limit of 10^{-3} at %. Tammann and Hinnüber (3) also could not detect the dissolution of tungsten in mercury. Raub and Plate (4) observed that there was no interaction between the two metals at 1273 K. Similarly, Nejedlik and Vargo (5) found that tungsten was inert to mercury after contact for more than 30 days at 719 K, thus indicating that the solubility of tungsten is very low. It is probable that the solubility of tungsten is of the same magnitude, or lower, as that of tantalum.

The low solubility of tungsten also is suggested by the semiempirical estimates of Kozin who reported values of 4.8×10^{-33} (6) and 6.8×10^{-20} at % (7) at 298 K.

References

1. Irvin, N.M.; Russell, A.S. *J. Chem. Soc.* 1932, 891.
2. Strachan, J.F.; Harris, N.L. *J. Inst. Metals* 1956-57, *85*, 17.
3. Tammann, G.; Hinnüber, J. *Z. Anorg. Chem.* 1927, *160*, 249.
4. Raub, E.; Plate, W. *Z. Metallk.* 1951, *42*, 76.
5. Nejedlik, J.F.; Vargo, E.J. *Electrochem. Technol.* 1965, *3*, 250.
6. Kozin, L.R. *Tr. Inst. Khim. Nauk Akad. Nauk Kaz. SSR* 1962, *9*, 101.
7. Kozin, L.F. *Fiziko-Khimicheskie Osnovy Amalgamnoi Metallurgii*, Nauka, Alma-Ata, 1964.

COMPONENTS:	EVALUATOR:
(1) Manganese; Mn; [7439-96-5] (2) Mercury; Hg; [7439-97-6]	C. Guminski; Z. Galus Department of Chemistry University of Warsaw Warsaw, Poland July, 1985

CRITICAL EVALUATION:

A number of authors have reported on the solubility of manganese in mercury near room temperature. The recommended solubility of 4.4×10^{-3} at % at 298 K was reported in two separate works by Krasnova and Zebreva (1) and Hurlen and Smaaberg (2); both groups employed potentiometry. Three other results support the recommended value: 3.6×10^{-3} at % at 293 K by Irvin and Russell (3); 4.6×10^{-3} at % at 293 K by Kemula and Galus (4); and 5.2×10^{-3} at % at 303 K by deWet and Haul (5). Chemical analysis (3,5) and voltammetry (4) were employed in the latter determinations.

There are several higher results but there is no basis for rejection of these data. Royce and Kahlenberg (6) determined a solubility of 1.13×10^{-2} at % at 293 K by chemical analysis. Jangg and Kirchmayr (7) reported a value of 6.8×10^{-3} at % at 288 K from potentiometric measurements. The solubility of 6.2×10^{-3} at % at 293 K reported by Ettmayer and Jangg (8) is slightly higher than the recommended value. Dowgird and Galus (9) employed potentiometry and determined the solubility to be 1.28×10^{-2} at % at 298 K. Sagadieva and Kozlovskii (10) used polarography to determine the solubility of 9.6×10^{-3} at % at 290 K.

Kozin (11) predicted a low solubility of 6.5×10^{-4} at % at 298 K. Some of the reported determinations of the solubility are rejected because the values were clearly too high, probably due to incomplete filtration: 0.014 (12), 0.44, 0.47 and 0.56 (13) at % at 282, 303, 328 and 343 K, respectively. The value of 9.2×10^{-4} at % reported by (14) is too low, probably because of corrosion of the manganese. Strachan and Harris (15) could not detect any dissolution of manganese in mercury at room temperature where their detection limit was 7×10^{-3} at %. Hickling and Maxwell (16) reported the solubility of 1.1×10^{-2} at % at 293 K but the work is not compiled due to lack of experimental details.

Jangg and Palman (17) determined the solubilities at 358 to 833 K, while Lange and coworkers (18,19,20) determined the solubilities over a temperature range of 293 to 368 K. There was agreement between (17) and (19) only in the region of 358 K. It may be that the dissolution of solid during electro-oxidation of the homogeneous amalgam resulted in increased estimates of the solubilities in (18-20). This system needs further investigation, especially in the temperature range of 300-600 K.

At temperatures below 345 ± 3 K the liquid phase is reported to be in equilibrium with Mn_2Hg_5, while at 345 to 538 K the liquid is in equilibrium with MnHg (5,6,19-23).

Recommended (r) and tentative values of the solubility of Mn in Hg:

T/K	Soly/at %	Reference
293	3.6×10^{-3}	[3]
298	4.5×10^{-3} (r)[a]	[1,2,4]
357	8×10^{-2}	[17,19]
473	0.4	[17]
573	1.3	[17]
673	3	[17]
773	6	[17]

[a] Mean value from cited references.

(Continued next page)

COMPONENTS:	EVALUATOR:
(1) Manganese; Mn; [7439-96-5] (2) Mercury; Hg; [7439-97-6]	C. Guminski; Z. Galus Department of Chemistry University of Warsaw Warsaw, Poland July, 1985

CRITICAL EVALUATION: (Continued)

References

1. Krasnova, I.E.; Zebreva, A.I. *Elektrokhimia* 1966, *2*, 96.
2. Hurlen, T.; Smaaberg, R. *J. Electroanal. Chem.* 1976, *71*, 157.
3. Irvin, N.M.; Russell, A.S. *J. Chem. Soc.* 1932, 891.
4. Kemula, W.; Galus, Z. *Roczniki Chem.* 1962, *36*, 1223.
5. deWet, J.F.; Haul, R.A.W. *Z. Anorg. Chem.* 1954, *277*, 96.
6. Royce, H.D.; Kahlenberg, L. *Trans. Electrochem. Soc.* 1931, *59*, 126.
7. Jangg, G.; Kirchmayr, H.R. *Z. Chem.* 1963, *3*, 47.
8. Ettmayer, P.; Jangg, G. *Monatsh. Chem.* 1973, *104*, 1120.
9. Dowgird, A.; Galus, Z. *J. Electroanal. Chem.* 1972, *34*, 457.
10. Sagadieva, K.Zh.; Kozlovskii, M.T. *Vest. Akad. Nauk. Kaz. SSR* 1963, No. 5, 85.
11. Kozin, L.F. *Fiziko-Khimicheskie Osnovy Amalgamnoi Metallurgii*, Nauka, Alma-Ata, 1964.
12. Campbell, A.N. *J. Chem. Soc.* 1924, 1713.
13. Campbell, A.N.; Carter, H.D. *Trans. Faraday Soc.* 1933, *29*, 1295.
14. Tammann, G.; Hinnüber, J. *Z. Anorg. Chem.* 1927, *160*, 249.
15. Strachan, J.F.; Harris, N.L. *J. Inst. Metals* 1956-57, *85*, 17.
16. Hickling, A.; Maxwell, J. *Trans. Faraday Soc.* 1955, *57*, 44.
17. Jangg, G.; Palman, H. *Z. Metallk.* 1963, *54*, 364.
18. Lange, A.A.; Bukhman, S.P. *Izv. Akad. Nauk Kaz. SSR, Ser. Khim.* 1964, No. 3, 27.
19. Lange, A.A.; Bukhman, S.P.; Kozlovskii, M.T. *Tr. Inst. Khim. Nauk Akad. Nauk Kaz. SSR* 1969, *21*, 92.
20. Lange, A.A.; Bukhman, S.P. *Elektrokhimia* 1969, *5*, 553.
21. Lihl, F. *Monatsh. Chem.* 1955, *86*, 186.
22. Jangg, G.; Steppan, F. *Z. Metallk.* 1965, *56*, 172.
23. deWet, J.F. *Angew. Chem.* 1955, *67*, 208.

COMPONENTS:	ORIGINAL MEASUREMENTS:
(1) Manganese; Mn; [7439-96-5] (2) Mercury; Hg; [7439-97-6]	Royce, H.D.; Kahlenberg, L. *Trans. Electrochem. Soc.* 1931, *59*, 121-33.
VARIABLES: One temperature: 20°C	PREPARED BY: C. Guminski; Z. Galus

EXPERIMENTAL VALUES:

The 20°C solubility of manganese in mercury was reported to be $(3.1 \pm 0.1) \times 10^{-3}$ mass %. The corresponding atomic % solubility calculated by the compilers is 1.1×10^{-2} at %. The liquid phase was reported to be in equilibrium with solid Mn_2Hg_5 up to 86°C. In the region of 86 to 100°C the solid phase was reported to be MnHg.

AUXILIARY INFORMATION

METHOD/APPARATUS/PROCEDURE:	SOURCE AND PURITY OF MATERIALS:
Heterogeneous amalgam was prepared electrolytically, then filtered through chamois skin. The liquid amalgam was analyzed by two methods: 1. A weighed sample of the amalgam was heated in conc. HCl for several hours to dissolve the Mn. Mercury was then washed, dried and weighed. 2. The Mn which was dissolved in HCl, as in the first method, was determined by the Volhard method.	"Purest obtainable" materials were employed.
	ESTIMATED ERROR: Soly: accuracy $\pm$ 3%. Temp: not specified.
	REFERENCES:

COMPONENTS:	ORIGINAL MEASUREMENTS:
(1) Manganese; Mn; [7439-96-5] (2) Mercury; Hg; [7439-97-6]	Irvin, N.M.; Russell, A.S. *J. Chem. Soc.* <u>1932</u>, 891-8.
VARIABLES: One temperature: 293 K	PREPARED BY: C. Guminski; Z. Galus

EXPERIMENTAL VALUES:

The solubility of manganese in mercury at 293 K was reported to be 1.0×10^{-3} mass %. The corresponding atomic % solubility calculated by the compilers is 3.6×10^{-3} at %.

AUXILIARY INFORMATION

METHOD/APPARATUS/PROCEDURE:	SOURCE AND PURITY OF MATERIALS:
The heterogeneous amalgam was prepared by electrolysis. After equilibration the amalgam was filtered through a ground-glass filter. The separated liquid amalgam was shaken with acidified ferric sulfate to oxidize the Mn. Mercuric ions were then reduced by treatment with zinc amalgam, and manganese was determined volumetrically as permanganate after oxidation with sodium bismuthate and nitric acid.	Nothing specified.
	ESTIMATED ERROR: Soly: accuracy ± 10%. Temp: not specified.
	REFERENCES:

COMPONENTS:	ORIGINAL MEASUREMENTS:
(1) Manganese; Mn; [7439-96-5] (2) Mercury; Hg; [7439-97-6]	deWet, J.F.; Haul, R.A.W. Z. Anorg. Chem. 1954, 277, 96-112.
VARIABLES: One temperature: 303 K	PREPARED BY: C. Guminski; Z. Galus

EXPERIMENTAL VALUES:

The 303 K solubility of manganese in mercury was reported to be 1.7×10^{-3} mass %. The corresponding atomic % solubility calculated by the compilers is 6.2×10^{-3} at %. The intermetallic compound, $MnHg_4$, was found to be in equilibrium with the homogeneous amalgam.

AUXILIARY INFORMATION

METHOD/APPARATUS/PROCEDURE:	SOURCE AND PURITY OF MATERIALS:
Amalgam was obtained by electrolysis of Mn(II)-sulfate at the mercury cathode and by the rotation of Mn rod in mercury at 30°C in a hydrogen atmosphere. The electrochemically obtained amalgam was filtered through sintered glass in a centrifuge vessel and sealed off under vacuum. After equilibration and sufficiently long centrifuging, about 10-12 grams of homogeneous amalgam was taken for analysis. This amalgam was treated with dilute phosphoric and sulfuric acids and the dissolved Mn, after oxidation with potassium periodate, was colorimetrically determined.	$MnSO_4 \cdot 4H_2O$ was Hopkins and Williams Analar grade. Purified Hg was distilled under vacuum and was found spectrochemically free of Mn. All other chemicals and vessels used were carefully cleaned.
	ESTIMATED ERROR: Soly: accuracy approximately $\pm$ 10%. Temp: nothing specified.
	REFERENCES:

COMPONENTS:	ORIGINAL MEASUREMENTS:
(1) Manganese; Mn; [7439-96-5] (2) Mercury; Hg; [7439-97-6]	Kemula, W.; Galus, Z. *Roczniki Chem.* 1962, *36*, 1223-38.
VARIABLES: One temperature: 20°C	PREPARED BY: C. Guminski; Z. Galus

EXPERIMENTAL VALUES:

The solubility of manganese in mercury at 20°C was found to be 3.1×10^{-3} mol dm^3. The corresponding atomic % solubility calculated by the compilers is 4.6×10^{-3} at %.

AUXILIARY INFORMATION

METHOD/APPARATUS/PROCEDURE:	SOURCE AND PURITY OF MATERIALS:
The heterogeneous manganese amalgam was prepared by electroreduction of Mn(II) on the hanging mercury-drop electrode. Then the peak of oxidation of the homogeneous amalgam was recorded under voltammetric conditions. The solubility was determined from the charge corresponding to this current peak and the volume of the mercury-drop.	Analytically pure chemicals and doubly distilled water were used in the study.
	ESTIMATED ERROR: Soly: nothing specified; precision no higher than $\pm$ 10% (compilers). Temp: not specified.
	REFERENCES:

COMPONENTS:	ORIGINAL MEASUREMENTS:
(1) Manganese; Mn; [7439-96-5] (2) Mercury; Hg; [7439-97-6]	Jangg, G.; Palman, H. Z. Metallk. 1963, 54, 364-69.
VARIABLES: Temperature: 86-565°C	PREPARED BY: C. Guminski; Z. Galus

EXPERIMENTAL VALUES:

The mass % solubility of manganese in mercury was presented graphically as a function of temperature. The data points were read from the curve and the solubilities were converted to atomic % by the compilers.

$t/°C$	Soly/at %	$t/°C$	Soly/at %
86	0.087	300	1.3
100	0.10	330	1.9
114	0.12	350	2.2
125	0.17	370	2.6
148	0.26	400	3.1
166	0.31	418	3.6
198	0.36	450	4.6
225	0.51	470	5.6
246	0.69	500	6.3
270	0.87	552	7.6
		565	8.2

AUXILIARY INFORMATION

METHOD/APPARATUS/PROCEDURE:	SOURCE AND PURITY OF MATERIALS:
Amalgam preparation was not specified. At below 320°C the amalgam was equilibrated for 12 hr in a glass vessel, after which the amalgam was filtered and analyzed. Above 320°C the heterogeneous amalgam was introduced into specially constructed apparatus made of refractory Cr-steel. Such apparatus could be used because of very low solubility of Fe in Hg, and because the Cr(III)-oxide film inhibits the wetting of the steel by Hg. After 12 hr of equilibration at the temperature of the experiment, the amalgam was filtered through the sintered iron frit under a pressure of purified nitrogen. Usually 3- to 4-fold filtration was necessary. The metal content was then analytically determined in the filtered saturated amalgam. Analytical procedure not described.	Nothing specified.
	ESTIMATED ERROR: Soly: precision better than ± 5%. Temp: precision ± 2 K.
	REFERENCES:

COMPONENTS:	ORIGINAL MEASUREMENTS:
(1) Manganese; Mn; [7439-96-5] (2) Mercury; Hg; [7439-97-6]	Jangg, G.; Kirchmayr, H. Z. Chem. 1963, 3, 47-56.
VARIABLES: One temperature: 15°C	PREPARED BY: C. Guminski; Z. Galus

EXPERIMENTAL VALUES:

The data were reported graphically; a solubility of 4.6×10^{-3} mol dm^{-3} at 15°C was read from the curve by the compilers. The corresponding atomic % solubility calculated by the compilers is 6.8×10^{-3} at %.

AUXILIARY INFORMATION

METHOD/APPARATUS/PROCEDURE:	SOURCE AND PURITY OF MATERIALS:
The amalgams were obtained by electrolysis with 100% efficiency. Concentration of the amalgam was determined by coulometry. Potentials of the Mn-amalgam were measured against the SCE in the cell, Mn(Hg)$_x$ | 0.01-1.0 mol dm^{-3} MnSO$_4$ | KCl, Hg$_2$Cl$_2$, Hg The concentration of the saturated amalgam was evaluated from the breakpoint in the plot of the potential vs. logarithm of the amalgam concentration. The experiments were performed in an atmosphere of nitrogen.	Nothing specified.
	ESTIMATED ERROR: Soly: error may be as high as ± 50%. Temp: precision better than ± 1 K.
	REFERENCES:

COMPONENTS:	ORIGINAL MEASUREMENTS:
(1) Manganese; Mn; [7439-96-5] (2) Mercury; Hg; [7439-97-6]	Sagadieva, K.Zh.; Kozlovskii, M.T. *Vestn. Akad. Nauk Kaz. SSR* 1963, No. 5, 85-7.
VARIABLES: One temperature: 17°C	PREPARED BY: C. Guminski; Z. Galus

EXPERIMENTAL VALUES:

The solubility of manganese in mercury at 17°C was found to be $(6.5 \pm 0.1) \times 10^{-3}$ mol dm^{-3}. The corresponding atomic % solubility calculated by the compilers is 9.6×10^{-3} at %.

AUXILIARY INFORMATION

METHOD/APPARATUS/PROCEDURE:	SOURCE AND PURITY OF MATERIALS:
The amalgam was prepared by electrolytic deposition of Mn on Hg cathode with 100% current efficiency. The concentration of the amalgam was determined from the current and time of the electrolysis. The solubility was determined polarographically; the anodic current was practically independent of the Mn content when amalgam saturation was attained. Oxidation of the Mn amalgam was carried out in two background electrolyte: 0.1 mol dm^{-3} KNO$_3$ and 1 mol dm^{-3} ammonia buffer.	Nothing specified.
	ESTIMATED ERROR: Soly: error probably $\pm$ 3% (compilers). Temp: precision $\pm$ 1 K.
	REFERENCES:

COMPONENTS:	ORIGINAL MEASUREMENTS:
(1) Manganese; Mn; [7439-96-5] (2) Mercury; Hg; [7439-97-6]	Lange, A.A.; Bukhman, S.P. *Izv. Akad. Nauk Kaz. SSR, Ser. Khim.* <u>1964</u>, No. 3, 27-32.

VARIABLES:	PREPARED BY:
Temperature: 20-50°C	C. Guminski; Z. Galus

EXPERIMENTAL VALUES:

The mass % solubility of manganese in mercury was reported graphically; only the numerical value for 20°C was presented by the authors. The solubility at 30, 40, and 50°C was read from the curve by the compilers, and the corresponding atomic % solubilities were calculated for all temperatures.

$t/°C$	Soly/mass % $\times 10^3$	Soly/at % $\times 10^3$
20	3.42	12.5
30	6.7	24
40	11	40
50	13	48

These results were also presented in (1).

AUXILIARY INFORMATION

METHOD/APPARATUS/PROCEDURE:	SOURCE AND PURITY OF MATERIALS:
The amalgams were prepared electrolytically. The amalgams were oxidized under voltammetric conditions and current-potential curves were constructed. For amalgams with manganese content exceeding its solubility in mercury the limiting current was constant, while in the region of homogeneity the current was linearly dependent on concentration. The concentration of the amalgam corresponding to the change of the character of this dependence was taken as the concentration equal to the solubility of manganese in mercury.	Not specified.

	ESTIMATED ERROR:
	Soly: not specified; error probably less than ± 10% (compilers). Temp: nothing specified.

	REFERENCES:
	1. Lange, A.A.; Bukhman, S.P. *Elektrokhimia* <u>1969</u>, 5, 553.

COMPONENTS:	ORIGINAL MEASUREMENTS:
(1) Manganese, Mn; [7439-96-5] (2) Mercury; Hg; [7439-97-6]	Krasnova, I.E.; Zebreva, A.I. *Elektrokhimia* 1966, 2, 96-9.
VARIABLES: One temperature: 25°C	PREPARED BY: C. Guminski; Z. Galus

EXPERIMENTAL VALUES:

The solubility of manganese in mercury at 25°C was reported to be 1.2×10^{-3} mass %. The corresponding atomic % solubility calculated by the compilers is 4.4×10^{-3} at %.

AUXILIARY INFORMATION

METHOD/APPARATUS/PROCEDURE:	SOURCE AND PURITY OF MATERIALS:
The manganese amalgams of various concentrations were prepared by the electro-reduction of manganese (II) on the hanging mercury drop electrode. The oxidation current of manganese from these amalgams was then recorded under voltammetric conditions. By plotting the peak current value versus the amalgam concentration, the change of the character of this dependence at the saturation point was observed. This enabled the determination of the solubility.	Nothing specified.
	ESTIMATED ERROR: Soly: error ± 33%. Temp: precision ± 2 K.
	REFERENCES:

COMPONENTS:	ORIGINAL MEASUREMENTS:
(1) Manganese; Mn; [7439-96-5] (2) Mercury; Hg; [7439-97-6]	Lange, A.A.; Bukhman, S.P.; Kozlovski, M.T. *Tr. Inst. Khim. Nauk, Akad. Nauk Kaz. SSR* <u>1969</u>, *21*, 92-102.
VARIABLES: Temperature: 20-95°C	PREPARED BY: C. Guminski; Z. Galus

EXPERIMENTAL VALUES:

Solubility of manganese in mercury:

$t/°C$	Soly/mol dm^{-3}	Soly/at %[a]
20	8.46×10^{-3}	0.012
30	1.68×10^{-2}	0.025
40	2.65×10^{-2}	0.039
50	3.23×10^{-2}	0.047
70	4.84×10^{-2}	0.071
80	5.6×10^{-2}	0.082
82	7.00×10^{-2}	0.098
85	7.60×10^{-2}	0.11
88	9.90×10^{-2}	0.14
90	1.11×10^{-1}	0.16
95	1.23×10^{-1}	0.18

[a] by compilers.

AUXILIARY INFORMATION

METHOD/APPARATUS/PROCEDURE:	SOURCE AND PURITY OF MATERIALS:
The amalgams were prepared by the electro-reduction of Mn(II) on the Hg cathode. The solubilities were determined on the basis of the limiting current of the manganese amalgam oxidation. When the content of metal in mercury exceeded the solubility the current ceased to be dependent on the manganese content.	Nothing specified.
	ESTIMATED ERROR: Soly: nothing specified; error may be as high as 4% (compilers). Temp: nothing specified.
	REFERENCES:

COMPONENTS:	ORIGINAL MEASUREMENTS:
(1) Manganese; Mn; [7439-96-5] (2) Mercury; Hg; [7439-97-6]	Dowgird, A.; Galus, Z. *J. Electroanal. Chem.* 1972, *34*, 457-61.

VARIABLES:	PREPARED BY:
One temperature: 25°C	C. Guminski; Z. Galus

EXPERIMENTAL VALUES:

Solubility of manganese in mercury at 25°C was reported to be 8.7×10^{-3} mol dm^{-3}. The corresponding atomic % solubility calculated by the compilers is 1.28×10^{-2} at %.

AUXILIARY INFORMATION

METHOD/APPARATUS/PROCEDURE:	SOURCE AND PURITY OF MATERIALS:
The amalgam was prepared by reduction of Mn(II) at constant current density on the hanging mercury-drop electrode. Then the potential was measured with respect to SCE over a period of 12 min. At higher concentrations the potential changes were only observed in 60 sec. The experiments were performed in hydrogen atmosphere. The inflection point of the plot of potential vs. logarithm of concentration corresponded to the saturation point. It is probable that the amalgams were slightly supersaturated at the highest concentrations.	All chemicals were of reagent grade. Mercury was chemically purified by prolonged shaking with acidified solution of $Hg_2(NO_3)_2$ then double-distilled at reduced pressure. All solutions prepared with triply-distilled water.
	ESTIMATED ERROR: Soly: nothing specified; precision no better than $\pm$ 10% (compilers). Temp: precision $\pm$ 0.2 K.
	REFERENCES:

COMPONENTS:	ORIGINAL MEASUREMENTS:
(1) Manganese; Mn; [7439-96-5] (2) Mercury; Hg; [7439-97-6]	Ettmayer, P.; Jangg, G. *Monatsh. Chem.* 1973, *104*, 1120-30.
VARIABLES: One temperature: 293 K	PREPARED BY: C. Guminski; Z. Galus

EXPERIMENTAL VALUES:

The solubility of manganese in mercury at 293 K was reported to be 1.7×10^{-3} mass %. The corresponding atomic % solubility calculated by the compilers is 6.2×10^{-3} at %.

AUXILIARY INFORMATION

METHOD/APPARATUS/PROCEDURE:	SOURCE AND PURITY OF MATERIALS:
The amalgam was prepared by electro-reduction of Mn(II) on a Hg cathode. The electrolyte contained $MnSO_4$ and $(NH_4)_2SO_4$ as a buffer. The amalgam was separated from the electrolyte, dried, filtered, and analyzed by an unknown method.	Nothing specified.
	ESTIMATED ERROR: Soly: nothing specified; accuracy probably no better than $\pm$ 10% (compilers). Temp: nothing specified.
	REFERENCES:

COMPONENTS:	ORIGINAL MEASUREMENTS:
(1) Manganese; Mn; [7439-96-5] (2) Mercury; Hg; [7439-97-6]	Hurlen, T.; Smaaberg, R. *J. Electroanal. Chem.* 1976, *71*, 157-68.

VARIABLES:	PREPARED BY:
One temperature: 25°C	C. Guminski; Z. Galus

EXPERIMENTAL VALUES:

The solubility of manganese in mercury at 25°C was reported to be 1.2×10^{-3} mass %. The corresponding atomic % solubility calculated by the compilers is 4.4×10^{-3} at %.

AUXILIARY INFORMATION

METHOD/APPARATUS/PROCEDURE:	SOURCE AND PURITY OF MATERIALS:
Amalgams of various concentrations were prepared electrolytically. The potentials of such amalgams, in the presence of a constant Mn(II) concentration, were measured in the cell, $Mn(Hg)_x \mid n$ mol dm^{-3} MnCl$_2$, (0.5-n) mol dm^{-3} MgCl$_2$, pH = 4.3-4.9 $\mid$ KCl, AgCl, Ag Up to the saturation point the potential of the amalgam was dependent on its concentration; at higher concentrations the potential was constant. The solubility was determined from the inflection point of the potential-concentration dependence.	Analytically pure reagents and doubly-distilled water were used.

	ESTIMATED ERROR: Soly: nothing specified; precision better than ± 10% (compilers). Temp: not specified.
	REFERENCES:

COMPONENTS:	EVALUATOR:
(1) Rhenium; Re; [7440-15-5] (2) Mercury; Hg; [7439-97-6]	C. Guminski; Z. Galus Department of Chemistry University of Warsaw Warsaw, Poland July, 1985

CRITICAL EVALUATION:

No specified data on the solubility of rhenium in mercury has been published. It has been reported (1) that rhenium powder is not attacked by mercury when the metals are heated in a reducing atmosphere at 573 K. Also, Jangg and Dörtbudak (2) equilibrated the two metals at 773 K and could not detect any dissolution of rhenium; their analytical detection limit was 10^{-5} at %. Kozin (3) estimated that the solubility of rhenium in mercury at 298 K is 5.9×10^{-18} at %; a previously predicted value of 3.5×10^{-29} at % at 298 K appears to be less probable (4).

The solid phase in equilibrium with the saturated rhenium amalgam should contain pure rhenium because no Re-Hg compound was found (2).

References

1. Heyne, R.; Moers, K.   *Z. Anorg. Chem.* 1931, *196*, 151.
2. Jangg, G.; Dörtbudak, T.   *Z. Metallk.* 1973, *64*, 715.
3. Kozin, L.F.   *Tr. Inst. Khim. Nauk Akad. Nauk Kaz. SSR* 1962, *9*, 101.
4. Kozin, L.F.   *Fiziko-Khimicheskie Osnovy Amalgamnoi Metallurgii*, Nauka, Alma-Ata, 1964.

COMPONENTS:	EVALUATOR:
(1) Iron; Fe; [7439-89-6] (2) Mercury; Hg; [7439-97-6]	C. Guminski; Z. Galus Department of Chemistry University of Warsaw Warsaw, Poland July, 1985

CRITICAL EVALUATION:

Early reports of the solubility of iron in mercury in the region of room temperature varied over a range from 4.1×10^{-17} to 6.39 at % (1-4). These results are all rejected because they are either much too low or too high as compared to recent more precise measurements. In some instances only the solubility limits were stated because the analytical methods could not detect the low solubility of iron near room temperature; the solubility limits reported varied from 10^{-6} to 10^{-3} at % in this temperature region (5-8). Palmaer (9) employed analytical methods and reported that the iron content in saturated iron amalgams remained nearly constant at about 2.5×10^{-4} at % between 293 and 484 K; this result is too high and is rejected. Kozin's (28) calculated solubility of 1.4×10^{-4} at % at 298 K is too high.

Marshall and coworkers (10) determined the solubility of iron between 298 and 973 K, and these authors observed an increase from 5.4×10^{-6} to 3.4×10^{-4} at %, respectively, in this temperature range. The data of (10) at temperatures below 700 K are clearly overstated, while the data at temperatures higher than 700 K are in good agreement with the subsequent works of Weeks and coworkers (11-14).

Weeks (11) graphically summarized the iron solubility determinations made at the Brookhaven National Laboratories by he and his coworkers (12-14). Numerical data were reported only at 873 and 923 K where the solubilities were 1.8×10^{-4} (12) and 2.7×10^{-4} at % (14), respectively. Earlier, preliminary results by these workers (15,16) are rejected because of the large scatter in the data. Nerad (17), without giving any experimental details, reported iron solubilities of 6.1×10^{-5} and 1.5×10^{-4} at % at 755 and 856 K, respectively; these solubilities are in good agreement with (10) and (11).

Wang (18) reported a solubility of 2.0×10^{-4} at % at 644 K, and Bowersox and Leary (19) determined a value of 5×10^{-5} at % at 623 K. Both these results are higher than the solubilities reported by (10); the result of (18) is rejected because it is too high as compared to the other measurements.

Parkman (20), using iron from two different sources, determined the iron solubility at several temperatures and at different equilibration times, but no definite conclusions may be made from the results of this study. Jangg and coworkers (21) reported that the iron content in saturated amalgams between 973 and 1073 K was less than 2×10^{-4} at %.

Because the scatter in the iron solubility data is large, it is difficult to make clear recommendations for the solubilities of this metal in mercury. There is an especial need for more precise measurements at temperatures below 573 K. Luborsky (22) found that a gel-like iron amalgam, which contained 1% Fe, was stable for long periods at room temperature, even though the apparent solubility was exceeded by more than a millionfold. In this instance, the particle size of the iron is about 2 nm in diameter and filtration through sintered glass does not appreciably change the composition. This formation of very fine crystallites of iron in the amalgam is the reason why almost all solubility determinations at the lower temperatures are strongly overstated.

Horsley (23) analyzed the data of (10) and reported iron solubility of $(0.27-6.8) \times 10^{-4}$ at % between 673 and 1073 K. This author also calculated grain boundary solubilities of $(1.5-13.6) \times 10^{-4}$ at % in this temperature range.

Gudtsov and Gavze (24, 25) investigated the solubility of steels in mercury, and they reported the content of iron in the mercury phase after hundreds of hours of contact at 673 to 1023 K. The authors found no evident dependence of the solubility on temperature, time of contact, or the composition of the steel; they reported solubilities ranging from $(0.109-8.4) \times 10^{-4}$ at %. Similar experiments were performed by Smith and Thompson (26) and by Parkman (20, 27). The solubilities obtained by (24, 25) are significantly higher than the solubility of pure iron; e.g., 6.2×10^{-2} at % for technical iron at 923 K as compared to 1.7×10^{-4} at % (11). On the other hand, (20, 27) found the solubility of technical iron to be of similar magnitude as that for pure iron.

Iron does not form any intermetallic compounds with mercury, and pure iron is in equilibrium with the liquid phase (8, 21, 24, 29).

(Continued next page)

COMPONENTS:	EVALUATOR:
(1) Iron; Fe; [7439-89-6] (2) Mercury; Hg; [7439-97-6]	C. Guminski; Z. Galus Department of Chemistry University of Warsaw Warsaw, Poland July, 1985

CRITICAL EVALUATION: (Continued)

Tentative values of iron solubility in mercury:

T/K	Soly/at %	Reference
673	4×10^{-5}	[10]
773	9×10^{-5} [a]	[10,11,12]
873	2×10^{-4}	[11]
973	3.5×10^{-4}	[10]

[a] Interpolated value from data of cited references.

References

1. Tammann, G.; Kollmann, K. *Z. Anorg. Chem.* 1927, 160, 242.
2. Richards, T.W.; Garrod-Thomas, R.N. *Z. Phys. Chem.* 1910, 72, 181.
3. Tammann, G.; Oelsen, W. *Z. Anorg. Chem.* 1930, 186, 257.
4. Nagaoka, H. *Ann. Phys. Chem.* 1896, 59, 66.
5. Gouy, M. *J. Phys.* 1895, 4, 320.
6. Irvin, N.M.; Russell, A.S. *J. Chem. Soc.* 1932, 891.
7. Strachan, J.F.; Harris, N.L. *J. Inst. Metals* 1956-57, 85, 17.
8. de Wet, J.F.; Haul, R.A.W. *Z. Anorg. Chem.* 1954, 277, 96.
9. Palmaer, E. *Z. Elektrochem.* 1932, 38, 70.
10. Marshall, A.L.; Epstein, L.F.; Norton, F.J. *J. Am. Chem. Soc.* 1950, 72, 3514.
11. Weeks, J.R. *Corrosion* 1967, 23, 98.
12. Weeks, J.R. *U.S. At. Ener. Comm. Rep.*, NASA-SP-41, 1963, p. 21; *U.S. At. Ener. Comm. Rep.*, BNL-7553, 1963.
13. Weeks, J.R.; Minardi, A.; Fink, S. *U.S. At. Ener. Comm. Rep.*, BNL-759, 1962, p. 63.
14. Weeks, J.R.; Fink, S. *U.S. At. Ener. Comm. Rep.*, BNL-900, 1964, p. 136.
15. Fleitman, A.H.; Romano, A.; Klamut, C. *U.S. At. Ener. Comm. Rep.*, TID-7626, 1962, Pt. I, p. 24.
16. Weeks, J.R.; Fink, S.; Minardi, A. *U.S. At. Ener. Comm. Rep.*, BNL-705, 1961, p. 56.
17. Nerad, A.J., as cited by A. H. Fleitman, J.R. Weeks, *Nucl. Eng. Des.* 1971, 16, 166.
18. Wang, J.Y.N. *Nucl. Sci. Eng.* 1964, 18, 18.
19. Bowersox, D.F.; Leary, J.A. *U.S. At. Ener. Comm. Rep.*, LAMS-2518, 1961.
20. Parkman, M.F. *Extended Abst., Electrothermics and Metallurgy Div.*, Vol. 2, No. 2, The Electrochemical Soc., New York, NY, 1964, pp. 16-21.
21. Jangg, G.; Fitzer, E.; Adlhart, O.; Hohn, H. *Z. Metallk.* 1958, 49, 557.
22. Luborsky, E. *J. Phys. Chem.* 1958, 61, 1336.
23. Horsley, G.W. *J. Nucl. Energy, Part B* 1959, 1, 84.
24. Gudtsov, N.T.; Gavze, M.N. *Izv. Akad. Nauk SSSR, Otd. Tekhn. Nauk* 1952, 67.
25. Gudtsov, N.T.; Gavze, M.N. *Vozdeistvie rtuti kak teplonositelya na stal v energeticheskikh ustanovkakh*, Izdatelstvo AN SSSR, Moskva, 1956, p. 16; 2nd Edition, 1963, p. 59.
26. Smith, A.R.; Thompson, E.S. *Trans. ASME* 1942, 64, 625.
27. Parkman, M.F.; Whaley, D.K. *Aerojet-General Nucleonics*, Rep. AN-957, 1963.
28. Kozin, L.F. *Fiziko-Khimicheskie Osnovy Amalgamnoi Metallurgii*, Nauka, Alma-Ata, 1964.
29. Lihl, F. *Z. Metallk.* 1953, 44, 160.

COMPONENTS:	ORIGINAL MEASUREMENTS:
(1) Iron; Fe; [7439-89-6] (2) Mercury; Hg; [7439-97-6]	Marshall, A.L.; Epstein, L.F.; Norton, F.J. J. Am. Chem. Soc. 1950, 72, 3514-16.
VARIABLES: Temperature: 25-700°C	PREPARED BY: C. Guminski; Z. Galus

EXPERIMENTAL VALUES:

Experimental Solubility of Iron in Mercury

$t/°C$	25	25	25	250	250	425	500	500	500	700	700
g Fe/10^6 g Hg	0.013	0.015	0.017	0.037	0.066	0.105	0.105	0.225	0.270	1.0	1.2

Smoothed Solubility of Iron in Mercury

$t/°C$	25	100	200	300	400	500	600	700
g Fe/10^6 Hg	0.015	0.019	0.030	0.054	0.11	0.21	0.45	0.96
[a]Soly/at % x 10^5	0.54	0.68	1.1	1.9	3.9	7.5	16	34

[a] by compilers.

AUXILIARY INFORMATION

METHOD/APPARATUS/PROCEDURE:	SOURCE AND PURITY OF MATERIALS:
Carefully deoxygenated iron cylinder was equilibrated with Hg in evacuated quartz bulbs for several hours to a month. Care was taken to assure wetting of iron. Bulb was sealed in a steel bomb with Hg to equalize pressure at high temperatures, and the bomb was rocked in the furnace to assure equilibrium. Finally, the bomb was tilted at temperature to separate the amalgam from the iron, and then cooled to remove the sample for analysis. After opening the bulb the iron was removed and the Hg distilled, collected and weighed. The iron was determined colorimetrically by complex formation with KCNS.	Redistilled Hg of original high purity, and pure Swedish iron were used.
	ESTIMATED ERROR: Soly: precision as high as ± 50%. Temp: nothing specified.
	REFERENCES:

COMPONENTS:	ORIGINAL MEASUREMENTS:

(1) Iron; Fe; [7439-89-6]
(2) Mercury; Hg; [7439-97-6]
	1. Weeks, J.R.
	  Corrosion 1967, *23*, 98-106.
	2. Weeks, J.R.; Minardi, A.; Fink, S.
	  U.S. At. Ener. Comm. Rep., BNL-759, 1962, p. 63.

VARIABLES: Temperature: 500-650°C	PREPARED BY: C. Guminski; Z. Galus

EXPERIMENTAL VALUES:

The mass % solubility was presented graphically as a function of temperature. The data points from the plot were read off and converted to atomic % by the compilers.

$t/°C$	Soly/mass % $\times 10^5$	Soly/at % $\times 10^4$
500	2.0	0.72
525	7.0	2.5
550	2.2	0.79
575	2.6	0.93
600	5.5	2.0
625	6.2	2.2
650	4.8	1.7

AUXILIARY INFORMATION

METHOD/APPARATUS/PROCEDURE:	SOURCE AND PURITY OF MATERIALS:
Hg and Fe were placed into the larger chamber of a fused quartz capsule which was constructed so that a coarse quartz filter separated the two chambers. The capsule containing the metals was sealed under vacuum and placed in a stainless-steel capsule. Hg was also placed in the steel capsule before it was welded shut by tungsten-inert-gas welding. The capsule was placed in a furnace of a high temperature centrifuge and the sample was equilibrated for 72 hours. The sample was centrifuged after this period and the sat. amalgam was collected in the smaller quartz chamber. Hg from the weighed amount of amalgam was distilled off and the residue dissolved in HNO_3-HF or aqua regia. Co or Y was added to the solution as internal standard and the Fe was determined spectrographically.	Mercury was triple-distilled, reagent grade. Iron source and purity not specified, but specimens were first irradiated in the Brookhaven Graphite Research Reactor. **ESTIMATED ERROR:** Soly: nothing specified. Temp: precision $\pm$ 2 K. **REFERENCES:**

COMPONENTS:	ORIGINAL MEASUREMENTS:
(1) Iron; Fe; [7439-89-6] (2) Mercury; Hg; [7439-97-6]	Bowersox, D.F.; Leary, J.A. *U.S. At. Ener. Comm. Rep.*, LAMS-2518, 1961.
VARIABLES: One temperature: 350°C	PREPARED BY: C. Guminski; Z. Galus

EXPERIMENTAL VALUES:

The solubility of iron in mercury at 350°C was reported to be 2×10^{-3} g Fe/dm^3 Hg. The corresponding atomic % solubility calculated by the compilers is 5×10^{-5} at %.

AUXILIARY INFORMATION

METHOD/APPARATUS/PROCEDURE:	SOURCE AND PURITY OF MATERIALS:
The solubility was determined by immersing a weighed coupon of Fe into a known amount of boiling Hg and periodically measuring the coupon weight. The solubility of Fe was determined from the weight loss of the coupon.	Mercury was triple-distilled material. Iron purity not specified.
	ESTIMATED ERROR: Soly: detection limit of method was 1×10^{-3} g; precision may be no better than $\pm$ 50%. Temp: nothing specified.
	REFERENCES:

COMPONENTS:	ORIGINAL MEASUREMENTS:
(1) Iron; Fe; [7439-89-6] (2) Mercury; Hg; [7439-97-6]	Parkman, M.F. *Extended Abst., Electrothermics and Metallurgy Div.*, Vol. 2, No. 2, The Electrochemical Soc., New York, NY <u>1964</u>, pp. 16-21.
VARIABLES:	PREPARED BY:
Temperature: 855-896 K	C. Guminski; Z. Galus

EXPERIMENTAL VALUES:

The mass % solubility data were presented graphically; the solubilities were read off the curve and converted to atomic % by the compilers.

T/K	Fe Source	Contact time, hrs.	Soly/mass %	Soly/at %
855	Armco	16	9×10^{-6}	3.6×10^{-5}
855	Armco	16	1.5×10^{-5}	6.0×10^{-5}
866	Westinghouse	64	2.8×10^{-5}	1.1×10^{-4}
896	Westinghouse	132	2.2×10^{-5}	8.8×10^{-5}

AUXILIARY INFORMATION

METHOD/APPARATUS/PROCEDURE:	SOURCE AND PURITY OF MATERIALS:
Specimen of Fe was placed in contact with Hg in a glass capsule and the Hg in the capsule was outgassed for at least 16 hours. The capsule was then sealed under vacuum. The capsule was placed in a copper block in a pressurized furnace and heated to the desired temperature and held there for 16 to 132 hours. A sample of the solution was then collected at temperature and cooled, and the Hg was separated from the sample by molecular distillation. The residue was dissolved into an acid solution and the Fe was determined by spectrophotometry.	Iron from Armco contained 0.01% C, 0.03% Mn, 0.02% Si, 0.007% P, 0.04% S, 0.0012% O. Iron from Westinghouse designated as "Puron". Mercury was probably triple-distilled.
	ESTIMATED ERROR:
	Soly: nothing specified. Temp: precision $\pm$ 3 K.
	REFERENCES:

COMPONENTS:	EVALUATOR:
(1) Ruthenium; Ru; [7440-18-8] (2) Mercury; Hg; [7439-97-6]	C. Guminski; Z. Galus Department of Chemistry University of Warsaw Warsaw, Poland July, 1985

CRITICAL EVALUATION:

The few reports on the experimental determinations of the solubility of ruthenium in mercury are at wide variance. Strachan and Harris (1) reported a solubility of 0.694 at % at room temperature, but this result is clearly much too high. Jangg and Dörtbudak (2) could not detect any dissolution at 773 K; the detection limit for ruthenium by the latter authors was 2×10^{-5} at %. Bowersox and Leary (3) equilibrated the two metals at 293 and at 523 K, and they could not detect any dissolution of ruthenium at these temperatures. These authors concluded that the solubility was below their detection limit of 3×10^{-5} at %. It also was reported (4) that there was no attack of ruthenium by mercury at 823 K.

Dean (5) reported that the solubility of ruthenium is of the order of 2×10^{-7} at %, but the temperature and other experimental details were not specified. Messing and Dean (6) reported that the solubility of ruthenium in a saturated uranium amalgam varied from 2.4×10^{-3} at % at 323 K to 2.2×10^{-2} at % at 629 K.

Kozin predicted a solubility of 1.2×10^{-11} at % at 298 K (7); he previously predicted 9.3×10^{-17} at % at 298 K (8). The first value appears to be more reliable to the evaluators.

It is clear that there is no dependable solubility data in this system; it only may be stated that the solubility of ruthenium at 298 K is less than 2×10^{-5} at %.

The saturated amalgam is in equilibrium with pure ruthenium (2).

References

1. Strachan, J.F.; Harris, N.L. *J. Inst. Metals* 1956-57, *85*, 17.
2. Jangg, G.; Dörtbudak, T. *Z. Metallk.* 1973, *64*, 715.
3. Bowersox, D.F.; Leary, J.A. *U.S. At. Ener. Comm. Rep.*, LAMS-2518, 1961.
4. Rhys, D.W.; Price, E.G. *Metal Ind.* 1964, *105*, 243.
5. Dean, O.C. *Unpublished data, Oak Ridge National Laboratory*, 1957; cited in ref. (6).
6. Messing, A.F.; Dean, O.C. *U.S. At. Ener. Comm. Rep.*, ORNL-2871, 1960.
7. Kozin, L.F. *Fiziko Khimicheskie Osnovy Amalgamnoi Metallurgii*, Nauka, Alma-Ata, 1964.
8. Kozin, L.F. *Tr. Inst. Khim. Nauk Akad. Nauk Kaz. SSR* 1962, *9*, 101.

COMPONENTS:	ORIGINAL MEASUREMENTS:
(1) Ruthenium; Ru; [7440-18-8] (2) Mercury; Hg; [7439-97-6]	Bowersox, D.F.; Leary, J.A. *U.S. At. Ener. Comm. Rep.*, LAMS-2518, 1961.
VARIABLES:	PREPARED BY:
Temperature: 20-250°C	C. Guminski; Z. Galus

EXPERIMENTAL VALUES:

The solubility of ruthenium in mercury at 20 and 250°C was reported to be less than the detection limit of 2×10^{-3} g of Ru in 1 dm^3 of Hg. The corresponding atomic % detection limit calculated by the compilers is 3×10^{-5} at %. Although Ru apparently dissolved in Hg at 350°C, it did not pass through a coarse Pyrex frit at either 30 or 350°C. Therefore, since the "solubility is considered to be the quantity that passes through such a frit, Ru would, by definition, be insoluble in Hg."

AUXILIARY INFORMATION

METHOD/APPARATUS/PROCEDURE:	SOURCE AND PURITY OF MATERIALS:
The solubility was determined by immersing a weighed coupon of Ru into a definite amount of Hg at specified temperatures. The coupon weight was measured periodically to determine the solubility from the weight loss.	Triple-distilled Hg was used. Ruthenium purity was not specified.
	ESTIMATED ERROR: Soly: detection limit as specified above. Temp: not specified.
	REFERENCES:

COMPONENTS:	EVALUATOR:
(1) Osmium; Os; [7440-04-2] (2) Mercury; Hg; [7439-97-6]	C. Guminski; Z. Galus Department of Chemistry University of Warsaw Warsaw, Poland July, 1985

CRITICAL EVALUATION:

Jangg and Dörtbudak (1), in an equilibration study at 773 K, could not detect any dissolution of osmium in mercury at their analytical detection limit of 10^{-5} at %. The low solubility of osmium is also suggested by the estimate reported by Kozin (2) of 1.1×10^{-14} at % at 298 K. The latter value appears to the evaluators to be more reliable than one predicted previously, i.e., 1.8×10^{-22} at % at 298 K (3).

The saturated osmium amalgam is in equilibrium with pure osmium; no Os-Hg compounds were found (1).

References

1. Jangg, G.; Dörtbudak, T. *Z. Metallk.* 1973, *64*, 715.
2. Kozin, L.F. *Fiziko-Khimicheskie Osnovy Amalgamnoi Metallurgii*, Nauka, Alma-Ata, 1964.
3. Kozin, L.F. *Tr. Inst. Khim. Nauk Akad. Nauk Kaz. SSR* 1962, *9*, 101.

COMPONENTS:	EVALUATOR:
(1) Cobalt; Co; [7440-48-4] (2) Mercury; Hg; [7439-97-6]	C. Guminski; Z. Galus Department of Chemistry University of Warsaw Warsaw, Poland July, 1985

CRITICAL EVALUATION:

The solubility of cobalt near room temperature is very low, and there are no reliable experimental determinations in this range; only the highest limit of the solubility based on the sensitivity of the analytical method has been reported. For example, Irvin and Russell (1) reported that the solubility is below 3×10^{-4} at % at 293 K, while Strachan and Harris (2) reported the solubility to be below 9×10^{-3} at % at room temperature, and deWet and Haul (3) reported that the solubility is below 3×10^{-6} at % at 303 K. Jangg and Palman (4) determined the solubility at 773 and 823 K and they extrapolated the two experimental points to 293 K to obtain a solubility of 2.4×10^{-7} at % at the latter temperature. However, this extrapolation is most likely erroneous because the temperature dependence of the solubility at high temperatures, reported by Weeks and coworkers (5-7), shows a steeper slope than the two measured points of Jangg and Palman. More recently, Speranskaya and Panina (8) reported that the solubility of cobalt at 363 K should be much lower than 10^{-5} at %, while Babkin and Omarova (9) reported that cobalt is insoluble in mercury at room temperature.

Kozin's (10) predicted solubility of 1.8×10^{-4} at % at 298 K is too high. Tammann and coworkers (11, 12) reported solubilities of 0.56 and 0.21 at % at 290 K and room temperature, respectively, while Nagaoka (13) reported a solubility of 1.7 at % at room temperature; the values reported by these authors are clearly too high and are rejected. The high values by the latter authors could be the result of the tendency for cobalt to form supersaturated amalgams.

At high temperatures, Weeks and coworkers (5-7, 14,15) determined solubilities of 6.5×10^{-5} to 1.1×10^{-3} at % in the temperature range of 798 to 1023 K; Weeks and Fink (7) reported a value of 3.7×10^{-4} at % for the solubility of Co at 923 K.

No intermetallic compounds have been found in this system; the amalgam is in equilibrium with solid cobalt (3, 16).

Parkman (17-19) determined the cobalt content in mercury after Co alloys were equilibrated with mercury at different temperatures; it was reported that the solubility of cobalt from the alloys was of similar magnitude as with pure cobalt.

Tentative values of Co solubility in Hg:

T/K	Soly/at %	Reference
773	7×10^{-5} [a]	[4,5]
873	2×10^{-4}	[5]
973	7×10^{-4}	[5]

[a] mean value from cited references.

References

1. Irvin, N.M.; Russell, A.S. *J. Chem. Soc.* 1932, 891.
2. Strachan, J.F.; Harris, N.L. *J. Inst. Metals* 1956-57, 85, 17.
3. deWet, J.F.; Haul, R.A.W. *Z. Anorg. Chem.* 1954, 277, 96.
4. Jangg, G.; Palman, H. *Z. Metallk.* 1963, 54, 364.
5. Weeks, J.R. *Corrosion*, 1967, 23, 98.
6. Weeks, J.R.; Minardi, A.; Fink, S. *U.S. At. Ener. Comm. Rep.*, BNL-841, 1963, p. 76.
7. Weeks, J.R.; Fink, S. *U.S. At. Ener. Comm. Rep.*, BNL-900, 1964, p. 136.
8. Speranskaya, E.F.; Panina, L.S. *Elektrokhimia* 1969, 5, 557.
9. Babkin, G.N.; Omarova, A.F. *Izv. Vyssh. Ucheb. Zaved., Khim. Khim. Tekhnol.* 1973, 16, 158.
10. Kozin, L.F. *Fiziko Khimicheskie Osnovy Amalgamnoi Metallurgii*, Nauka, Alma-Ata, 1964.
11. Tammann, G.; Kollmann, K. *Z. Anorg. Chem.* 1927, 160, 242.
12. Tammann, G.; Oelsen, W. *Z. Anorg. Chem.* 1930, 186, 280.
13. Nagaoka, H. *Ann. Phys. Chem.* 1896, 59, 66.

(Continued next page)

COMPONENTS:	EVALUATOR:
(1) Cobalt; Co; [7440-48-4] (2) Mercury; Hg; [7439-97-6]	C. Guminski; Z. Galus Department of Chemistry University of Warsaw Warsaw, Poland July, 1985

CRITICAL EVALUATION: (continued)

14. Weeks, J.R.; Fink, S. *U.S. At. Ener. Comm. Rep.*, *BNL-799*, 1963, p. 85.
15. Weeks, J.R.; Fink, S. *U.S. At. Ener. Comm. Rep.*, *BNL-823*, 1963, p. 80.
16. Lihl, F. *Z. Metallk.* 1953, *44*, 160.
17. Parkman, M.F. *Ext. Abst.*, *Electrothermics and Metallurgy Div.*, Vol. 2, No. 2, The Electrochemical Soc., 1964, pp. 16-21.
18. Parkman, M.F. *U.S. At. Ener. Comm. Rep.*, *TID-7626*, 1962, Pt. I, p. 35.
19. Parkman, M.F.; Whaley, D.K. *Aerojet-General Nucleonics*, Rep. AN-957, 1963; as cited in 6.

COMPONENTS:	ORIGINAL MEASUREMENTS:
(1) Cobalt; Co; [7440-48-4] (2) Mercury; Hg; [7439-97-6]	Jangg, G.; Palman, H. *Z. Metallk.* 1963, *54*, 364-9.
VARIABLES: Temperature: 20-550°C	PREPARED BY: C. Guminski; Z. Galus

EXPERIMENTAL VALUES:

The solubility of cobalt in mercury:

t/°C	Soly/mass %	Soly/at %
550	2.4×10^{-5}	8.2×10^{-5}
500	2.0×10^{-5}	6.8×10^{-5}
400[a]	1.2×10^{-5}	4.1×10^{-5}
300[a]	6×10^{-6}	2.0×10^{-5}
200[a]	2.2×10^{-6}	7.5×10^{-6}
100[a]	5×10^{-7}	1.7×10^{-6}
20[a]	7×10^{-8}	2.4×10^{-7}

[a] Only the 500 and 550°C solubilities were experimentally determined. The values at the lower temperatures were estimated by extrapolation from the two experimental points.

AUXILIARY INFORMATION

METHOD/APPARATUS/PROCEDURE:	SOURCE AND PURITY OF MATERIALS:
The heterogeneous amalgams were introduced into a specially constructed apparatus of refractory chromium steel. This type of apparatus could be used because the solubility of iron in mercury is very low and the chromous oxide film on the surface inhibits the wetting of the steel by mercury. After 12 hr of equilibration at the temperature of the experiment the amalgam was filtered through the sintered iron frit under a pressure of purified nitrogen; 3- to 4-fold filtration was usually necessary. The metal content in the filtered amalgam was determined analytically by an unspecified procedure.	Nothing specified.
	ESTIMATED ERROR: Soly: accuracy ± 5%. Temp: precision ± 2 K.
	REFERENCES:

COMPONENTS:	ORIGINAL MEASUREMENTS:
(1) Cobalt; Co; [7440-48-4] (2) Mercury; Hg; [7439-97-6]	Weeks, J.R. *Corrosion* 1967, *23*, 98-106.
VARIABLES: Temperature: 525-750°C	PREPARED BY: C. Guminski; Z. Galus

EXPERIMENTAL VALUES:

The solubility of cobalt in mercury was presented graphically as a function of temperature. The data points were read off the curve by the compilers.

t/°C	Soly/mass % $\times 10^5$	Soly/at % $\times 10^{4}$ [a]
750	32	11
725	18	6.1
700	21	7.1
675	7.0	2.4
650	12	4.1
625	4.1	1.4
600	4.0	1.4
575	8.0	2.7
550	5.4	1.8
525	1.9	0.65

[a] by compilers.

The data for this paper were also reported in refs. (1) and (2).

AUXILIARY INFORMATION

METHOD/APPARATUS/PROCEDURE:	SOURCE AND PURITY OF MATERIALS:
Cobalt was immersed in Hg which was contained in the upper part of a two-chambered quartz tube; a coarse quartz filter separated the two chambers. After loading, the tube was sealed under vacuum, then placed in a steel bomb which contained Hg to equalize the pressures inside and outside the quartz tube at high temperatures. The bomb was then placed in a centrifuge which was contained in a furnace, and the sample was equilibrated for 72 hr at the desired temperature. After this time the sample was centrifuged at temperature to force the amalgam through the filter. After cooling, the tube was opened and the Hg content in the amalgam was determined by the evaporation method. Cobalt was dissolved in $HF-HNO_3$ and determined spectrographically.	Mercury was triple-distilled, reagent grade. Cobalt purity and source not specified.
	ESTIMATED ERROR: Soly: nothing specified. Temp: precision $\pm$ 2 K.
	REFERENCES: 1. Weeks, J.R.; Minardi, A.; Fink, S. *U.S. At. Ener. Comm. Rep.*, BNL-841, 1963, p. 76. 2. Weeks, J.R.; Fink, S. *U.S. At. Ener. Comm. Rep.*, BNL-900, 1964, p. 136.

COMPONENTS:	EVALUATOR:
(1) Rhodium; Rh; [7440-16-6] (2) Mercury; Hg; [7439-97-6]	C. Guminski; Z. Galus Department of Chemistry University of Warsaw Warsaw, Poland July, 1985

CRITICAL EVALUATION:

The solubility of rhodium in mercury is very low at 298 K; Kozin predicted this solubility as 1.1×10^{-11} (1) and 1.0×10^{-8} at % (2). Jangg and Dörtbudak (3) reported an experimental solubility of 1.2×10^{-4} at % at 773 K. Kozin's second estimate at 298 K appears to be of the correct order of magnitude by comparison with the high temperature determination of Jangg and Dörtbudak. Strachan and Harris (4) reported a solubility of 0.31 at % at room temperature, but this value is much too high to be acceptable.

The saturated rhodium amalgams are in equilibrium with the compounds, $RhHg_2$, $RhHg_{4.63}$ and $RhHg_5$; the respective decomposition temperatures of the compounds are 593, 689 and 833 K (5).

References

1. Kozin, L.F. *Tr. Inst. Khim. Nauk Akad. Nauk Kaz. SSR* 1962, *9*, 101.
2. Kozin, L.F. *Fiziko-Khimicheskie Osnovy Amalgamnoi Metallurgii*, Nauka, Alma-Ata, 1964.
3. Jangg, G.; Dörtbudak, T. *Z. Metallk.* 1973, *64*, 715.
4. Strachan, J.F.; Harris, N.L. *J. Inst. Metals* 1956-57, *85*, 17.
5. Jangg, G.; Kirchmayr, H.R.; Mathis, H.B. *Z. Metallk.* 1967, *58*, 724.

COMPONENTS:	ORIGINAL MEASUREMENTS:
(1) Rhodium; Rh; [7440-16-6] (2) Mercury; Hg; [7439-97-6]	Jangg, G.; Dörtbudak, T. *Z. Metallk.* 1973, *64*, 715-9.
VARIABLES: One temperature: 773 K	PREPARED BY: C. Guminski; Z. Galus

EXPERIMENTAL VALUES:

The solubility of rhodium in mercury at 773 K was found to be 6×10^{-5} mass %. The corresponding atomic % solubility calculated by the compilers is 1.2×10^{-4} at %.

AUXILIARY INFORMATION

METHOD/APPARATUS/PROCEDURE:	SOURCE AND PURITY OF MATERIALS:
The amalgam was equilibrated in a quartz tube which was contained in a pressurized bomb. One end of the tube consisted of a fused quartz filter through which the amalgam was filtered at the equilibration temperature. Subsequently, tin was added to the amalgam and the mercury was removed by evaporation. The rhodium, which was alloyed into the tin, was then analyzed spectroscopically; the tin served as an internal standard.	Rhodium: powder material supplied by Degussa. Hg purity not specified.
	ESTIMATED ERROR: Soly: nothing specified. Temp: nothing specified.
	REFERENCES:

COMPONENTS:	EVALUATOR:
(1) Iridium; Ir; [7439-88-5] (2) Mercury; Hg; [7439-97-6]	C. Guminski; Z. Galus Department of Chemistry University of Warsaw Warsaw, Poland July, 1985

CRITICAL EVALUATION:

Strachan and Harris (1) equilibrated iridium and mercury, but they could not detect any solubility of iridium in mercury at room temperature; their analytical detection limit for iridium was 10^{-3} at %. Jangg and Dörtbudak (2) equilibrated the two metals at 773 K and could not detect any dissolution of iridium at this higher temperature; the detection limit by the latter authors was 10^{-5} at %. Exposure of iridium to mercury at 823 K showed no corrosion of the iridium (3). The extremely low solubility of iridium in mercury is suggested by the estimates of 6.6×10^{-18} (5) and 2.9×10^{-12} at % (4) at 298 K by Kozin; the second value appears to be more reliable to the evaluators.

Pure iridium should be in equilibrium with its saturated amalgam (2).

References

1. Strachan, J.F.; Harris, N.L. *J. Inst. Metals* 1956-57, *85*, 17.
2. Jangg, G.; Dörtbudak, T. *Z. Metallk.* 1973, *64*, 715.
3. Rhys, D.W.; Price, E.G. *Metal. Ind.* 1964, *105*, 243.
4. Kozin, L.F. *Fiziko-Khimicheskie Osnovy Amalgamnoi Metallurgii*, Nauka, Alma-Ata, 1964.
5. Kozin, L.F. *Tr. Inst. Khim. Nauk Akad. Nauk Kaz. SSR* 1962, *9*, 101.

COMPONENTS:	EVALUATOR:
(1) Nickel; Ni; [7440-02-0] (2) Mercury; Hg; [7439-97-6]	C. Guminski; Z. Galus Department of Chemistry University of Warsaw Warsaw, Poland July, 1985

CRITICAL EVALUATION:

The solubility of nickel in mercury is very low in the region of room temperature. Although there have been many determinations, the following reported solubilities, expressed in atomic %, have varied over five orders of magnitude: 2×10^{-3} at 290 K (1), 0.5 at 291 K (2), 4.8×10^{-4} at 293 K (3), less than 7×10^{-5} at 293 K (4), less than 7×10^{-6} at 303 K (5), 7×10^{-3} at room temperature (6), 1.7×10^{-5} at room temperature (7), 4.8×10^{-4} at 298 K (8), 4.8×10^{-5} at 293 K (9), 1.6×10^{-4} at 298 K (10,11), 1.0×10^{-4} at 290 K (12,13), 1.4×10^{-5} at room temperature (14), 7×10^{-6} probably at 303 K (15), and 6.7×10^{-5} at 293 K (16). The above determinations were made by various methods: EMF (1), magnetic susceptibility (2), chemical analysis (3-6, 8,9), coulometry (7), voltammetry (10,11), chronoamperometry (12,13,16), chronopotentiometry (14), and pulse polarography (15). The wide variation in the reported solubilities shows that the system is very susceptible to oversaturation. All results, except that of (5) and (15), are too high and are rejected. Kozin's prediction of 1.0×10^{-5} at 298 K is also too high.

There was better agreement of the solubility data at higher temperatures. Epstein (8) reported a nickel solubility of 4.8×10^{-3} at % at 573 K, but no details of the analytical procedure were given. Toner (18) reported that the solubility of nickel increased from 4.3×10^{-5} to 1.09×10^{-2} at % in the temperature range of 401 to 605 K; a break on the solubility vs. temperature curve was observed at 520 K. Jangg and Palman (9) reported that the solubility increased from 4.8×10^{-5} to 2.9×10^{-2} at % in the range of 293-826 K. These authors observed a break in the solubility curve at 498 K. Weeks (19-21) determined the solubility at 773-1023 K and agreed with the high temperature values of Jangg and Palman; at 1023 K the solubility was found to be 0.24 at %. A single determination of the Ni solubility by Parkman and Whaley (22,23), 3.8×10^{-2} at % at 866 K, agrees well with the results of (9, 19-21). The results in (9, 19-23) were presented graphically; the only numerical values presented were 6.1×10^{-2} at % at 923 K (20) and 3.4×10^{-2} at % at 873 K (19).

Barański and Galus (24) explained part of the discrepancies in the nickel solubilities reported by various authors at temperatures below 500 K. The disparities are attributed to the formation of $NiHg_2$, $NiHg_3$ and $NiHg_4$, and to the differences in solubilities of these compounds and nickel. The authors state that true equilibrium is attained only for $NiHg_4$; this compound is formed in the last step in the reaction between electrolytically introduced nickel and mercury. The solubilities were determined from potentiometric measurements, and unit activity coefficient of Ni was assumed. Because the activity coefficient in the homogeneous amalgam is most probably less than unity, the nickel solubility would be higher than it would be with the above assumption. The solubilities of the compounds increase in the order $NiHg_4 < NiHg_3 < NiHg_2 < Ni$, but it should be indicated that the equilibrium with the last two species is unstable. Below 500 K there is good agreement of the $NiHg_4$ solubility values of (24) and (18).

The existence of $NiHg_4$ up to 493 K and $NiHg_3$ up to 483 K have been confirmed, and $NiHg_2$ is stable to approximately 458 K (4, 24-26). Above 493 K the saturated amalgams are in equilibrium with pure nickel.

(Continued next page)

COMPONENTS:	EVALUATOR:
(1) Nickel; Ni; [7440-02-0] (2) Mercury; Hg; [7439-97-6]	C. Guminski; Z. Galus Department of Chemistry University of Warsaw Warsaw, Poland July, 1985

CRITICAL EVALUATION: (continued)

Recommended (r) and tentative values of nickel solubility in mercury:

T/K	Soly/at %	Reference
293	1×10^{-7}	[24]
298	2×20^{-7}	[24]
323	2×10^{-6}	[24]
373	4×10^{-5}	[24]
473	2×10^{-3a}	[18,9]
573	7×10^{-3a}	[18,9]
673	1.5×10^{-2}	[9]
773	2.5×10^{-2}	[9]
873	3.5×10^{-2} (r)^a	[19,22]
973	5×10^{-2b}	[21], [18,9,19,22]

[a] Mean value from cited references.

[b] Extrapolated value from data of cited references.

References

1. Tammann, G.; Kollmann, K. *Z. Anorg. Chem.* 1927, *160*, 242.
2. Tammann, G.; Oelsen, W. *Z. Anorg. Chem.* 1930, *186*, 257.
3. Palmaer, E. *Z. Elektrochem.* 1932, *38*, 70.
4. Irvin, N.M.; Russell, A.S. *J. Chem. Soc.* 1932, 891.
5. deWet, J.F.; Haul, R.A.W. *Z. Anorg. Chem.* 1954, *277*, 96.
6. Strachan, J.F.; Harris, N.L. *J. Inst. Metals* 1956-57, *85*, 17.
7. Liebl, G.; quoted by H. Spengler *Metall.* 1958, *12*, 105.
8. Epstein, L.F. *Proc. Inter. Conf. on the Peaceful Uses of At. Ener.*, Geneva, 1956, *9*, 311.
9. Jangg, G.; Palman, H. *Z. Metallk.* 1963, *54*, 364.
10. Krasnova, I.E.; Zebreva, A.I.; Kozlovskii, M.T. *Dokl. Akad. Nauk SSSR* 1964, *156*, 415.
11. Krasnova, I.E.; Zebreva, A.I. *Elektrokhimia* 1966, *2*, 247.
12. Podkorytova, N.V.; Nazarov, B.F. *Zh. Anal. Khim.* 1973, *28*, 1535.
13. Nazarov, B.F.; Podkorytova, N.V. *Uspekhi Polarografii s Nakopleniem*, Tomsk, 1973, p. 103.
14. Babkin, G.N.; Omarova, A.F. *Izv. Vyssh. Ucheb. Zaved., Khim. Khim. Tekhnol.* 1973, *16*, 158.
15. Zutić, V.; Batel, R.; Chevalet, J. *J. Electroanal. Chem.* 1979, *105*, 115.
16. Nazarov, B.F.; Podkorytova, N.V. *Elektrokhimia* 1980, *16*, 1847.
17. Kozin, L.F. *Fiziko Khimicheskie Osnovy Amalgamnoi Metallurgii*, Nauka, Alma-Ata, 1964.
18. Toner, D.F. *U.S. At. Ener. Comm. Rep.*, ORNL-2839, 1959, p. 187.
19. Weeks, J.R. *U.S. At. Ener. Comm. Rep.*, NASA-SP-41, 1963, p. 21; *U.S. At. Ener. Comm. Rep.*, BNL-7553, 1963.
20. Weeks, J.R.; Fink, S. *U.S. At. Ener. Comm. Rep.*, BNL-900, 1964, p. 136.
21. Weeks, J.R. *Corrosion* 23
22. Parkman, M.F. *Extended Abst., Electrothermics and Metallurgy Div.*, Vol. 2, No. 2, The Electrochemical Soc., 1964, pp. 16-21.
23. Parkman, M.F.; Whaley, D.K. *Aerojet-General Nucleonics*, Rep. AN-957, 1963; as cited in (21).
24. Barański, A.; Galus, Z. *J. Electroanal. Chem.* 1973, *46*, 289.
25. Lihl, F. *Z. Metallk.* 1953, *44*, 160.
26. Jangg, G.; Steppan, F. *Z. Metallk.* 1965, *56*, 172.

COMPONENTS:	ORIGINAL MEASUREMENTS:
(1) Nickel; Ni; [7440-02-0] (2) Mercury; Hg; [7439-97-6]	deWet, J.F.; Haul, R.A.W. *Z. Anorg. Chem.* 1954, *277*, 96-112.
VARIABLES: One temperature: 303 K	PREPARED BY: C. Guminski; Z. Galus

EXPERIMENTAL VALUES:

Solubility of nickel in mercury at 303 K was found to be less than 2×10^{-6} mass %. The corresponding atomic % solubility limit calculated by the compilers is 7×10^{-6} at %. Solid $NiHg_4$ was reported to be in equilibrium with the saturated amalgam.

AUXILIARY INFORMATION

METHOD/APPARATUS/PROCEDURE:	SOURCE AND PURITY OF MATERIALS:
Heterogenous nickel amalgam was obtained by electrolysis of 1 mol dm^{-3} $NiSO_4$ solution at the mercury-pool cathode. The amalgam was washed with water, dried with acetone and sealed off under vacuum in a glass tube. A portion of the amalgam was introduced into a centrifuge vessel and after sufficiently long centrifuging the homogenous part of the amalgam was taken for analysis. Mercury was carefully distilled off and the residue was dissolved in 2.8 mol dm^{-3} HCl. The resulting solution was analyzed spectrochemically for nickel content.	$NiSO_4 \cdot 6-7H_2O$ was Analar grade from Hopkins and Williams. Purified mercury was distilled under vacuum and was found to be spectrochemically free from traces of nickel. Water was triply distilled.
	ESTIMATED ERROR: Soly: analytical detection limit was 2×20^{-6} mass % Ni in the amalgam. Temp: nothing specified.
	REFERENCES:

COMPONENTS:	ORIGINAL MEASUREMENTS:
(1) Nickel; Ni; [7440-02-0] (2) Mercury; Hg; [7439-97-6]	Toner, D.F. *U.S. At. Ener. Comm. Rep.*, *ORNL-2839*, 1959, pp. 187-191.
VARIABLES: Temperature: 128-332°C	PREPARED BY: C. Guminski; Z. Galus

EXPERIMENTAL VALUES:

The mass % solubility of nickel in mercury was reported graphically as a function of temperature; the solubility values were read from the plotted data and the corresponding atomic % conversion was calculated by the compilers.

$t/°C$	Soly/mass % $\times 10^4$	Soly/at % $\times 10^3$
128	0.125	0.043
162	0.90	0.31
173	2.2	0.75
181	1.3	0.44
205	4.2	1.43
227	9.0	3.1
243	12	4.1
245	14	4.8
258	19	6.5
278	36	12.2
308	25	8.5
327	26	8.9
332	32	10.9

AUXILIARY INFORMATION

METHOD/APPARATUS/PROCEDURE:	SOURCE AND PURITY OF MATERIALS:
A Ni specimen was inserted into the isothermal hot zone of a Hg thermal convection loop made of quartz. The sample was mechanically polished with a grit paper or electropolished. The system was operated under a hydrogen atmosphere. A series of thermocouples indicated the temperature profile in the loop. A Hg sample was extracted through a fritted disk and then chemically analyzed. The measurements were performed at various times of Hg circulation in the loop. Constant values of the Ni solubility in Hg were obtained after over 10 hours of equilibration.	Nothing specified.
	ESTIMATED ERROR: Soly: nothing specified; precision probably better than ± 20%. Temp: nothing specified.
	REFERENCES:

COMPONENTS:	ORIGINAL MEASUREMENTS:
(1) Nickel; Ni; [7440-02-0] (2) Mercury; Hg; [7439-97-6]	Jangg, G.; Palman, H. *Z. Metallk.* 1963, *54*, 364-69.
VARIABLES:	PREPARED BY:
Temperature: 20-553°C	C. Guminski; Z. Galus

EXPERIMENTAL VALUES:

The mass % solubility of nickel in mercury was presented graphically as a function of temperature. The data points were read from the curve and converted to atomic % by the compilers.

	Soly				Soly	
$t/°C$	mass % × 10^4	at % × 10^4	$t/°C$	mass % × 10^3	at % × 10^3
20	0.14	0.48	236	1.1	3.8
50	0.36	1.2	243	1.2	4.1
100	1.1	3.7	252	1.3	4.5
150	2.6	8.8	302	2.1	7.2
200	5.0	17	353	3.3	11
225	6.2	21	402	4.3	15
230	8.5	29	454	5.7	20
232	9.4	32	503	7.1	24
234	10	34	553	8.5	29

It was reported that $NiHg_4$ is in equilibrium with the liquid below 225°C.

The results below 225°C appear to be overstated.

AUXILIARY INFORMATION

METHOD/APPARATUS/PROCEDURE:	SOURCE AND PURITY OF MATERIALS:
The heterogeneous amalgam was introduced into specially constructed apparatus made of refractory chromium steel. Such steel apparatus could be used because the solubility of iron in mercury is very low and the chromium (III) oxide film inhibits the wetting of the steel by mercury. After twelve hours of equilibration at the temperature of the experiment, the amalgam was filtered through the sintered iron frit under a pressure of purified nitrogen. Usually 3- to 4-fold filtration was necessary. The nickel content was then analytically determined in the filtered amalgam. For experiments carried out below 320°C, amalgam was equilibrated in a glass vessel. The analytical procedures are not described in the paper.	Nothing specified.
	ESTIMATED ERROR: Soly: precision ± 5%. Temp: precision ± 2 K.
	REFERENCES:

COMPONENTS:	ORIGINAL MEASUREMENTS:
(1) Nickel; Ni; [7440-02-0] (2) Mercury; Hg; [7439-97-6]	Parkman, M.F. *Extended Abst., Electrothermics and Metallurgy Div.*, Vol. 2, No. 2, The Electrochemical Soc., 1964, pp. 16-21.
VARIABLES: One temperature: 866 K	PREPARED BY: C. Guminski; Z. Galus

EXPERIMENTAL VALUES:

The mass % solubility of nickel in mercury was presented graphically as a function of temperature; the compilers read off a value of 1.1×10^{-2} mass % at 866 K from the curve. The corresponding atomic % solubility calculated by the compilers is 3.8×10^{-2} at %.

AUXILIARY INFORMATION

METHOD/APPARATUS/PROCEDURE:	SOURCE AND PURITY OF MATERIALS:
Specimen of Ni was placed in contact with Hg in a glass capsule. The capsule was sealed under vacuum after at least 16 h outgassing of Hg. The capsule was heated to the desired temperature and held for 16 hr. A sample of the solution was then collected and cooled. Hg was separated from the sample by molecular distillation, and the residue was taken into acid solution, dried and analyzed by emission spectroscopy.	Nothing specified.
	ESTIMATED ERROR: Soly: nothing specified. Temp: precision ± 3 K.
	REFERENCES:

COMPONENTS:	ORIGINAL MEASUREMENTS:
(1) Nickel; Ni; [7440-02-0] (2) Mercury; Hg; [7439-97-6]	1. Weeks, J.R. *Corrosion* 1967, *23*, 98-106. 2. Weeks, J.R.; Fink, S. *U.S. At. Ener. Comm. Rep.*, BNL-900, 1964, p. 136.
VARIABLES: Temperature: 500-755°C	PREPARED BY: C. Guminski; Z. Galus

EXPERIMENTAL VALUES:

The solubility of nickel in mercury as a function of temperature was presented graphically. The data points were read off the plot and recalculated to at % by compilers.

$t/°C$	Soly/at % $\times 10^2$
665	43
750	24
500	13
725	12
700	9.7
755	7.3
655	6.5
700	5.5
605	4.5
550	3.8
500	2.9
625	2.4

AUXILIARY INFORMATION

METHOD/APPARATUS/PROCEDURE:	SOURCE AND PURITY OF MATERIALS:
Hg and Ni were equilibrated in a quartz capsule consisting of two chambers separated by a sintered quartz filter. The capsule was sealed with the metals in the larger chamber, then the capsule was placed inside of a stainless steel bomb which contained some Hg to equalize the pressure inside of the quartz capsule. The bomb was sealed and placed inside of an electric oven which was mounted on a centrifuge. The sample was equilibrated for 72 hr under stationary condition at the desired temperature; subsequently, the equilibrated amalgam was centrifuged at temperature to filter and separate the liquid phase. After cooling the capsule a known quantity of filtrate was analyzed by distilling off the mercury, the residue dissolved in HF-HNO_3 or aqua regia and the Ni determined spectrographically.	Triple-distilled, reagent grade mercury was used. ESTIMATED ERROR: Soly: nothing specified. Temp: precision $\pm$ 2 K. REFERENCES:

COMPONENTS:	ORIGINAL MEASUREMENTS:
(1) Nickel; Ni; [7440-02-0] (2) Mercury; Hg; [7439-97-6]	Barański, A.; Galus, Z. J. Electroanal. Chem. 1973, 46, 289-305.
VARIABLES: Temperature: 20-500°C	PREPARED BY: C. Guminski; Z. Galus

EXPERIMENTAL VALUES:

Solubilities of Ni and the compounds $NiHg_4$, $NiHg_3$ and $NiHg_2$ in mercury at various temperatures.

Solute	Soly/at %							
	20°C	50°C	100°C	150°C	200°C	300°C	400°C	500°C
$NiHg_4$	1.5×10^{-7}	1.8×10^{-6}	4.1×10^{-5}	4.7×10^{-4}	3.0×10^{-3}	-	-	-
$NiHg_3$	3.4×10^{-6}	1.9×10^{-5}	1.6×10^{-4}	9.3×10^{-4}	3.6×10^{-3}	-	-	-
$NiHg_2$	2.3×10^{-5}	8.5×10^{-5}	$\underline{4.7 \times 10^{-4}}$	$\underline{1.8 \times 10^{-3}}$	-	-	-	-
Ni	1.1×10^{-3}	1.6×10^{-3}	2.4×10^{-3}	3.3×10^{-3}	4.2×10^{-3}	$\underline{6.1 \times 10^{-3}}$	$\underline{8.1 \times 10^{-3}}$	9.8×10^{-3}

It was assumed that the nickel activity in the homogeneous amalgam was unity; this assumption may not have been valid. The underlined results were obtained by long extrapolation, so that the solubility of pure nickel at higher temperatures was understated.

AUXILIARY INFORMATION

METHOD/APPARATUS/PROCEDURE:	SOURCE AND PURITY OF MATERIALS:
The intermetallic compounds of Ni and Hg were prepared by electrolysis on various cathodes and the amalgam was used as an electrode for the cell: $Ni(Hg)_x$ | 6 mol dm^{-3} $CaCl_2$, X mol dm^{-3} $NiCl_2$ || 6 mol dm^{-3} $CaCl_2$ | Hg_2Cl_2, Hg at different temperatures. The solubilities were calculated from the measured EMF.	All chemicals of reagent grade from Ciech were additionally purified by crystallization. Mercury was chemically purified by shaking with acidic solution of $Hg_2(NO_3)_2$, washed and then distilled under reduced pressure.
	ESTIMATED ERROR: Soly: large because equilibrium was not attained in case of Ni, $NiHg_2$ and $NiHg_3$; potential reproducibility was 2-3 mV. Temp: nothing specified.
	REFERENCES:

COMPONENTS:	ORIGINAL MEASUREMENTS:
(1) Nickel; Ni; [7440-02-0] (2) Mercury; Hg; [7439-97-6]	Zutić, V.; Batel, R.; Chevalet, J. *J. Electroanal. Chem.* 1979, *105*, 115-25.
VARIABLES: One temperature: 30°C	PREPARED BY: C. Guminski; Z. Galus

EXPERIMENTAL VALUES:

Solubility of Ni in Hg, probably at 30°C, was reported to be 5×10^{-6} mol dm^{-3}. The corresponding atomic % solubility calculated by the compilers is 7×10^{-6} at %.

NiHg$_3$ is assumed to crystallize in the amalgam so that the solubility is referred to this compound.

Kinetics of the crystallization was also investigated.

AUXILIARY INFORMATION

METHOD/APPARATUS/PROCEDURE:	SOURCE AND PURITY OF MATERIALS:
The reduction of Ni(II) (5×10^{-4} – 1×10^{-2} mol dm^{-3}) in 10 mol dm^{-3} LiCl and subsequent oxidation of Ni amalgam were performed at the dropping mercury electrode. The generation potential was scanned along the whole NI(II) reduction wave and oxidation of Ni from the amalgam formed was carried out at plateau of the polarographic anodic wave (-0.15 V vs. SCE). The Ni solubility was estimated from a relation of critical concentration of Ni on generation time.	NiCl$_2$ was "Prolabo" from Rhône-Poulenc (Co content below 5×10^{-3}%). LiCl was analytical grade and was heated several hours at 500°C. Mercury was double distilled. Water was triple distilled.
	ESTIMATED ERROR: Soly: nothing specified. Temp: precision ± 2 K.
	REFERENCES:

COMPONENTS:	EVALUATOR:
(1) Palladium; Pd; [7440-05-3] (2) Mercury; Hg; [7439-97-6]	C. Guminski; Z. Galus Department of Chemistry University of Warsaw Warsaw, Poland July, 1985

CRITICAL EVALUATION:

There have been only three reports of the experimental determination of palladium in mercury. Jangg and Gröll (1) determined the solubility over a temperature range of 298 to 573 K, and reported a solubility of 5.1×10^{-3} at % at 298 K; a smooth curve was plotted through the data in this work. The room temperature value of Butler and Makrides (2), 5.5×10^{-3} at %, is in good agreement with that of Jangg and Gröll. The room temperature solubility of 1.2×10^{-2} at % reported by Strachan and Harris (3) is twofold higher than those of (1) and (2). The data of Jangg and Gröll appear to be the most accurate, and these authors reported their experimental procedure in some detail.

Kozin (4,5) predicted the solubility of Pd in Hg at 298 K, but his values are more than a hundredfold too low.

Messing and Dean (6) reported that the solubility of palladium in saturated uranium amalgam is nearly a hundredfold lower than in pure mercury.

Palladium forms the intermediate compounds, $PdHg_4$ (stable up to 363 K), Pg_2Hg_5 (stable up to 511 K) and PdHg (1,7); Pd_2Hg_3 also has been reported (8), but this compound was shown not to exist in this system (1,7).

Tentative solubilities of Pd in Hg:

T/K	Soly/at % $\times 10^3$	Reference
298	5.1	[1,2]
323	5.4	[1]
373	11	[1]
473	71	[1]
573	370	[1]

References

1. Jangg, G.; Gröll, W. *Z. Metallk.* 1965, *56*, 232.
2. Butler, J.N.; Makrides, A.C. *Trans. Faraday Soc.* 1964, *60*, 938.
3. Strachan, J.F.; Harris, N.L. *J. Inst. Metals* 1956-57, *85*, 17.
4. Kozin, L.F. *Tr. Inst. Khim. Nauk Akad. Nauk Kaz. SSR* 1962, *9*, 101.
5. Kozin, L.F. *Fiziko Khimicheskie Osnovy Amalgamnoi Metallurgii*, Nauka, Alma-Ata, 1964.
6. Messing, A.F.; Dean, O.C. *U.S. At. Energy Comm. Rep.*, ORNL-2871, 1960.
7. Galus, Z. *Crit. Rev. Anal. Chem.* 1975, *5*, 359.
8. Bittner, H.; Novotny, H. *Monatsh. Chem.* 1953, *84*, 211.

COMPONENTS:	ORIGINAL MEASUREMENTS:
(1) Palladium; Pd; [7440-05-3] (2) Mercury; Hg; [7439-97-6]	Strachan, J.F.; Harris, N.L. *J. Inst. Metals* <u>1956-57</u>, *85*, 17-24.
VARIABLES:	PREPARED BY:
Room temperature	C. Guminski; Z. Galus

EXPERIMENTAL VALUES:

The solubility of palladium in mercury at room temperature was reported to be 0.012 at %.

AUXILIARY INFORMATION

METHOD/APPARATUS/PROCEDURE:	SOURCE AND PURITY OF MATERIALS:
Mercury and palladium were equilibrated in evacuated glass tubes and maintained either at room temperature or at 773 K for times lasting for many hours. The solubility was determined from the change in weight of the specimens after equilibration, and by chemical analysis of the amalgam after filtration through a sintered glass filter. The analytical method was not specified.	99.997% pure mercury was submitted to cleaning, filtration, drying and distillation before use. Palladium was 99 to 99.99% pure.
	ESTIMATED ERROR: Soly: precision ± 20%. Temp: not specified.
	REFERENCES:

COMPONENTS:	ORIGINAL MEASUREMENTS:
(1) Palladium; Pd; [7440-05-3] (2) Mercury; Hg [7439-97-6]	Butler, J.N.; Makrides, A.C. *Trans. Faraday Soc.* 1964, *60*, 938-946.
VARIABLES: Room temperature measurement	PREPARED BY: C. Guminski; Z. Galus

EXPERIMENTAL VALUES:

The solubility of palladium in mercury at room temperature was reported to be 5.5×10^{-3} at %.

AUXILIARY INFORMATION

METHOD/APPARATUS/PROCEDURE:	SOURCE AND PURITY OF MATERIALS:
Mercury was saturated with palladium in a sealed glass tube for few days at 523 K. The amalgam was then cooled down to room temperature and filtered through a sintered glass filter. The filtrate was analyzed spectroscopically.	Palladium purity not specified. Triply-distilled mercury was used. Content of other metals in the amalgam was below 10^{-4} %.
	ESTIMATED ERROR: Nothing specified.
	REFERENCES:

COMPONENTS:	ORIGINAL MEASUREMENTS:
(1) Palladium; Pd; [7440-05-3] (2) Mercury; Hg; [7439-97-6]	Jangg, G.; Gröll, W. Z. Metallk. 1965, 56, 232-34.

VARIABLES:	PREPARED BY:
Temperature: 25-300°C	C. Guminski; Z. Galus

EXPERIMENTAL VALUES:

The solubility of palladium in mercury:

t/°C	Soly/mass %	Soly/at %[a]
25	0.0027	0.0051
50	0.0029	0.0054
90	0.0047	0.0089
100	0.0060	0.011
150	0.015	0.031
200	0.038	0.071
238	0.089	0.17
250	0.099	0.19
300	0.20	0.37

[a] by compilers.

Additional data were presented graphically. There were two breaks in the reciprocal temperature-solubility plot, at 90 and 240°C; these corresponded to the decomposition temperatures of $PdHg_4$ and Pd_2Hg_5, respectively.

AUXILIARY INFORMATION

METHOD/APPARATUS/PROCEDURE:	SOURCE AND PURITY OF MATERIALS:
The heterogeneous amalgam was introduced into a specially constructed apparatus made of glass. After twelve hours of equilibration at the temperature of the experiment, the amalgam was filtered through the sintered-glass frit under a pressure of purified nitrogen. The palladium content in the filtered, saturated amalgam was determined by an unspecified analytical method.	Not specified in detail.
	ESTIMATED ERROR: Soly: nothing specified; precision better than ± 10% (compilers). Temp: nothing specified.
	REFERENCES:

COMPONENTS:	EVALUATOR:
(1) Platinum; Pt; [7440-06-4] (2) Mercury; Hg; [7439-97-6]	C. Guminski; Z. Galus Department of Chemistry University of Warsaw Warsaw, Poland July, 1985

CRITICAL EVALUATION:

There is a large variation on the reported solubilities of platinum in mercury. The solubility is very low, and the presence of platinum oxides on the surface of the metal further inhibits its dissolution in mercury. Plaksin and Suvorovskaya (1,2) determined the solubility in the range of 289-473 K by filtration and analyses of the saturated solution, and they found that the solubility increased from 2.05×10^{-2} to 1.77 at % in this temperature range. At room temperature, Strachan and Harris (3) reported a solubility of 0.002 at %, while Butler and Makrides (4) obtained a solubility of 0.028 at %; the latter solubility is in agreement with that by ref. (2). Yoshida (5) reported that the solubility is less than 0.01 at %.

Kozin, with the use of his semiempirical equations at 298 K, predicted solubilities of 2.6×10^{-9} (6) and 3.1×10^{-7} at % (7); these results are too low because interaction of the metals was neglected.

Jangg and Dörtbudak (8) determined the solubilities at 374-593 K, and they found that the solubility increased from 3.4×10^{-5} to 9.0×10^{-4} at % in this temperature range. These solubilities are significantly lower than the experimental determination of the previous authors. Based on the work of Barlow and Planting (9), the evaluators are of the opinion that the solubility of platinum at 573 K should be $\gtrsim 2 \times 10^{-2}$ at %.

There is a variation of approximately 10^3 in the reported solubilities, and though the experimental procedures were similar there was a lack of detailed description by the different authors. This has resulted in some difficulty in assigning the more accurate measurements. Although the data in refs. (1,2,4) are similar, the results appear to be too high. Recent, careful electroanalytical measurements in the evaluators' laboratory (10) resulted in a solubility of 5×10^{-4} at % at 298 K. In this work, it was found that the platinum had to be equilibrated with mercury at 600 K for at least a week, followed by an equilibration at 298 K for at least two weeks. Shorter equilibration times resulted in erroneous solubilities; the discrepancies in previously reported data probably are the result of incomplete equilibration. In view of the evaluators' 298 K determination, it appears that the data of Jangg and Dörtbudak (8) should be rejected.

During investigations of the corrosion of pure metals in refluxing mercury at 756 K it was found that the solubility of platinum is similar to, or lower than, those of aluminum and manganese, but higher than those of nickel, titanium and zirconium (11); this observation adds further proof that the results of (8) are too low.

The saturated platinum amalgams are in equilibrium with the intermediate solid phases (2,5,8,9,12), $PtHg_4$, $PtHg_2$ and $PtHg$; the latter two compounds are stable to 523 K and the first compound is stable to 873 K.

The tentative value for the solubility of platinum in mercury at 298 K is 5×10^{-4} at % (10).

References

1. Plaksin, I.N.; Suvorovskaya, N.A. *Zh. Fiz. Khim.* 1941, *15*, 978.
2. Plaksin, I.N.; Suvorovskaya, N.A. *Izv. Sekt. Platiny* 1945, *18*, 67.
3. Strachan, J.F.; Harris, N.L. *J. Inst. Metals* 1956-57, *85*, 17.
4. Butler, J.N.; Makrides, A.C. *Trans. Faraday Soc.* 1964, *60*, 938.
5. Yoshida, Z. *Bull. Chem. Soc. Jap.* 1981, *54*, 556.
6. Kozin, L.F. *Tr. Inst. Khim. Nauk Akad. Nauk Kaz. SSR* 1962, *9*, 101.
7. Kozin, L.F. *Fiziko-Khimicheskie Osnovy Amalgamnoi Metallurgii*, Nauka, Alma-Ata, 1964.
8. Jangg, G.; Dörtbudak, T. *Z. Metallk.* 1973, *64*, 715.
9. Barlow, M.; Planting, P.J. *Z. Metallk.* 1969, *60*, 292.
10. Guminski, C.; Roslonek, H.; Galus, Z. *J. Electroanal. Chem.* 1983, *158*, 357.
11. Fleitman, A.H.; Weeks, J.R. *Nucl. Eng. Des.* 1971, *16*, 266.
12. Jangg, G.; Steppan, F. *Z. Metallk.* 1965, *56*, 172.

COMPONENTS:	ORIGINAL MEASUREMENTS:
(1) Platinum; Pt; [7440-06-4] (2) Mercury; Hg; [7439-97-6]	Plaksin, I.N.; Suvorovskaya, N.A. Izv. Sekt. Platiny 1945, 18, 67-76.
VARIABLES: Temperature: 16-200°C	PREPARED BY: C. Guminski; Z. Galus

EXPERIMENTAL VALUES:

Solubility of platinum in mercury:

t/°C	Soly/at %	t/°C	Soly/at %
16.5	0.0205	86.0	0.904
17.5	0.0513	101.0	0.980
20.0	0.0021	131.5	1.040
24.0	0.102	144.0	1.080
39.5	0.151	167.0	1.12
54.0	0.202	171.0	1.20
71.0	0.910	200.0	1.77

Pt-Hg phase diagram was presented; Pt_3Hg, Pt_2Hg and $PtHg$ are in equilibrium with the saturated amalgams.

Substantially the same results of the solubility were reported in the previous work by the same authors (1).

AUXILIARY INFORMATION

METHOD/APPARATUS/PROCEDURE:	SOURCE AND PURITY OF MATERIALS:
Purified platinum was dissolved in mercury during cathodic polarization of Pt while in contact with Hg and H_2SO_4 solution. Prepared amalgams were kept in a thermostat for 2 hours and filtered through a capillary of 0.24-0.4 mm diameter. The filtration and storing of amalgam were done with the use of a special air-free glass apparatus. The mercury was evaporated from the filtrate and platinum was analyzed by colorimetric, cuppellation, or gravimetric method, depending upon the metal content.	Mercury purified with HNO_3 and distilled under vacuum. Platinum, 99.84% pure, was dissolved in aqua regia, then transformed into $(NH_4)_2PtCl_4$ and reduced to the metallic state with HCOOH.
	ESTIMATED ERROR: Soly: precision no better than several percent (compilers). Temp: precision no better than $\pm$ 0.5 K (compilers).
	REFERENCES: 1. Plaskin, I.N.; Suvorovskaya, N.A. Zh. Fiz. Khim. 1941, 15, 978.

COMPONENTS:	ORIGINAL MEASUREMENTS:
(1) Platinum; Pt; [7440-06-4] (2) Mercury; Hg; [7439-97-6]	Strachan, J.F.; Harris, N.L. *J. Inst. Metals* 1956-57, *85*, 17-24.
VARIABLES:	PREPARED BY:
Room temperature	C. Guminski; Z. Galus

EXPERIMENTAL VALUES:

Solubility of platinum in mercury at room temperature was reported to be 0.002 at %.

AUXILIARY INFORMATION

METHOD/APPARATUS/PROCEDURE:	SOURCE AND PURITY OF MATERIALS:
A preweighed piece of platinum was equilibrated with Hg in an evacuated glass tube over long periods of time. After equilibration, the amalgam was filtered through a sintered glass filter of 90-150 μm pore size. The solubility was determined from: (1) the weight loss of the platinum; and (2) analysis of the filtrate by an unspecified method.	Mercury: 99.997% pure was filtered, dried and distilled before use. Platinum: minimum 99.99% pure.
	ESTIMATED ERROR: Soly: precision ± 50%. Temp: nothing specified.
	REFERENCES:

COMPONENTS:	ORIGINAL MEASUREMENTS:
(1) Platinum; Pt; [7440-06-4] (2) Mercury; Hg; [7439-97-6]	Butler, J.N.; Makrides, A.C. *Trans. Faraday Soc.* 1964 60, 938-46.
VARIABLES:	PREPARED BY:
Room temperature	C. Guminski; Z. Galus

EXPERIMENTAL VALUES:

The solubility of platinum in mercury at room temperature was reported to be 0.028 at %.

AUXILIARY INFORMATION

METHOD/APPARATUS/PROCEDURE:	SOURCE AND PURITY OF MATERIALS:
Mercury was saturated with platinum in a sealed glass tube for few days at 523 K, then the amalgam was cooled down to room temperature and filtered through a sintered glass filter. The filtrate was analyzed spectroscopically.	Purity of platinum not specified. Triply distilled mercury was used. Content of other metals in the amalgam was below 10^{-4}%.
	ESTIMATED ERROR:
	Nothing specified.
	REFERENCES:

COMPONENTS:	ORIGINAL MEASUREMENTS:
(1) Platinum; Pt; [7440-06-4] (2) Mercury; Hg; [7439-97-6]	Guminski, C.; Roslonek, H.; Galus, Z. *J. Electroanal. Chem.* 1983, *158*, 357-68.
VARIABLES: One temperature: 298 K	PREPARED BY: C. Guminski; Z. Galus

EXPERIMENTAL VALUES:

Solubility of platinum in mercury at 298 K was reported to be $(5 \pm 1) \times 10^{-4}$ at %.

When the dissolution time was shorter than a week, or the conditioning shorter than 2 weeks, the solubilities determined were lower and higher, respectively.

AUXILIARY INFORMATION

METHOD/APPARATUS/PROCEDURE:	SOURCE AND PURITY OF MATERIALS:
A clean and degreased Pt foil was placed in a closed glass vessel with mercury and heated at 600 K for 8 hours per day during a period of 2 months. After a subsequent month of conditioning at 298 K the heterogeneous amalgam was filtered through a sintered-glass crucible under vacuum. Employing a hanging-drop electrode filled with the saturated amalgam, cyclic chronopotentiometric curves were recorded in 0.01 mol dm^{-3} ZnCl$_2$ + 1 mol dm^{-3} NaCl solution. The current density was varied between 10.8 and 1080 μA cm^{-2} and reduction time from 5 to 100 s. The differences of oxidation times observed were due to bonding of Zn by Pt to form PtZn$_2$, a stable compound which is rapidly formed and is insoluble in mercury. The calculations of the solubility were based on these differences.	Mercury (from Ciech) was chemically purified with acidic Hg$_2$(NO$_3$)$_2$ and then twice distilled in vacuum. Platinum (from Polish Mint) was 99.9999% pure. ZnCl$_2$ and NaCl (from Ciech) were analytically pure; their solutions were additionally refined by a cathodic electrolysis at -0.9 V vs. SCE.
	ESTIMATED ERROR: Soly: precision $\pm$ 20%. Temp: $\pm$ 1 K.
	REFERENCES:

COMPONENTS:	EVALUATOR:
(1) Copper; Cu; [7440-50-8] (2) Mercury; Hg; [7439-97-6]	C. Guminski; Z. Galus Department of Chemistry University of Warsaw Warsaw, Poland July, 1985

CRITICAL EVALUATION:

Although readily wetted by mercury, the solubility of copper in mercury at room temperature is low. There have been numerous reports of solubility measurements near room temperature. Gouy (1) reported the first solubility determination, but this author's value of 3 x 10^{-3} at % at 288 to 291 K is too low and therefore rejected. Humphreys (2), from his investigation of the solution and diffusion of copper in mercury, estimated a solubility of 8.5 x 10^{-3} at % at 299.4 K. Richards and Garrod-Thomas (3) determined a solubility of 7.42 x 10^{-3} at % at 293 K; these authors' result was based on colorimetric and potentiometric analyses of the equilibrated liquid amalgam. Tammann and Kollmann (4) utilized potentiometry to determine a solubility of 1.02 x 10^{-2} at % at 288 K. Irvin and Russell (5) performed careful analytical measurements of the solubility at 293 K, but their value of 6.3 x 10^{-3} at % is lower than those of the earlier authors. Without presenting experimental details, Hickling and Maxwell (6) reported a solubility of 6 x 10^{-3} at % at 293 K; these authors had studied the electrochemical oxidation of copper amalgams. Liebl (7) employed coulometry to determine a solubility of 1.0 x 10^{-2} at % at room temperature, but no experimental details for this determination were presented. Sagadieva and Kozlovskii (8) employed amalgam polarography and also determined a solubility of 1.0 x 10^{-2} at % at 293 K. From potentiometric measurements, Schupp and coworkers (9) determined a solubility of 8.5 x 10^{-3} at % at 298 K. Jangg and Kirchmayr (10) measured the potentials of various copper amalgams at 288 K, and though numerical data were not presented, a solubility of 8.9 x 10^{-3} at % was estimated from the graphical presentation; employing polarographic oxidation of the copper amalgam they estimated the solubility of 8.4 x 10^{-3} at % at 293 K. Jangg and Palman (11), from measurements based on a filtration method, reported solubilities over a temperature range of 293 to 823 K; the solubility at 293 K was 6.3 x 10^{-3} at % and it increased to 3.7 at % at 823 K. The experimental results of the latter authors at temperatures above 298 K were much higher than the predicted liquidus from the phase diagrams which were proposed earlier by Tammann and Strassfurth (12) and by Schmidt (13).

Levitskaya and Zebreva (14) employed potentiometric measurements at 293 to 323 K and obtained solubilities which increased from 1.07 x 10^{-2} to 2.4 x 10^{-2} at % in this temperature range. Chao and Costa (15) also employed potentiometry and obtained a solubility of 9.3 x 10^{-3} at % at room temperature.

Baletskaya and coworkers (16), from voltammetric measurements on the amalgam, determined a solubility of 1.02 x 10^{-2} at % at room temperature. Dragavtseva and Bukhman (17) performed chronoamperometric oxidation of heterogeneous copper amalgams and determined a solubility of 9 x 10^{-3} at % at 293 K. Lange and coworkers (18) also carried out similar measurements as the previous authors, but at 313 to 363 K, and determined that the solubilities increased from 1.7 x 10^{-2} to 4.3 x 10^{-2} at % in this temperature range. Ostapczuk and Kublik (19) employed voltammetric oxidation of a saturated copper amalgam at 298 K and found a solubility of 1.09 x 10^{-2} at %. Hurlen and coworkers (20) employed potentiometry and determined a solubility of 6.2 x 10^{-3} at % at 298 K. Sasim and coworkers (21) also used potentiometric measurements and determined a solubility of 1.1 x 10^{-2} at % at 298 K; the latter result is in agreement with the most dependable determinations at this temperature.

Grønlund and Kristensen (22) performed precise concentration cell measurements at 283.4 to 298.3 K and reported solubilities of 7.46 x 10^{-3} to 13.9 x 10^{-3} at % in this temperature range. The latter measurements were made in evacuated cells; probably because of the reduced pressure the results obtained are higher than those determined under normal conditions. Ignateva and coworkers determined Cu solubility at 298.2 K of 1.1 x 10^{-2} at % (23), 1.0 x 10^{-2} at % (24), and 1.25 x 10^{-2} at % (25); chronoamperometric oxidation of the Cu saturated amalgams was applied in these works.

There were a number of other solubilities reported near room temperature, but these are rejected because the results are too high (26-29), or the data were presented without sufficient definition of experimental conditions (30-32). Kozin predicted a solubility of 5.7 x 10^{-3} at % at 298 K (33).

It is clear that there is wide scatter in the solubility data near room temperature. As indicated by Chao and Costa (15), the crystallization of copper attains equilibrium very slowly, so that the determinations based on electrochemical methods in which the copper is crystallized from an amalgam may result in solubilities that may be too high.

COMPONENTS:	EVALUATOR:
(1) Copper; Cu; [7440-50-8] (2) Mercury; Hg; [7439-97-6]	C. Guminski; Z. Galus Department of Chemistry University of Warsaw Warsaw, Poland July, 1985

CRITICAL EVALUATION:

The solubility measurements at 293 to 824 K, by Jangg and Palman (11), were extended to higher temperatures by Lugscheider and Jangg (34). The latter authors analyzed samples of the immiscible, mutually saturated copper and mercury phases in the temperature range of 883 to 1073 K, and they established the corresponding values of maximum and minimum solubility in this temperature range. The solubility of 0.67 at % at 644 K, reported by Wang (35), is more than an order of magnitude too low, and is rejected.

The solid phases in equilibrium with the saturated amalgam consist of unstable Cu-Hg compounds (15,34,36,37); the peritectic temperature has been reported to be 369.4 (12), 371 (37) and 401 K (34). The phase diagram is shown in Figure 1 (34). The solid compounds, CuHg, Cu_4Hg_3, Cu_3Hg_2, and Cu_7Hg_6 have been reported, but the existence of the last of these is the most probable.

Tentative values of the solubility of Cu in Hg:

T/K	Soly/at %	Source
293	9.2×10^{-3} [a]	[3,7,8,10,17]
298	1.00×10^{-2} [a]	[2,9,15,19,23,21,24]
373	4×10^{-2}	[11]
473	0.2	[11]
573	0.6	[11]
673	1.1	[11]
773	2.1	[11]
873	5.5[b]	[11,34]
933	8	[34]
973	14	[34]
1073	40	[34]

a. Mean value from cited references.
b. Interpolated value from cited references

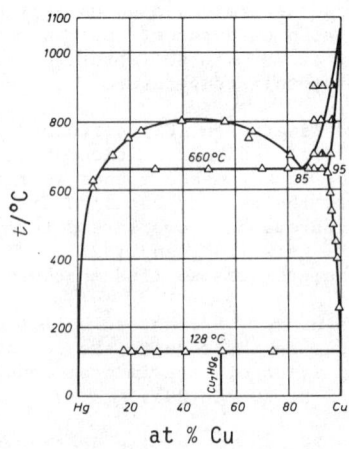

Fig. 1. The Cu-Hg system (34).

COMPONENTS:	EVALUATOR:
(1) Copper; Cu; [7440-50-8] (2) Mercury; Hg; [7439-97-6]	C. Guminski; Z. Galus Department of Chemistry University of Warsaw Warsaw, Poland July, 1985

CRITICAL EVALUATION:

REFERENCES

1. Gouy, M. *J. Phys.* 1895, *4*, 320.
2. Humphreys, W.J. *J. Chem. Soc.*, 1896, 243.
3. Richards, T.W.; Garrod-Thomas, R.N. *Z. Phys. Chem.*, 1910, *72*, 165.
4. Tammann, G.; Kollmann, K. *Z. Anorg. Chem.*, 1927, *160*, 242.
5. Irvin, N.M.; Russell, A.S. *J. Chem. Soc.*, 1932, 891.
6. Hickling, A.; Maxwell, J. *Trans. Faraday Soc.*, 1955, *51*, 44.
7. Liebl, G.; cited by H. Spengler, *Metall*, 1958, *12*, 105.
8. Sagadieva, K.Zh.; Kozlovskii, M.T. *Izv. Akad. Nauk Kaz. SSR, Ser. Khim*, 1959, *1 (15)*, 22.
9. Schupp, O.E.; Youness, T.; Watters, J.I. *J. Am. Chem. Soc.*, 1962, *84*, 505.
10. Jangg, G.; Kirchmayr, H. *Z. Chem.*, 1963, *3*, 47.
11. Jangg, G.; Palman, H. *Z. Metallk.*, 1963, *54*, 364.
12. Tammann, G.; Strassfurth, T. *Z. Anorg. Chem.*, 1925, *143*, 357.
13. Schmidt, W. *Metall.*, 1949, *3*, 10.
14. Levitskaya, S.A.; Zebreva, A.I. *Tr. Inst. Khim. Nauk Akad. Nauk Kaz. SSR*, 1967, *15*, 66.
15. Chao, F.; Costa, M. *C.R. Acad. Sci., Ser. 2* 1965, *261*, 990; *Bull. Soc. Chim. Fr.* 1968, 549.
16. Baletskaya, L.G.; Zakharova, Z.A.; Zakharov, M.S. *Izv. Akad. Nauk Kaz. SSR, Ser. Khim.*, 1969, *19*, No. 1, 34.
17. Dragavtseva, N.A.; Bukhman, S.P. *Izv. Akad. Nauk Kaz. SSR, Ser. Khim.*, 1970, *20*, No. 6, 33.
18. Lange, A.A.; Bukhman, S.P.; Kairbaeva, A.A. *Izv. Akad. Nauk Kaz. SSR, Ser. Khim.*, 1974, *24*, No. 5, 37.
19. Ostapczuk, P.; Kublik, Z. *J. Electroanal. Chem.*, 1977, *83*, 1.
20. Hurlen, T.; Staurset, A.; Eriksrud, E. *J. Electroanal. Chem.*, 1977, *83*, 263.
21. Sasim, D.; Srudka, M.; Guminski, C. *Monatsh. Chem.*, 1984, *115*, 45.
22. Grønlund, F.; Kristensen, B. *Acta Chem. Scand., Ser. A* 1984, *38*, 229.
23. Ignateva, L.A.; Zakharova, E.A.; Nazarov, B.F. *Dep. ONITEKhim*, 1731-78, 1978.
24. Kataev, G.A.; Zakharova, E.A.; Ignateva, L.A. *ONITEKhim*, 3092-79, 1979.
25. Zakharova, E.A.; Kataev, G.A.; Ignateva, L.A.; Morosova, V.E. *Tr. Tomsk. Gos. Univ.*, 1973, *249*, 103.
26. Strachan, J.F.; Harris, N.L. *J. Inst. Metals.*, 1956-57, *85*, 17.
27. Igolinskii, V.A.; Shalaevskaya, V.N. *Elektrokhimia*, 1974, *10*, 536.
28. Igolinskii, V.A.; Shalaevskaya, V.N.; Guryanova, O.N.; Igolinskaya, I.M.; Kotova, N.A. *Sovremiennye Problemy Polarografii Nakopleniem*, Tomsk, 1975, p. 150.
29. Shalaevskaya, V.N.; Igolinskii, V.A.; Kataev, G.A. *Dep. VINITI*, 588-75, 1975.
30. Spencer, J.F. *Z. Elektrochem.* 1905, *11*, 681.
31. Zebreva, A.I.; Filippova, L.M.; Omarova, N.D. *Izv. Vyssh. Ucheb. Zaved., Khim. Khim. Tekhnol.*, 1977, *20*, 19.
32. Zebreva, A.I.; Filippova, L.M.; Omarova, N.D.; Gayfullin, A.Sh. *Izv. Vyssh. Ucheb. Zaved., Khim. Khim. Tekhnol.* 1976, *19*, 1043.
33. Kozin, L.F. *Fiziko-Khimicheskie Osnovy Amalgamnoi Metallurgii*, Nauka, Alma-Ata, 1964.
34. Lugscheider, E.; Jangg, G. *Z. Metallk.*, 1971, *62*, 548.
35. Wang, J.Y.N. *Nucl. Sci. Eng.*, 1964, *18*, 18.
36. Lihl, F. *Z. Metallk.*, 1953, *44*, 160.
37. Noer, S.; Grønlund, F. *Ber. Bunsenges. Phys. Chem.*, 1975, *79*, 517.

COMPONENTS:	ORIGINAL MEASUREMENTS:
(1) Copper; Cu; [7440-50-8] (2) Mercury; Hg; [7439-97-6]	Humphreys, W.J. J. Chem. Soc. 1896, 243-53.
VARIABLES: One temperature: 26°C	PREPARED BY: C. Guminski; Z. Galus

EXPERIMENTAL VALUES:

The solubility of copper in mercury at 26.2°C was reported as $(2.7 \pm 0.1) \times 10^{-3}$ mass %. Converting to atomic %, the compilers calculated 8.5×10^{-3} at %.

AUXILIARY INFORMATION

METHOD/APPARATUS/PROCEDURE:	SOURCE AND PURITY OF MATERIALS:
Disc of Cu was placed on the surface of a column of Hg contained in a glass or wooden vessel, and the liquid sampled for analysis after several days of equilibration. The Hg was evaporated from the amalgam and the residual copper was determined as the oxide after treatment with HNO_3 and ignition.	Nothing specified.
	ESTIMATED ERROR: Soly: precision probably no better than $\pm$ 3% (compilers). Temp: nothing specified.
	REFERENCES:

COMPONENTS:	ORIGINAL MEASUREMENTS:
(1) Copper; Cu; [7440-50-8] (2) Mercury; Hg; [7439-97-6]	Irvin, N.M.; Russell, A.S. *J. Chem. Soc.* 1932, 891-8.
VARIABLES: One temperature: 20°C	PREPARED BY: C. Guminski; Z. Galus

EXPERIMENTAL VALUES:

The solubility of copper in mercury at 20°C was found to be $(2.0 \pm 0.1) \times 10^{-3}$ mass %.

Converting to atomic %, the compilers calculate 6.3×10^{-3} at %.

AUXILIARY INFORMATION

METHOD/APPARATUS/PROCEDURE:	SOURCE AND PURITY OF MATERIALS:
Amalgams were prepared by electrolysis and by chemical reduction of copper (II) with V(II) solutions. The prepared amalgams were filtered through their own paste on a ground-glass filter. Copper was removed from the homogeneous amalgam by oxidation with $KMnO_4$. In the final determination of copper the iodide-thiosulphate volumetric method was applied. When chamois leather was used instead of the sintered glass for filtration of the amalgam, some irregular and higher solubilities were obtained.	Not given
	ESTIMATED ERROR: Soly: accuracy $\pm$ 5%. Temp: nothing specified.
	REFERENCES:

COMPONENTS:	ORIGINAL MEASUREMENTS:
(1) Copper; Cu; [7440-50-8] (2) Mercury; Hg; [7439-97-6]	Richards, T.W.; Garrod-Thomas, R.N. Z. Phys. Chem. 1910, 72, 165-201.
VARIABLES: One temperature: 20°C	PREPARED BY: C. Guminski; Z. Galus

EXPERIMENTAL VALUES:

Solubility of copper in mercury at 20°C was found to be $(2.35 \pm 0.35) \times 10^{-3}$ mass %.

Converting to atomic %, the compilers calculate 7.42×10^{-3} at %.

The result was confirmed also by potentiometric measurements.

AUXILIARY INFORMATION

METHOD/APPARATUS/PROCEDURE:	SOURCE AND PURITY OF MATERIALS:
Copper amalgam was prepared by the electroreduction of a $CuSO_4$ solution on a mercury cathode. The amount of the reduced copper was determined coulometrically. The amalgams were filtered through chamois leather after equilibration. The solid residue was analyzed by volatilizing the mercury and determination of copper by colorimetric method.	Mercury was purified with $Hg_2(NO_3)_2$ and doubly distilled. Pure $CuSO_4$ was recrystallized.
	ESTIMATED ERROR: Soly: precision no better than $\pm$ 15%. Temp: nothing specified.
	REFERENCES:

COMPONENTS:	ORIGINAL MEASUREMENTS:
(1) Copper; Cu; [7440-50-8] (2) Mercury; Hg; [7439-97-6]	Tammann, G.; Kollmann, K. Z. Anorg. Chem. 1927, 160, 242-8.
VARIABLES: One temperature: 15°C	PREPARED BY: C. Guminski; Z. Galus

EXPERIMENTAL VALUES:

Solubility of copper in mercury at 15°C was reported to be $(3.23 \pm 0.07) \times 10^{-3}$ mass %.
Converting to atomic %, the compilers calculate 1.02×10^{-2} at %.

AUXILIARY INFORMATION

METHOD/APPARATUS/PROCEDURE:	SOURCE AND PURITY OF MATERIALS:
The amalgams of various concentrations were obtained by electrolysis of saturated $CuSO_4$ solutions by varying the current and the time of the electrolysis. Subsequently, the steady-state potentials of the cell, $Cu(Hg)_x \mid CuSO_4, Hg_2SO_4 \mid Hg$, were measured.	Nothing specified.
	ESTIMATED ERROR: Soly: precision $\pm$ 5%. Temp: nothing specified.
	REFERENCES:

COMPONENTS:	ORIGINAL MEASUREMENTS:
(1) Copper; Cu; [7440-50-8] (2) Mercury; Hg; [7439-97-6]	Sagadieva, K.Zh.; Kozlovski, M.T. *Izv. Akad. Nauk Kaz. SSR, Ser. Khim.* 1959, No. 1 (15), 22-5.
VARIABLES: One temperature: 20°C	PREPARED BY: C. Guminski; Z. Galus

EXPERIMENTAL VALUES:

Solubility of copper in mercury at 20°C was reported to be 6.8×10^{-3} mol dm^{-3}.

Converting to atomic %, the compilers calculate 1.01×10^{-2} at %.

AUXILIARY INFORMATION

METHOD/APPARATUS/PROCEDURE:	SOURCE AND PURITY OF MATERIALS:
The amalgams were obtained by the exhaustive electrolysis of copper (II) from solution. The polarograms of amalgam dissolution were determined in 0.1 mol dm^{-3} solution of KNO_3. All operations were performed in a hydrogen atmosphere. Estimation of the copper concentration was based on the polarograms.	Nothing specified.
	ESTIMATED ERROR: Soly: nothing specified; ± 10% (compilers). Temp: ± 1 K.
	REFERENCES:

COMPONENTS:	ORIGINAL MEASUREMENTS:
(1) Copper; Cu; [7440-50-8] (2) Mercury; Hg; [7439-97-6]	Schupp, O.E.; Youness, T.; Watters, J.I. *J. Am. Chem. Soc.* <u>1962</u>, *84*, 505-13.
VARIABLES:	PREPARED BY:
One temperature: 25°C	C. Guminski; Z. Galus

EXPERIMENTAL VALUES:

Solubility of copper in mercury at 25°C was found to be 2.7×10^{-3} mass %.

Converting to atomic %, the compilers calculate 8.5×10^{-3} at %.

AUXILIARY INFORMATION

METHOD/APPARATUS/PROCEDURE:	SOURCE AND PURITY OF MATERIALS:
Potentials of the amalgams versus pure copper immersed in solution of ethylenediamine complex of Cu(II) were measured. The copper amalgams were prepared electrolytically, and the solubility was determined from EMF measurements on the cell, $Cu(Hg)_x	Cu(II)$ in ethylenediamine $| Cu$.	Nothing specified.
	ESTIMATED ERROR:
	Soly: nothing specified; accuracy probably better than a few percent (compilers). Temp: not specified.
	REFERENCES:

COMPONENTS:	ORIGINAL MEASUREMENTS:
(1) Copper; Cu; [7440-50-8] (2) Mercury; Hg; [7439-97-6]	Jangg, G.; Palman, H. Z. Metallk. 1963, 54, 364-369.
VARIABLES: Temperature: 293-823 K	PREPARED BY: C. Guminski; Z. Galus

EXPERIMENTAL VALUES:

The solubility of copper in mercury at 20 to 550°C was presented graphically as a plot of log N against temperature, where N was in mass %. The data points were read off the curve and converted to atom % by the compilers.

T/K	Soly/at %	T/K	Soly/at %	T/K	Soly/at %	T/K	Soly/at %
293	6.3×10^{-3}	508	0.28	663	1.0	823	3.7
323	1.3×10^{-2}	523	0.35	673	1.1		
373	4.0×10^{-2}	548	0.40	698	1.2		
383	5.0×10^{-2}	563	0.57	723	1.6		
423	9.6×10^{-2}	573	0.62	743	1.7		
438	0.14	608	0.78	763	2.1		
463	0.17	623	0.86	773	2.3		
473	0.19	643	0.93	798	2.9		

AUXILIARY INFORMATION

METHOD/APPARATUS/PROCEDURE:	SOURCE AND PURITY OF MATERIALS:
Method of the amalgam preparation is not given. The amalgams were mixed and kept for 12 hours in thermostated glass cylinders and then filtered in an atmosphere of pure nitrogen. For temperatures above 600 K a pressure apparatus made of hard chromium steel was used. No method of analyzing of the filtrate is given.	Nothing specified.
	ESTIMATED ERROR: Soly: precision ± 5%. Temp: nothing specified.
	REFERENCES:

COMPONENTS:	ORIGINAL MEASUREMENTS:
(1) Copper; Cu; [7440-50-8] (2) Mercury; Hg; [7439-97-6]	Jangg, G.; Kirchmayr, H. Z. Chem. 1963, 3, 47-56.
VARIABLES: Temperature: 15-20°C	PREPARED BY: C. Guminski; Z. Galus

EXPERIMENTAL VALUES:

Solubility of copper in mercury.

$t/°C$	$mol\ cm^{-3}$	at %[a]
15	$(6.0 \pm 0.2) \times 10^{-3}$	8.9×10^{-3}
20	5.7×10^{-3}	8.4×10^{-3}

[a] by compilers

15°C data based on EMF measurements, and 20°C data based on polarographic measurements (see below).

AUXILIARY INFORMATION

METHOD/APPARATUS/PROCEDURE:	SOURCE AND PURITY OF MATERIALS:
The amalgams were obtained electrolytically and the potentials of the cell, $Cu(Hg)_x\|CuSO_4 + CH_3COOH\|KCl, Hg_2Cl_2, Hg$, were measured at 15°C. The concentration of the saturated amalgam was determined from the breakpoint of the plot of EMF against the logarithm of the copper concentration in the amalgam. Anodic polarographic currents of various copper amalgams were determined at 20°C, and the current was plotted against the Cu concentration in the amalgam. The solubility was determined from the breakpoint in the plot. All experiments were performed in an inert gas atmosphere.	Nothing specified.
	ESTIMATED ERROR: Soly: accuracy better than $\pm$ 3%. Temp: nothing specified.
	REFERENCES:

COMPONENTS:	ORIGINAL MEASUREMENTS:
(1) Copper; Cu; [7440-50-8] (2) Mercury; Hg; [7439-97-6]	Chao, F.; Costa, M. *C.R. Acad. Sci.*, Ser. 2 <u>1965</u>, *261*, 990-3.
VARIABLES: One temperature: 295 K	PREPARED BY: C. Guminski; Z. Galus

EXPERIMENTAL VALUES:

Solubility of copper in mercury at 295 K was reported to be $(6.2 \pm 0.2) \times 10^{-3}$ mol dm^{-3}. The corresponding atomic % solubility calculated by the compilers is 9.3×10^{-3} at %.

This work is part of an extensive study of copper amalgams by the same authors (1). The authors investigated the tendency toward the formation of semi-stable, supersaturated amalgams.

AUXILIARY INFORMATION

METHOD/APPARATUS/PROCEDURE:	SOURCE AND PURITY OF MATERIALS:
The copper amalgams were prepared electrolytically at constant current from 0.5 mol dm^{-3} CuSO$_4$ + 0.25 mol dm^{-3} H$_2$SO$_4$. Dilution was made by adding Hg to the amalgam in a nitrogen atmosphere. Potentials of the amalgams were measured in the cell of the type, Cu(Hg)$_x$ | 0.5 mol dm^{-3} CuSO$_4$ +0.36 mol dm^{-3} H$_2$SO$_4$ | Hg$_2$SO$_4$, Hg Concentration of the saturated amalgam was determined from the breakpoint of the plot of EMF against the logarithm of the copper concentration.	Reagents were specified as "pure for analysis", and salts were recrystallized. Hg purity was specified as 99.99999%.
	ESTIMATED ERROR: Soly: precision $\pm$ 3%. Temp: precision $\pm$ 0.5 K.
	REFERENCES: 1. Chao, F.; Costa, M. *Bull. Soc. Chim. Fr.*, <u>1968</u>, 549-55.

COMPONENTS:	ORIGINAL MEASUREMENTS:
(1) Copper; Cu; [7440-50-8] (2) Mercury; Hg; [7439-97-6]	Levitskaya, S.A.; Zebreva, A.I. *Tr. Inst. Khim. Nauk Akad. Nauk Kaz. SSR* 1967, *15*, 66-8.
VARIABLES:	PREPARED BY:
Temperature	C. Guminski; Z. Galus

EXPERIMENTAL VALUES:

Solubility of copper in mercury.

$t/°C$	Soly/mol dm^{-3}	Soly/at %[a]
20	7.2×10^{-3}	1.07×10^{-2}
40	1.2×10^{-2}	1.8×10^{-2}
50	1.6×10^{-2}	2.4×10^{-2}

[a] by compilers

AUXILIARY INFORMATION

METHOD/APPARATUS/PROCEDURE:	SOURCE AND PURITY OF MATERIALS:
Potentials of the galvanic cell, $Cu(Hg)	CuSO_4, H_2SO_4	Cu(Hg)_x$ were measured at various temperatures. The solubilities were determined from the breakpoint in the plot of cell EMF against the logarithm of the copper concentration in the amalgam.	Nothing specified.
	ESTIMATED ERROR:
	Nothing specified.
	REFERENCES:

COMPONENTS:	ORIGINAL MEASUREMENTS:
(1) Copper; Cu; [7440-50-8] (2) Mercury; Hg; [7439-97-6]	Dragavtseva, N.A.; Bukhman, S.P. *Izv. Akad. Nauk Kaz. SSR, Ser. Khim.* <u>1970</u>, No. 6, 33-7.
VARIABLES: One temperature: 20°C	PREPARED BY: C. Guminski; Z. Galus

EXPERIMENTAL VALUES:

Solubility of copper in mercury at 20°C was found to be 9×10^{-3} at %.

AUXILIARY INFORMATION

METHOD/APPARATUS/PROCEDURE:	SOURCE AND PURITY OF MATERIALS:
The amalgams were obtained by exhaustive electrolysis of solutions containing copper ions. Oxidation of the prepared amalgams were performed after 2 hours under chrono-amperometric conditions. Estimation of the copper concentrations was based on analysis of the current-time curves.	Nothing specified.
	ESTIMATED ERROR: Soly: nothing specified; precision better than $\pm$ 10% (compilers). Temp: not specified.
	REFERENCES:

COMPONENTS:	ORIGINAL MEASUREMENTS:
(1) Copper; Cu; [7440-50-8] (2) Mercury; Hg; [7439-97-6]	Lugscheider, E.; Jangg, G. *Z. Metallk.* 1971, *62*, 548-551.
VARIABLES: Temperature: 610-800°C	PREPARED BY: C. Guminski; Z. Galus

EXPERIMENTAL VALUES:

The solubility of copper in mercury in the two regions at 610-800°C.

	Copper solubility, at %	
$t/°C$	Hg-rich region	Cu-rich region
610	5.9	
630	6.1	
700	14.1	$\geq$ 80.8
750	19.9	$\geq$ 65.1
770	24.4	$\geq$ 66.6
800	40.0	$\geq$ 56.5

The experiments were performed in the range of miscibility gap so that the solubility of copper in mercury may be higher than values given in the third column. When copper content in the amalgam is higher than 85 at % at 660°C the liquidus curve goes up to 100 at % of Cu at 1070°C. The experimental data have large scatter in the Cu-rich region.

AUXILIARY INFORMATION

METHOD/APPARATUS/PROCEDURE:	SOURCE AND PURITY OF MATERIALS:
The amalgams were obtained by dissolution of copper in mercury (1:1) and heating up to 850°C. After annealing for 170 hours at the experimental temperature, the samples were quickly cooled and solidified as two separate phases. The samples were analyzed in hydrogen atmosphere after the mercury was removed by distillation.	Nothing specified
	ESTIMATED ERROR: Soly: nothing specified; precision no better than $\pm$ 10% (compilers). Temp: nothing specified.
	REFERENCES:

COMPONENTS:	ORIGINAL MEASUREMENTS:
(1) Copper; Cu; [7440-50-8] (2) Mercury; Hg; [7439-97-6]	Zakharova, E.A.; Kataev, G.A.; Ignateva, L.A.; Morozova, V.E. *Tr. Tomsk. Univ.* 1973, *249*, 103-9.
VARIABLES:	PREPARED BY:
One temperature: 25°C	C. Guminski; Z. Galus

EXPERIMENTAL VALUES:

The solubility of Cu in Hg at 25°C was reported to be: 8.4×10^{-3} mol dm^{-3} from concentration dependence of Cu(II) vs. Cu(Hg), and 8.6×10^{-3} mol dm^3 from dependence of diffusion coefficient of Cu in Hg vs. concentration of Cu in Hg (see below).

Converting to atomic %, the compilers calculate 1.25×10^{-2} and 1.28×10^{-2} at %, respectively.

AUXILIARY INFORMATION

METHOD/APPARATUS/PROCEDURE:	SOURCE AND PURITY OF MATERIALS:
The experiments were carried out on a semispherical Hg electrode on a Ag base. The electrode was prepared by electrochemical reduction of Hg(II); the electrode was then transferred to a cell which contained the reference electrode of "manganese half element" and an auxiliary Pt net electrode. Cu was introduced into the Hg by electro-reduction from solution: 0.1 mol dm^{-3} of $(NH_4)_2C_2H_4O_6$ and Cu(II) of concentration ranging between 2.9×10^{-4} and 3.92×10^{-3} mol dm^{-3}. After various periods of waiting the chronoamperometric curves of oxidation were recorded. The results were analyzed from the curves: Concentrations of Cu(II) in the solution vs. Cu in the amalgam, and diffusion coefficient of Cu in Hg vs. concentration of Cu in Hg. Breakpoints on the curves correspond to saturation of Hg with Cu.	Nothing specified
	ESTIMATED ERROR:
	Temperature: $\pm$ 0.2 K. Diffusion coefficients are accurate to $\pm$ 2% but the waiting times up to 15 min are too short to reach a true equilibrium in the system so the results may be overstated.
	REFERENCES:

COMPONENTS:	ORIGINAL MEASUREMENTS:
(1) Copper; Cu; [7440-50-8] (2) Mercury; Hg; [7439-97-6]	1. Ignateva, L.A.; Zakharova, E.A.; Nazarov, B.F. *Dep. ONITEKhim.* 1731-78, <u>1978</u>. 2. Kataev, G.A.; Zakharova, E.A.; Ignateva, L.A. *Dep. ONITEKhim.* 3092-79, <u>1979</u>.
VARIABLES:	PREPARED BY:
One temperature: 298 K	C. Guminski; Z. Galus

EXPERIMENTAL VALUES:

Solubility of copper in mercury at 298.2 K.

mol dm^{-3}	at %[a]	Ref.
$(7.5 \pm 0.2) \times 10^{-3}$	1.1×10^{-2}	(1)
6.9×10^{-3}	1.0×10^{-2}	(2)

[a] by compilers

AUXILIARY INFORMATION

METHOD/APPARATUS/PROCEDURE:	SOURCE AND PURITY OF MATERIALS:
The experiments were performed in a cell containing 3 electrodes: (Ref. 1) Working - half spherical Hg; reference - the SCE; auxiliary - Pt net. The Cu amalgams were obtained by electroreduction of Cu(II) from 0.1 mol dm^{-3} solutions of $(NH_4)_2C_4H_4O_6$ at -1.20 V. The amalgams were conditioned for 30 min. at -0.40 V and then oxidized at 0.05-0.10 V in chronoamperometric conditions. The solubility was calculated from analysis of $it^{1/2}$ vs. $t^{1/2}$ curves, where i is the limiting current and t the time. (Ref. 2) Working - half sperhical Hg on Ag base; reference - manganese oxide electrode; auxiliary - Pt net. Cu amalgams were obtained by electroreduction of Cu(II) from 0.1 mol dm^{-3} solutions of $(NH_4)_2C_4H_4O_6$ at -1.8 V. The amalgams were conditioned for 15 min and then oxidized in chronoamperometric conditions at -0.46 V. When the amalgams contained more than 6.9×10^{-3} mol dm^{-3} Cu the curves obtained were irregular, indicating that the amalgams were not homogeneous.	Nothing specified.
	ESTIMATED ERROR:
	Soly: precision $\pm$ 3% in (1). Temp: $\pm$ 0.2 K.
	REFERENCES:

COMPONENTS:	ORIGINAL MEASUREMENTS:
(1) Copper; Cu; [7440-50-8] (2) Mercury; Hg; [7439-97-6]	Lange, A.A.; Bukhman, S.P.; Kairbaeva, A.A. *Izv. Akad. Nauk Kaz. SSR, Ser. Khim.* **1974**, No. 5, 37-41.
VARIABLES:	PREPARED BY:
Temperature: 40-90°C	C. Guminski; Z. Galus

EXPERIMENTAL VALUES:

Solubility of copper in mercury:

$t/°C$	Soly/mass %	Soly/at %[a]
40	5.4×10^{-3}	1.7×10^{-2}
50	7.1×10^{-3}	2.2×10^{-2}
60	8.0×10^{-3}	2.5×10^{-2}
70	9.6×10^{-3}	3.0×10^{-2}
80	1.16×10^{-2}	3.7×10^{-2}
90	1.36×10^{-2}	4.3×10^{-2}

[a] by compilers

AUXILIARY INFORMATION

METHOD/APPARATUS/PROCEDURE:	SOURCE AND PURITY OF MATERIALS:
The copper amalgams were obtained by the exhaustive electrolysis of $CuSO_4$ solutions in 0.1 mol dm^{-3} H_2SO_4. Concentrations of Cu(II) were determined colorimetrically. Amalgams were kept 6-8 hours at a chosen temperature, then oxidation currents were recorded potentiostatically at +0.3 V (vs. SCE). The amalgam and the solution were mixed at a constant velocity during the measurements.	$CuSO_4$ was of reagent grade. Mercury purity not specified.
	ESTIMATED ERROR: Soly: no better than ± 3% (compilers). Temp: nothing specified.
	REFERENCES:

COMPONENTS:	ORIGINAL MEASUREMENTS:
(1) Copper; Cu; [7440-50-8] (2) Mercury; Hg; [7439-97-6]	Ostapczuk, P.; Kublik, Z. *J. Electroanal. Chem.* 1977, *83*, 1-17.
VARIABLES: One temperature: 25°C	PREPARED BY: C. Guminski; Z. Galus

EXPERIMENTAL VALUES:

Solubility of copper in mercury at 25°C was found to be 7.4×10^{-3} mol dm^{-3}.

Converting to atomic %, the compilers calculate 1.1×10^{-2} at %.

AUXILIARY INFORMATION

METHOD/APPARATUS/PROCEDURE:	SOURCE AND PURITY OF MATERIALS:
A piece of copper was introduced into the mercury of the hanging drop electrode. The electrode was conditioned for 5 days with occasional shaking to ascertain saturation of the amalgam, then the chronovoltammetric oxidation peak currents were recorded. Constancy of such peaks was taken as evidence of amalgam saturation.	Copper: spectroscopically pure. Mercury: purified with $Hg_2(NO_3)_2$, then twice distilled under vacuum.
	ESTIMATED ERROR: Soly: accuracy no better than ± 5% (compilers). Temp: nothing specified.
	REFERENCES:

COMPONENTS:	ORIGINAL MEASUREMENTS:
(1) Copper; Cu; [7440-50-8] (2) Mercury; Hg; [7439-97-6]	Hurlen, T.; Staurset, A.; Eriksrud, E. *J. Electroanal. Chem.* <u>1977</u>, *83*, 263-72.
VARIABLES: One temperature: 25°C	PREPARED BY: C. Guminski; Z. Galus

EXPERIMENTAL VALUES:

Solubility of copper in mercury at 25°C was reported to be $(4.2 \pm 0.3) \times 10^{-3}$ mol dm^{-3}.
Converting to atomic %, the compilers calculate 6.2×10^{-3} at %.

AUXILIARY INFORMATION

METHOD/APPARATUS/PROCEDURE:	SOURCE AND PURITY OF MATERIALS:
The amalgams were prepared by controlled electrolytic deposition of copper into a weighed amount of mercury, and the amalgams were used to determine the EMF of the cell, $Cu(Hg)_x	CuSO_4, MgSO_4	Hg_2SO_4, Hg$. The solubility was determined from the breakpoint in the plot of EMF against the logarithm of copper concentration in the amalgam.	Nothing specified.
	ESTIMATED ERROR: Soly: precision better than ± 7%. Temp: not specified.
	REFERENCES:

COMPONENTS:	ORIGINAL MEASUREMENTS:
(1) Copper; Cu; [7440-50-8] (2) Mercury; Hg; [7439-97-6]	Sasim, D.; Srudka, M.; Guminski, C. *Monatsh. Chem.* 1984, *115*, 45-56.
VARIABLES: One temperature: 298 K	PREPARED BY: C. Guminski; Z. Galus

EXPERIMENTAL VALUES:

The solubility of copper at 298 K was reported to be $(1.1 \pm 0.1) \times 10^{-2}$ at %.

AUXILIARY INFORMATION

METHOD/APPARATUS/PROCEDURE:	SOURCE AND PURITY OF MATERIALS:
The experiments were performed with the use of a hanging mercury-drop electrode in a solution of 0.10 mol dm^{-3} CuSO$_4$ at pH = 2. Controlled amounts of Cu were introduced into the electrode by electrolysis at constant current, and the concentration of Cu was varied over a range of 7.7×10^{-4} - 2.3×10^{-2} mol dm^{-3}. Potentials of the electrodes were recorded for 1000 s after the electrolysis; the potentials were practically constant after 600 s. A breakpoint on the curve relating potential to logarithm of Cu concentration corresponds to the saturation of the amalgam. A +6 mV correction was applied to the potentials of the heterogeneous amalgams because of very slow attainment of true equilibrium in the system [1,2].	Mercury was purified with acidified solution of Hg$_2$(NO$_3$)$_2$ and then twice distilled. All reagents were analytically pure (Ciech) and solutions were prepared with triply-distilled water. The solution of CuSO$_4$ was cathodically electrolyzed.
	ESTIMATED ERROR: Soly: precision $\pm$ 10%. Temp: precision $\pm$ 0.2 K.
	REFERENCES: 1. Chao, F.; Costa, M.; *C.R. Acad. Sci., Ser. 2* 1965, *261*, 990; *Bull. Soc. Chim. Fr.* 1968, 549. 2. Hurlen, T.; Staurset, A.; Eriksrud, E. *J. Electroanal. Chem.*, 1977, *83*, 263.

COMPONENTS:	ORIGINAL MEASUREMENTS:
(1) Copper; Cu; [7440-50-8] (2) Mercury; Hg; [7439-97-6]	Grønlund, F.; Kristensen, B. *Acta Chem. Scand., Ser. A* 1984, *38*, 229-32.
VARIABLES: Temperature: 10-25°C	PREPARED BY: C. Guminski; Z. Galus

EXPERIMENTAL VALUES:

The solubility of Cu in Hg:

t/°C	Soly/10^4 mol kg^{-1}	Stand. Dev.	Soly/10^3 at %[a]
10.2	3.72	0.05	7.46
15.8	5.00	0.10	10.0
22.1	6.08	0.07	12.2
25.1	6.95	0.05	13.9

[a] by compilers

The measurements are of high precision but it is not clear whether equilibrium was reached in the time span of the experiments (compilers).

AUXILIARY INFORMATION

METHOD/APPARATUS/PROCEDURE:	SOURCE AND PURITY OF MATERIALS:
The Cu amalgams were obtained by coulometric addition of Cu to the measuring Hg electrode. Cu anode served as a Cu source. Potentials (E) of the cell, $Cu(Hg)_{sat.} \vert CuSO_4 \vert Cu(Hg)_x$, were monitored during 6 to 12 h. The electrolyte contained 0.7 mol dm^{-3} of $CuSO_4$ at pH = 2. The results were placed on a plot of E vs. log N_{Cu}. A line with the Nernstian slope was fitted numerically to the experimental points; its intersection with E = 0 gives the saturated concentration. The measurements were performed under vacuum.	99.99999% pure Hg from Mercure-Industrie; 99.999% pure Cu from ASARCO; $CuSO_4$ analytically pure from Merck; H_2SO_4 analytically pure from BDH, and low-conductivity H_2O were used.
	ESTIMATED ERROR: Standard deviation is lower than 2% but see the comments. Temp: probably ± 0.1 K (compilers).
	REFERENCES:

COMPONENTS:	EVALUATOR:
(1) Silver; Ag; [7440-22-4] (2) Mercury; Hg; [7439-97-6]	C. Guminski; Z. Galus Department of Chemistry University of Warsaw Warsaw, Poland July, 1985

CRITICAL EVALUATION:

The solubility of silver in mercury is rather low near room temperature. Gouy (1), by using a filtration method, was the first to report a measured solubility of 0.06 at % at 288-291 K. Humphreys (2) determined a solubility of 0.086 at % at 301 K while studying the diffusion of silver into mercury. Reinders (3) electrolytically saturated the amalgam with silver, and determined a solubility of 0.076 at % at 298 K; the latter solubility appears to be an acceptable determination. The determination by Strachan and Harris (4), 7.2×10^{-2} at % Ag at room temperature, is in good agreement with that of Reinders. Kozin (5) predicted a solubility of 4.3×10^{-2} at % at 298 K.

Several authors determined the solubility of silver in mercury over a range of temperatures. Joyner (6) reported that the silver concentration in the saturated amalgam varied from 0.07 to 1.13 at %, respectively, over the equilibration temperature range of 287 to 436 K. The results agree with precise measurements reported more recently by others.

Very careful determinations of silver solubility in mercury were made over a temperature range of 278 to 486 K in the same laboratory by Sunier and Hess (7), DeRight (8) and Maurer (9). These authors equilibrated and filtered the liquid amalgam at the various temperatures, and chemically analyzed the amalgams to determine the solubility. Smoothing equations were fitted to the data in all of the measurements.

Murphy (10) determined the liquidus over the temperature range of 651 to 1201 K from thermoanalysis, and he utilized previously published data to draw a complete phase diagram. Tammann and Strassfurth (11) earlier reported a phase diagram based on their thermal analyses and potentiometry, but their data do not agree with the accepted phase diagram (12) shown in Fig. 1.

Hudson (13a) equilibrated known weights of silver and mercury at temperatures varying from 289 to 718 K, and determined the solubility of silver from the loss in weight of the silver which was immersed in the liquid. These results agreed very well with those of refs. (7-9). Hudson combined his data with those of others and derived three equations of the form, $\log N = A - B/(T/K)$, where A and B are constants and N is the at % solubility of silver. These equations were derived for the temperature ranges: 290-603 K, 603-723 K, and 723-1234 K. These equations fitted the experimental solubilities with good agreement.

Jangg and Palman (14) equilibrated silver and mercury at various temperatures between 293 and 823 K, and found that the solubility varied from 0.071 to 44 at % Ag in this temperature range. Although the method of analysis for the solubility determination was not described, the solubilities from the latter work are in good agreement with the earlier reliable measurements (7-9, 13).

Other solubility determinations of silver are rejected in this evaluation because they are too high (15,16) or too low (17).

The kinetics of dissolution of silver in mercury was reported by Hinzner and Stevenson (18).

The saturated amalgams are in equilibrium with intermediate solid phases, and various compounds have been proposed. However, only the ε and γ phases have been confirmed (12,19).

The solubility of silver in saturated tin amalgam was reported by Joyner (6).

(Continued next page)

COMPONENTS:	EVALUATOR:
(1) Silver; Ag; [7440-22-4] (2) Mercury; Hg; [7439-97-6]	C. Guminski; Z. Galus Department of Chemistry University of Warsaw Warsaw, Poland July, 1985

CRITICAL EVALUATION: (continued)

Recommended (r) and tentative values of the solubility of silver in mercury:

T/K	Soly/at %	Reference
273	3.5×10^{-2} [a]	[9]
293.2	6.5×10^{-2} (r)[b]	[8,9,14]
298.2	7.6×10^{-2} (r)[b]	[2,3,8]
323	0.15 (r)	[8,14]
373	0.42 (r)	[7,13,13a,14]
473	1.8 (r)	[7,14]
573	5.1 (r)	[13,13a,14]
673	19[c]	[13,13a,14]
773	39	[14]
873	49[b]	[10,14]
973	61[b]	[10]
1073	76	[10]
1173	91[b]	[10]

[a] Extrapolated value from data of cited references.
[b] Interpolated value from data of cited references.
[c] Mean value from cited references.

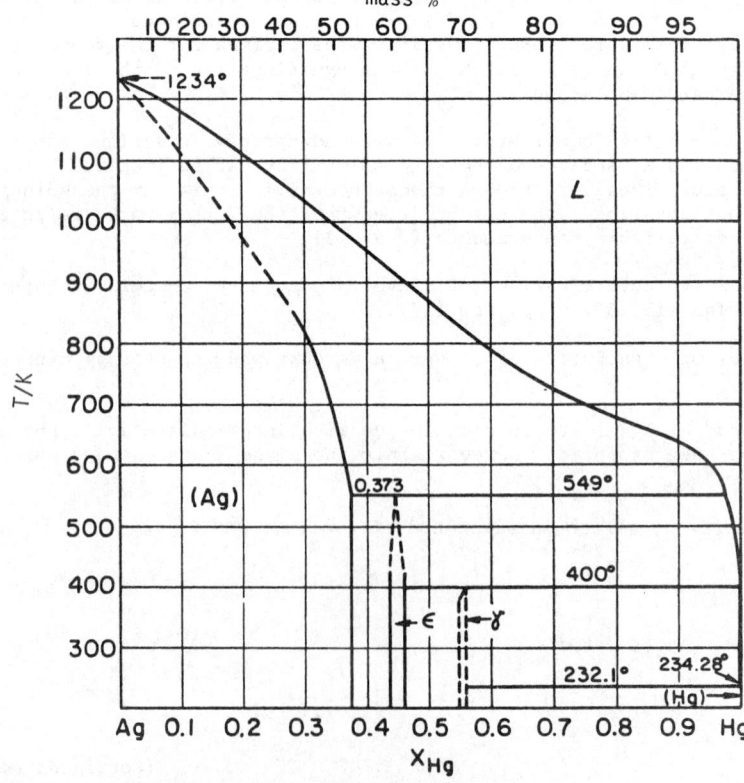

Fig. 1. Silver-Mercury System (12).

COMPONENTS:	EVALUATOR:
(1) Silver; Ag; [7440-22-4] (2) Mercury; Hg; [7439-97-6]	C. Guminski; Z. Galus Department of Chemistry University of Warsaw Warsaw, Poland July, 1985

CRITICAL EVALUATION:

References

1. Gouy, M. *J. Phys.* 1895, *4*, 320.
2. Humphreys, W.J. *J. Chem. Soc.* 1896, 243.
3. Reinders, W. *Z. Phys. Chem.* 1906, *54*, 609.
4. Strachan, J.F.; Harris, N.L. *J. Inst. Metals* 1956-57, *85*, 17.
5. Kozin, L.F. *Fiziko-Khimicheskie Osnovy Amalgamnoi Metallurgii*, Nauka, Alma-Ata, 1964.
6. Joyner, R.A. *J. Chem. Soc.* 1911, 195.
7. Sunier, A.A.; Hess, C.B. *J. Am. Chem. Soc.* 1928, *50*, 662.
8. DeRight, R. *J. Phys. Chem.* 1933, *37*, 405.
9. Maurer, R.J. *J. Phys. Chem.* 1938, *42*, 515.
10. Murphy, A.J. *J. Inst. Metals* 1931, *46*, 507, 522.
11. Tammann, G.; Strassfurth, T. *Z. Anorg. Chem.* 1925, *143*, 357.
12. Hultgren, R.; Desai, P.D.; Hawkins, D.T.; Gleiser, M.; Kelley, K.K. *Selected Values of the Thermodynamic Properties of Binary Alloys*, Am. Soc. Metals, Metals Park, OH, 1973, p. 62.
13. Hudson, D.R. *J. Phys. Chem.* 1945, *49*, 483.
13a. Hudson, D.R. *Metallurgia* 1943, *28*, 203.
14. Jangg, G.; Palman, H. *Z. Metallk.* 1963, *54*, 364.
15. Ogg, A. *Z. Phys. Chem.* 1898, *27*, 285.
16. Maey, E. *Z. Phys. Chem.* 1905, *50*, 200.
17. Ostapczuk, P.; Kublik, Z. *J. Electroanal. Chem.* 1977, *83*, 1.
18. Hinzner, F.W.; Stevenson, D.A. *J. Phys. Chem.* 1963, *67*, 2424.
19. Trebukhov, A.A., as cited by Kozin, L.F.; Nigmetova, R.Sh.; Dergacheva, M.B. *Termodinamika Binarnykh Amalgamnykh Sistem*, Nauka, Alma-Ata, 1977, p. 136.

COMPONENTS:	ORIGINAL MEASUREMENTS:
(1) Silver; Ag; [7440-22-4] (2) Mercury; Hg; [7439-97-6]	Humphreys, W.J. *J. Chem. Soc.* 1896, 243-53.
VARIABLES: Temperature: 26-28°C	PREPARED BY: C. Guminski; Z. Galus

EXPERIMENTAL VALUES:

Solubility of silver in mercury at 26.4 and 28.2°C was reported to be 0.043 ± 0.002 and 0.046 ± 0.002 mass %, respectively. The corresponding atomic % solubilities calculated by the compilers are 0.080 ± 0.004 and 0.086 ± 0.004 at %, respectively.

AUXILIARY INFORMATION

METHOD/APPARATUS/PROCEDURE:	SOURCE AND PURITY OF MATERIALS:
Disc of Ag was placed on the surface of a column of Hg contained in a glass or a wooden vessel, and the liquid was sampled for analysis after 10 days. Silver was determined as the metal by evaporating the Hg from the known weight of the amalgam.	Nothing specified.
	ESTIMATED ERROR: Soly: precision better than ± 5%. Temp: nothing specified.
	REFERENCES:

COMPONENTS:	ORIGINAL MEASUREMENTS:
(1) Silver; Ag; [7440-22-4] (2) Mercury; Hg; [7439-97-6]	Reinders, W. *Z. Phys. Chem.* 1906, *54*, 609-27.
VARIABLES: One temperature: 25°C	PREPARED BY: C. Guminski; Z. Galus

EXPERIMENTAL VALUES:

The solubility of silver in mercury at 25°C was reported to be 7.6×10^{-2} at %.

AUXILIARY INFORMATION

METHOD/APPARATUS/PROCEDURE:	SOURCE AND PURITY OF MATERIALS:
Aqueous solutions of $AgNO_3$ and $Hg_2(NO_3)_2$ were shaken with metallic silver in a thermostat. The cell was then opened and the potential difference between pure silver and the metallic residue was determined. Both the metallic residue and the solution were analyzed; the amalgams were analyzed gravimetrically after mercury was distilled off.	Nothing specified.
	ESTIMATED ERROR: Nothing specified.
	REFERENCES:

COMPONENTS:	ORIGINAL MEASUREMENTS:
(1) Silver; Ag; [7440-22-4] (2) Mercury; Hg; [7439-97-6]	Joyner, R.A. *J. Am. Chem. Soc.* 1928, *50*, 662-68.
VARIABLES: Temperature: 287-403 K	PREPARED BY: C. Guminski; Z. Galus

EXPERIMENTAL VALUES:

Solubility of silver in mercury:

T/K	Soly/at %
287	0.07
298	0.082
303	0.086
336	0.19
363	0.34
403	1.13

The results obtained at 287 and 403 K are too high; the other values agree with most precise reported works.

AUXILIARY INFORMATION

METHOD/APPARATUS/PROCEDURE:	SOURCE AND PURITY OF MATERIALS:
Ag was carefully purified by chemical treatments, and filings of this metal were equilibrated with excess Hg in sealed, hydrogen-filled tubes at different temperatures. After opening the tubes, the amalgams were pipetted through a glass wool plug and dissolved in HNO_3. The solution was then treated with NH_4Cl to precipitate AgCl. The precipitate was redissolved in ammonium hydroxide, then the AgCl was reprecipitated by acidifying the solution with HNO_3. The AgCl was then estimated in the "usual way".	Hg purity was not specified. Ag was dissolved in HNO_3, then precipitated as AgCl. The AgCl was dissolved in NH_4OH and reprecipitated as AgCl after filtration of the ammoniacal solution. The AgCl was then fused with Na_2CO_3, and the molten Ag was successively treated with KNO_3, NH_4Cl, and borax, with a bone ash support being employed.
	ESTIMATED ERROR: Soly: nothing specified. Temp: nothing specified.
	REFERENCES:

COMPONENTS:	ORIGINAL MEASUREMENTS:
(1) Silver; Ag; [7440-22-4] (2) Mercury; Hg; [7439-97-6]	Sunier, A.A.; Hess, C.B. J. Am. Chem. Soc. 1928, 50, 662-68.
VARIABLES: Temperature: 80-213°C	PREPARED BY: C. Guminski; Z. Galus

EXPERIMENTAL VALUES:

Solubility of silver in mercury:

$t/°C$	Soly/at %	Ave. Deviation x 1000	$t/°C$	Soly/at %[a]
80.2	0.286	3.5	181.8	1.365
98.2	0.411	4.9	193.3	1.573
121.9	0.612	1.6	212.7	1.953
144.5	0.849	1.2		
160.6	1.057	7.6		
177.9	1.346	2.2		
198.9	1.746	5.2		

[a] Unpublished data of G. H. Reed from the same laboratory.

AUXILIARY INFORMATION

METHOD/APPARATUS/PROCEDURE:	SOURCE AND PURITY OF MATERIALS:
An excess of Ag was equilibrated with Hg in one of the bulbs of a Pyrex apparatus which was immersed and shaken in a thermostated bath. The amalgam was then filtered through an integral capillary by inverting the apparatus in the bath. The filtrate was analyzed by distilling the Hg, dissolving the Ag with HNO_3, then gravimetrically determining the Ag as the chloride.	Ag was "999 fine" from the U.S. Mint and "1000 fine foil" from the Philadelphia Mint. Mercury was first washed by dropping through a column of $Hg_2(NO_3)_2$, then triply distilling after drying.
	ESTIMATED ERROR: Soly: accuracy better than ± 0.7%. Temp: precision ± 0.1 K.
	REFERENCES:

COMPONENTS:	ORIGINAL MEASUREMENTS:
(1) Silver; Ag; [7440-22-4] (2) Mercury; Hg; [7439-97-6]	Murphy, A.J. J. Inst. Metals 1931, 46, 507-22.
VARIABLES: Temperature: 378-928°C	PREPARED BY: C. Guminski; Z. Galus

EXPERIMENTAL VALUES:

Temperatures of crystallization of silver amalgams:

	Silver Content	
t/°C	mass %	at %[a]
928	88.97	93.75
886	80.08	88.2
843	69.6	81.0
786	60.23	73.8
721	49.96	65.0
630	37.72	52.96
541	28.86	43.0
465	20.08	31.84
407	10.03	17.17
378	5.00	8.91

[a] by compilers

AUXILIARY INFORMATION

METHOD/APPARATUS/PROCEDURE:	SOURCE AND PURITY OF MATERIALS:
Amalgams were prepared by mixing the required amounts of Hg and precipitated Ag in a shaking apparatus. The specimens were transferred into silica tubes which were sealed and contained in a pressurized bomb for the high temperature measurements. Cooling curve temperatures were measured with a Chromel-Alumel thermocouple. Analytical method for the amalgam analysis was not specified.	Chemically precipitated silver was better than 99.9% pure. High purity mercury was redistilled.
	ESTIMATED ERROR: Temp: nothing specified. Amalgam composition: accuracy better than ± 0.2%.
	REFERENCES:

COMPONENTS:	ORIGINAL MEASUREMENTS:
(1) Silver; Ag; [7440-22-4] (2) Mercury; Hg; [7439-97-6]	DeRight, R. *J. Phys. Chem.* 1933, *37*, 405-16.

VARIABLES:	PREPARED BY:
Temperature: 9-81°C	C. Guminski; Z. Galus

EXPERIMENTAL VALUES:

Solubility of silver in mercury:

$t/°C$	Soly/at %	% Ave. Deviation
8.92	0.0641	6.4
18.17	0.0643	1.0
19.01	0.0636	0.6
25.28	0.0766	2.1
25.60	0.0792	1.4
29.93	0.0881	1.3
30.15	0.0965	9.1
40.11	0.1139	0.5
50.02	0.1450	0.4
60.26	0.1901	2.7
70.54	0.2404	3.2
80.94	0.2892	0.4

Ag_3Hg_4 was reported to be in equilibrium with the saturated amalgam.

AUXILIARY INFORMATION

METHOD/APPARATUS/PROCEDURE:	SOURCE AND PURITY OF MATERIALS:
The solubility apparatus consisted of two Pyrex bulbs connected by a capillary filter. Hg and excess Ag were sealed in one bulb under a pressure of hydrogen, and the system was equilibrated in a thermostat. The amalgam was then filtered through the capillary and the Ag analyzed gravimetrically as the metal after evaporation of the Hg.	Mercury was purified by dropping through a column of 6 mol dm^{-3} HNO$_3$, washed with H$_2$O, dried and distilled. No residue upon evaporation of this Hg. Silver was "1000 fine foil" from the Philadelphia Mint.
	ESTIMATED ERROR: Soly: ± 1% ave. deviation from mean, except ± 9% at 303 K. Temp: precision ± 0.02 K.
	REFERENCES:

COMPONENTS:	ORIGINAL MEASUREMENTS:
(1) Silver; Ag; [7440-22-4] (2) Mercury; Hg; [7439-97-6]	Maurer, R.J. *J. Phys. Chem.* 1938, *42*, 515-19.
VARIABLES: Temperature: 5-19°C	PREPARED BY: C. Guminski; Z. Galus

EXPERIMENTAL VALUES:

Solubility of silver in mercury:

t/°C	Soly/at % x 100	Ave. dev. from mean x 1000
5.72	4.03	1.3
9.71	4.74	1.3
12.39	5.19	1.5
16.12	5.86	4.1
18.98	6.25	1.4
19.24	6.52	7.7

AUXILIARY INFORMATION

METHOD/APPARATUS/PROCEDURE:	SOURCE AND PURITY OF MATERIALS:
The amalgams were prepared by dissolution of silver in mercury and equilibrated in a thermostat. The amalgams were filtered through a G-1 Schott-Jena filter, then analyzed gravimetrically after evaporation of mercury.	Silver was "1000 fine foil" from the Philadelphia Mint. Mercury was purified by dropping into HNO_3, then washed, dried and triply distilled.
	ESTIMATED ERROR: Soly: precision better than 0.6%. Temp: precision ± 0.02 K.
	REFERENCES:

COMPONENTS:	ORIGINAL MEASUREMENTS:

(1) Silver; Ag; [7440-22-4]
(2) Mercury; Hg; [7439-97-6]
	1. Hudson, D.R.
	   J. Phys. Chem. 1945, *49*, 483-506.
	2. Same author
	   Metallurgia 1943, *28*, 203-6.

VARIABLES: Temperature: 289-718 K	PREPARED BY: C. Guminski; Z. Galus

EXPERIMENTAL VALUES:

Solubility of silver in mercury:

T/K	Soly	
	g Ag/100 g Hg	at %
289.4	0.030	0.0558
372.8	0.222	0.4121
457.6	0.768	1.4192
533.2	1.885	3.450
579.2	2.823	5.251
611.2	3.816	6.872
629.9	5.22	9.294
678.2	10.59	18.053
717.7	17.35	28.081

AUXILIARY INFORMATION

METHOD/APPARATUS/PROCEDURE:	SOURCE AND PURITY OF MATERIALS:
Solubilities were determined by equilibrating a cube of Ag with Hg in a tube of refractory glass. The sealed tube with the known amounts of the metals was suspended in a constant temperature, vapor-bath for various periods. Knowing the total weight, the subsequent analyses were made by determining the loss in weight of the solid Ag core after Hg was removed by evaporation from the surface of the Ag.	Silver: 99.95% pure. Mercury: "analytical reagent" grade, 99.998% pure.
	ESTIMATED ERROR: Soly: not specified; precision better than ± 1% (compilers). Temp: precision ± 0.25 K.
	REFERENCES:

COMPONENTS:	ORIGINAL MEASUREMENTS:
(1) Silver; Ag; [7440-22-4] (2) Mercury; Hg; [7439-97-6]	Jangg, G.; Palman, H. Z. Metallk. 1963, 54, 364-69.
VARIABLES: Temperature: 293-823 K	PREPARED BY: C. Guminski; Z. Galus

EXPERIMENTAL VALUES:

Solubility of silver in mercury was presented graphically as a function of temperature. The mass % solubilities were read off the curve and converted to atomic % by the compilers.

T/K	Soly/at %
293	0.071
323	0.16
373	0.41
423	0.91
473	1.8
523	3.1
548	4.4
573	5.1
623	12
673	20
723	29
773	39
823	44

AUXILIARY INFORMATION

METHOD/APPARATUS/PROCEDURE:	SOURCE AND PURITY OF MATERIALS:
Method of the amalgam preparation was not specified. The amalgams were shaken and kept for 12 hours in thermostated glass cylinders and subsequently filtered under pure nitrogen pressure. For temperatures above 600 K a pressure apparatus of hard chromium steel was used. The method of analysis of the amalgam was not specified.	Nothing specified.
	ESTIMATED ERROR: Soly: precision ± 5%. Temp: nothing specified.
	REFERENCES:

COMPONENTS:	EVALUATOR:
(1) Gold; Au; [7440-57-5] (2) Mercury; Hg; [7439-97-6]	C. Guminski; A. Galus Department of Chemistry University of Warsaw Warsaw, Poland July, 1985

CRITICAL EVALUATION:

Because of the ready wetting of gold by mercury, there has been a mistaken belief by many scientists that gold has a relatively high solubility in mercury at room temperature. Contrary to this belief, it was shown as early as 1855, by Henry (1), that the solubility of gold in mercury was approximately 0.14 at %, presumably at room temperature. Kazantsev (2), in 1878, employed a filtration method and reported solubilities of 0.112, 0.128, and 0.662 at % at 273, 293 and 373 K, respectively; the results at 293 and 373 K are in remarkably good agreement with more precise measurements reported approximately fifty years later. Gouy (3) also reported a solubility of approximately 0.13 at % at 288 to 291 K. More recently, Strachan and Harris (4) equilibrated the two metals at room temperature and determined a solubility of 0.128 at %, while Kozin and coworkers (5) utilized a capillary phase separation technique and determined a solubility of 0.135 at % at 295 K; these solubilities near room temperature are in good agreement with the most accurate measurements of Sunier and White (see below).

Tammann (6) determined that the freezing point of mercury is elevated by 0.1, 0.1 and 0.2 K upon dissolution of 6×10^{-3}, 1.2×10^{-2} and 2.5×10^{-2} at %, respectively, of gold.

The most detailed and accurate determinations of the solubility of gold in mercury were made by Sunier and coworkers (7-11). These authors equilibrated the metals in a glass apparatus at various temperatures from 280 to 662 K, then the liquid phase was separated at equilibration temperatures by filtration through a capillary which was constructed into the apparatus. The filtrate was then chemically analyzed to determine the solubility. A total of nearly three hundred data points was obtained in this series of papers, and the data were fitted with a smooth curve to form the liquidus. Sunier and White (8) fitted a smoothing equation to their data at 280 to 357 K, and obtained a solubility of 0.1306 at % at 293.2 K.

From vapor pressure measurements of gold amalgams, Eastman and Hildebrand (12) reported a solubility of 16.5 at % gold in mercury at 590 K. This solubility is in good agreement with Anderson's data (11). Parravano (13), from freezing point determinations, reported the solubility of gold in mercury at 353 to 598 K, but his solubilities are in agreement with the more accurate measurements (7-11) only for temperatures above 553 K; at lower temperatures, probably because of supercooling, the solubilities were as much as two times higher than those of Sunier et al. (7,8). Britton and McBain (14) determined the solubility of gold at 291 to 683 K by equilibrating the metals at various temperatures, then separating the solid phase by filtration through a sintered glass filter, followed by chemical analysis of the filtrate. These authors reported solubilities of 0.212 to 55.33 at % over their temperature range; the solubilities in the lower temperature region are too high compared to those of Sunier et al. (7,8). Plaksin (15), by employing thermoanalysis, determined the liquidus of gold amalgams from 395 to 733 K; the solubilities at temperatures below 500 K were higher than those determined by chemical analysis (7-9). Rolfe and Hume-Rothery (16) determined the liquidus from 402 to 1324 K from measurements of cooling curves, and the data were used to construct a complete phase diagram of the Au-Hg system. The data of the latter authors were in general agreement with those of Anderson (11).

Other solubility data have been reported but these are rejected in the evaluation because they are either too high (17-19) or too low (20).

In a brief review, Brown (21) tabulated selected values of the solubility of gold in mercury in the lower temperature range.

The phase diagram (22) for the Au-Hg system is shown in Fig. 1. The identification of the following compounds has been made: Au_4Hg (16), Au_3Hg (11,13,15-17,23) and Au_2Hg (15-17). Other compounds suggested for this system are Au_5Hg (23), Au_2Hg_3 (11,13,23,24), Au_2Hg_5 (17), $AuHg_2$ (15,17,23,24), $AuHg_4$ (17) and $AuHg_6$ (14).

(Continued next page)

COMPONENTS:	EVALUATOR:
(1) Gold; Au; [7440-57-5] (2) Mercury; Hg; [7439-97-6]	C. Guminski; A. Galus Department of Chemistry University of Warsaw Warsaw, Poland July, 1985

CRITICAL EVALUATION: (Continued)

Recommended (r) and tentative values for the solubility of gold in mercury:

T/K	Soly/at %		References
273	0.08[a]		[8]
293.2	0.13	(r)	[2,4,5,8]
298.2	0.14[b]		[5,8]
323	0.25		[8]
373	0.68	(r)	[2,7,14]
473	3.0	(r)[c]	[7,9,10,14,16]
573	14	(r)[c]	[9,10,16]
673	44[b]		[16]
773	54[b]		[16]
873	62[b]		[16]
973	71[b]		[16]
1073	77		[16]
1173	82		[16]
1273	92		[16]

[a] Extrapolated from data of [8].
[b] Interpolated value from data of cited references.
[c] Mean value of data from cited references.

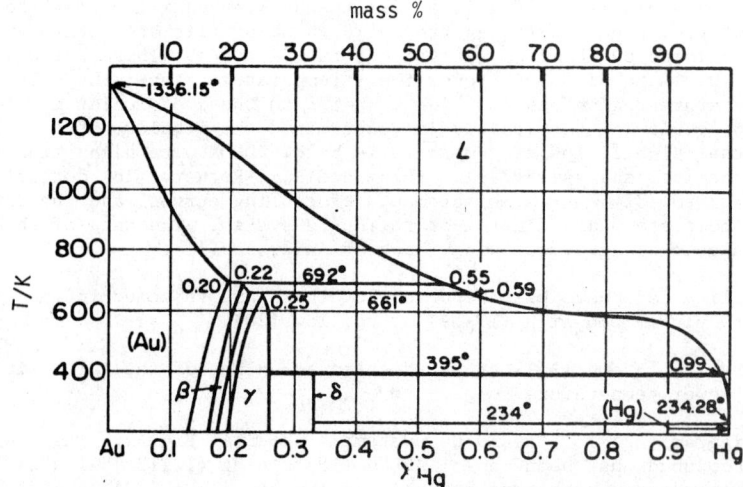

Fig. 1. The Au-Hg System (22).

(Continued next page)

COMPONENTS:	EVALUATOR:
(1) Gold; Au; [7440-57-5] (2) Mercury; Hg; [7439-97-6]	C. Guminski; A. Galus Department of Chemistry University of Warsaw Warsaw, Poland July, 1985

CRITICAL EVALUATION: (Continued)

REFERENCES

1. Henry, T.H. *Phil. Mag.*, Ser. 4, 1855, *9*, 458; cited by [21].
2. Kazantsev, M. *Zh. Russ. Fiz. Khim. Obshch.* 1878, *10*, 233; *Bull. Soc. Chim. Fr.* 1878, *30*, 20; *Ber.* 1878, *11*, 1255; *Brit. Abstr.* 1878, 937.
3. Gouy, M. *J. Phys.* 1895, *4*, 320.
4. Strachan, N.F.; Harris, N.L. *J. Inst. Metals* 1956-57, *85*, 17.
5. Kozin, L.F.; Dergacheva, M.B.; Nikushkina, N.L. *Tr. Inst. Khim. Nauk Akad. Nauk Kaz. SSR* 1976, *42*, 82.
6. Tammann, G. *Z. Phys. Chem.* 1889, *3*, 441.
7. Sunier, A.A.; Gramkee, B.E. *J. Am. Chem. Soc.* 1929, *51*, 1703.
8. Sunier, A.A.; White, C.M. *J. Am. Chem. Soc.* 1930, *52*, 1842.
9. Sunier, A.A.; Weiner, L.G. *J. Am. Chem. Soc.* 1931, *53*, 1714.
10. Mees, G. *J. Am. Chem. Soc.* 1938, *60*, 870.
11. Anderson, J.T. *J. Phys. Chem.* 1932, *36*, 2145.
12. Eastman, E.D.; Hildebrand, J.H. *J. Am. Chem. Soc.* 1914, *36*, 2020.
13. Parravano, N. *Gazz. Chim. Ital.* 1918, *48*, 123.
14. Britton, G.T.; McBain, J.W. *J. Am. Chem. Soc.* 1926, *48*, 593.
15. Plaksin, I.N. *Izv. Sekt. Fiz. Khim. Anal.* 1938, *10*, 129.
15a. Plaksin, I.N. *Zh. Russ. Fiz. Khim. Obshch., Ser. Khim.* 1929, *61*, 521; abstracted in *Z. Metallk.* 1932, *24*, 56, 89; *J. Inst. Metals* 1928, *39*, 498; 1929, *41*, 459.
16. Rolfe, C.; Hume-Rothery, W. *J. Less-Common Metals* 1967, *13*, 1.
17. Braley, S.A.; Schneider, R.F. *J. Am. Chem. Soc.* 1921, *43*, 740.
18. Loomis, A.G. *J. Am. Chem. Soc.* 1922, *44*, 8.
19. Williams, T.C.; Evans, E.J. *Phil. Mag.*, Ser. 7, 1928, *6*, 231.
20. Kozin, L.F. *Fiziko-Khimicheskie Osnovy Amalgamnoi Metallurgii*, Nauka, Alma-Ata, 1964.
21. Brown, J.B. *J. Chem. Educ.* 1960, *37*, 415.
22. Hultgren, R.; Desai, P.D.; Hawkins, D.T.; Gleiser, M.; Kelley, K.K. *Selected Values of the Thermodynamic Properties of Binary Alloys*, Am. Soc. Metals, Metals Park, OH, 1973, p. 279.
23. Aramian, K.T.; Shklar, R.Sh.; Kakovskii, I.A. *Izv. Vyssh. Ucheb. Zaved., Tsvet. Met.* 1972, *15*, No. 1, 57.
24. Winterhager, H.; Schlösser, W. *Metall* 1960, *14*, 1.

COMPONENTS:	ORIGINAL MEASUREMENTS:
(1) Gold; Au; [7440-57-5] (2) Mercury; Hg; [7439-97-6]	Kazantsev, M. *Zh. Russ. Fiz. Khim. Obshch.* 1878, *10*, 233-5.
VARIABLES:	PREPARED BY:
Temperature: 0-100°C	C. Guminski; Z. Galus

EXPERIMENTAL VALUES:

Solubility of gold in mercury:

t/°C	Soly/mass %	Soly/at %[a]
0	0.110	0.112
20	0.126	0.128
100	0.650	0.662

[a] by compilers.

These data were also reported in (1-3).

AUXILIARY INFORMATION

METHOD/APPARATUS/PROCEDURE:	SOURCE AND PURITY OF MATERIALS:
Mercury was saturated with gold by contact of the metals for minimum of 1 hour. The amalgams containing excess of gold were squeezed through a chamois leather or a capillary of 0.15-0.40 mm dia. The filtrate was analyzed by an unspecified method after amalgam was dissolved in nitric acid. Rate of filtration, pressure applied and source of gold had no influence on the solubility.	Nothing specified.
	ESTIMATED ERROR:
	Nothing specified.
	REFERENCES: 1. Kazantsev, M. *Bull. Soc. Chim. Fr.* 1878, *30*, 20. 2. Same. *Ber.* 1878, *11*, 1255. 3. Same. *Brit. Abstr.* 1878, 937.

COMPONENTS:	ORIGINAL MEASUREMENTS:
(1) Gold; Au; [7440-57-5] (2) Mercury; Hg; [7439-97-6]	Tammann, G. Z. Phys. Chem. 1889, 3, 441-9.
VARIABLES: Temperature	PREPARED BY: C. Guminski; Z. Galus

EXPERIMENTAL VALUES:

Elevation of the melting point of mercury, ΔT/K, upon addition of small amounts of gold:

g Au/100 g Hg	Soly/at %[a]	ΔT/K
0.006	0.006	0.1
0.012	0.012	0.1
0.025	0.025	0.2

[a] by compilers

AUXILIARY INFORMATION

METHOD/APPARATUS/PROCEDURE:	SOURCE AND PURITY OF MATERIALS:
The melting temperatures of the amalgams were measured, probably with the use of a thermometer. No further details are given. The melting temperature of mercury was reported to be 244 instead of 234 K, but one may assume that the experimental ΔT values are correct.	Nothing specified.
	ESTIMATED ERROR: Soly: nothing specified. Temp: precision $\pm$ 0.05 K.
	REFERENCES:

COMPONENTS:	ORIGINAL MEASUREMENTS:
(1) Gold; Au; [7440-57-5] (2) Mercury; Hg; [7439-97-6]	Eastman, E.D.; Hildebrand, J.H. *J. Am. Chem. Soc.* 1914, *36*, 2020-30.
VARIABLES:	PREPARED BY:
One temperature: 317°C	C. Guminski; Z. Galus

EXPERIMENTAL VALUES:

The solubility of gold in mercury at 317°C was reported to be 16.5 at %.

AUXILIARY INFORMATION

METHOD/APPARATUS/PROCEDURE:	SOURCE AND PURITY OF MATERIALS:
The metals were introduced into U-tubes in the desired proportions then they were heated and outgassed by boiling. The tubes, which were connected to a Hg manometer, were agitated in a thermostat. The vapor pressure of the amalgams was measured manometrically. The vapor pressure of pure Hg was concurrently determined in an identical apparatus, with the Hg tube immersed in the same thermostat. Temperature of sample was determined from the measured vapor pressure of Hg by correlating the pressure to the vapor pressure equation determined by earlier workers. The breakpoint in the relationship of amalgam vapor pressure to composition gave the solubility of gold.	Mercury was carefully purified by washing with dilute nitric acid, then distilled in a current of air. Gold stated to be purified by "usual methods".
	ESTIMATED ERROR: Soly: precision better than ± 1% (compilers). Temp: precision ± 2 K.
	REFERENCES:

COMPONENTS:	ORIGINAL MEASUREMENTS:
(1) Gold; Au; [7440-57-5] (2) Mercury; Hg; [7439-97-6]	Parravano, N. *Gazz. Chim. Ital.* 1918, *48*, 123-38.
VARIABLES: Temperature: 113-312°C	PREPARED BY: C. Guminski; Z. Galus

EXPERIMENTAL VALUES:

Solubility of gold in mercury:

t/°C	Soly/at %[a]	Soly/mass %	t/°C	Soly/at %[a]	Soly/mass %
113	1.63	1.60	305	17.47	17.21
168	3.63	3.57	308	18.07	17.80
220	6.36	6.25	306	18.93	18.65
270	10.16	10.00	308	19.32	19.04
288	11.12	10.94	310	19.92	19.63
293	13.74	13.53	312	25.55	25.20
302	16.71	16.46			

[a] by compilers

These solubilities are generally high compared to those reported by other workers, but the data at temperatures above 293°C are nearly the same as those for more precise measurements.

AUXILIARY INFORMATION

METHOD/APPARATUS/PROCEDURE:	SOURCE AND PURITY OF MATERIALS:
The amalgams were prepared by dissolution of gold in mercury accompanied by heating. The melting points of the known compositions were then determined.	Gold: 99.9% pure. Mercury purified with HNO_3, washed and distilled under reduced pressure.
	ESTIMATED ERROR: Soly: nothing specified. Temp: nothing specified.
	REFERENCES:

COMPONENTS:	ORIGINAL MEASUREMENTS:
(1) Gold; Au; [7440-57-5] (2) Mercury; Hg; [7439-97-6]	Britton, G.T.; McBain, J.W. *J. Am. Chem. Soc.* 1926, *48*, 593-598.
VARIABLES:	PREPARED BY:
Temperature: 18-410°C	C. Guminski; Z. Galus

EXPERIMENTAL VALUES:

Solubilities of gold in mercury determined with three different apparatus, as indicated by series numbers:

SERIES I		SERIES II			
$t/°C$	at % Au	$t/°C$	at % Au	$t/°C$	at % Au
18	0.212	64	0.379	143	1.591
18	0.287	65.4	0.378	147.5	1.628
47	0.388 ± 0.010	93.0	0.599 ± 0.002	149	1.630
52.5	0.293 ± 0.020	98	0.682 ± 0.006	153	1.785
77	0.538 ± 0.046	105	0.736	155	1.815
80.5	0.623 ± 0.012	106.5	0.753	158	1.920
92	0.721 ± 0.006	114.5	0.948	159	1.929
99.5	0.812 ± 0.073	115.5	0.944	163	2.028
103.5	0.906 ± 0.020	121	1.226 ± 0.010	163.5	2.052
108	0.952 ± 0.036	122.5	1.101	172	2.212
128	1.474 ± 0.014	133.5	1.421	173	2.212
132	1.462	136	1.409	174	2.260
145	1.667 ± 0.011	142.5	1.576	174.5	2.158

(Series II continued next page)

AUXILIARY INFORMATION

METHOD/APPARATUS/PROCEDURE:	SOURCE AND PURITY OF MATERIALS:
Three different apparatus, each compatible for the temperature ranges 18-150, 60-350, and 300-410°C, were used for the solubility measurements. In each case the amalgams were equilibrated in a glass bulb immersed in a thermostated system, and the liquid was drawn off through a capillary or glasswool filter for analysis. In the two higher temperature ranges the amalgams were equilibrated under an atmosphere of hydrogen at pressures up to 4 atm. Amalgams from the highest temperature range were analyzed by evaporation of the Hg in a stream of coal gas at 350°C in a Pyrex tube and weighing the Au residue. Amalgams from the other two ranges were analyzed by reduction of the dissolved $AuCl_3$ by standard $Fe(NH_4)_2(SO_4)_2$.	Chemically pure Au from Johnson-Matthey. Mercury was purified with $Hg_2(NO_3)_2$, then dried and distilled.
	ESTIMATED ERROR:
	Soly: precision better than ± 10%. Temp: nothing specified.
	REFERENCES:

COMPONENTS:	ORIGINAL MEASUREMENTS:
(1) Gold; Au; [7440-57-5] (2) Mercury; Hg; [7439-97-6]	Britton, G.T.; McBain, J.W. *J. Am. Chem. Soc.* 1926, *48*, 593-598.
VARIABLES: Temperature: 18-410°C	PREPARED BY: C. Guminski; Z. Galus

EXPERIMENTAL VALUES:

SERIES II (concluded)				SERIES III			
t/°C	at % Au	t/°C	at % Au	t/°C	at % Au	t/°C	at % Au
186.2	2.616	247	5.300	280	6.89	366	34.48
189.5	2.746	258.5	6.585	289	8.30	370	39.48
192.5	2.615	259	6.036	321	10.67	378	42.99
201.5	3.137	259.5	6.725	327	20.49	386	42.96
206	3.302	260	6.753	327	23.27	398	52.00
207	3.265	288	8.542	331	18.32	410.5	55.33
206.5	3.345	291	8.716	334	18.92		
207.5	3.732			339	16.14		
234	4.559			343.5	27.16		
234.5	4.601			354	32.96		
241	5.225			361	33.32		

Most reliable results are those in Series II (compilers).

AUXILIARY INFORMATION	
METHOD/APPARATUS/PROCEDURE:	SOURCE AND PURITY OF MATERIALS:
	ESTIMATED ERROR:
	REFERENCES:

COMPONENTS:	ORIGINAL MEASUREMENTS:

(1) Gold; Au; [7440-57-5]
(2) Mercury; Hg; [7439-97-6]
	1. Sunier, A.A.; Gramkee, B.E.
	   J. Am. Chem. Soc. 1929, *51*, 1703-8.
	2. Sunier, A.A.; White, C.M.
	   J. Am. Chem. Soc. 1930, *52*, 1842-52.

VARIABLES: Temperature: 6-201°C	PREPARED BY: C. Guminski; Z. Galus

EXPERIMENTAL VALUES:

Solubility of gold in mercury:

		Experimental Soly		Smoothed Soly	
	t/°C	at %	% Ave. Dev.	t/°C	at %
Ref. (1)	80.8	0.467	1.2	80	0.459
	101.2	0.697	0.6	100	0.684
	121.7	1.021	1.2	120	0.996
	142.1	1.482	0.1	140	1.385
	159.2	1.847	0.3	160	1.871
	182.3	2.434	0.2	180	2.380
	201.1	2.875	2.5	200	2.849
Ref. (2)	6.96	0.1006	0.82	0	(0.081)
	20.00	0.1290	0.77	10	0.1038
	29.68	0.1638	0.47	20	0.1306
	39.98	0.2045	0.54	30	0.1629
	49.50	0.2461	0.41	40	0.2014
	60.32	0.3152	0.52	50	0.2489
	70.36	0.3753	0.31	60	0.3076
	80.40	0.4647	0.43	70	0.3767
	69.2	0.375		80	0.4614
	83.8	0.498			

AUXILIARY INFORMATION

METHOD/APPARATUS/PROCEDURE:	SOURCE AND PURITY OF MATERIALS:
The solubility apparatus consisted of two Pyrex bulbs separated by a connecting capillary filter. Hg and Au were introduced into one bulb, evacuated, then sealed and equilibrated in a thermostat. Subsequently, the liquid was drawn through the capillary filter into the empty bulb. The weighed amalgam was analyzed by evaporating the mercury in a stream of hydrogen at temperatures up to 550°C for several hours. Equilibration of amalgams was approached from higher and from lower temperatures.	(1) "Thousand-fine" gold foil from the Philadelphia Mint. Mercury was passed through a column of $Hg_2(NO_3)_2$ then washed and distilled. (2) 99.95% pure gold. Mercury was purified with HNO_3 then distilled several times.
	ESTIMATED ERROR: Soly: precision better than ± 1% in (1) and ± 0.8% in (2). Temp: precision ± 0.1 in (1) and ± 0.02 K in (2).
	REFERENCES:

COMPONENTS:	ORIGINAL MEASUREMENTS:
(1) Gold; Au; [7440-57-5] (2) Mercury; Hg; [7439-97-6]	1. Plaksin, I.N. *Izv. Sekt. Fiz. Khim. Anal.* 1938, *10*, 129-59. 2. Same author *Zh. Russ. Fiz. Khim. Obshch.*, *Ser. Khim.* 1929, *61*, 521-34.
VARIABLES: Temperature: 122-515°C	PREPARED BY: C. Guminski; Z. Galus

EXPERIMENTAL VALUES:

Crystallization temperatures of gold amalgams:

$t/°C$	at % Au	$t/°C$	at % Au
122	1.3	376	40.1
180	3.1	390	44.8
272	9.2	387	45.0
285	10.0	403	50.6
307	12.8	412	53.0
310 (315)	15.2	430	57.0
323	17.0	460	60.2
327	18.2	487	63.0
335	21.0	515	66.8
341	24.0		
351	28.1		
357	32.3		
361	33.8		
369	37.0		

AUXILIARY INFORMATION

METHOD/APPARATUS/PROCEDURE:	SOURCE AND PURITY OF MATERIALS:
The alloys were prepared by mixing the metals in appropriate ratios in sealed, evacuated tubes. The samples were annealed for 10 hours at 300-400°C. The cooling and heating curves were recorded with the use of various thermocouples.	Mercury: chemically pure from Kahlbaum, as well as that which was double-distilled under vacuum. Gold was purified by dissolution in aqua regia, then reduced with oxalic acid or hydrazine; traces of silver were deposited upon treatment with HBr.
	ESTIMATED ERROR: Soly: nothing specified. Temp: precision ± 1 K.
	REFERENCES:

COMPONENTS:	ORIGINAL MEASUREMENTS:
(1) Gold; Au; [7440-57-5] (2) Mercury; Hg; [7439-97-6]	Anderson, J.T. *J. Phys. Chem.* 1932, *36*, 2145-65.
VARIABLES: Temperature: 286-390°C	PREPARED BY: C. Guminski; Z. Galus

EXPERIMENTAL VALUES:
Solubility of gold in mercury:

$t/°C$	Soly/at %	$t/°C$	Soly/at %	$t/°C$	Soly/at %
286.3	10.22	300.2	15.99	321.7	26.97
286.5	12.50	300.4	17.02	327.5	29.02
288.2	13.17	300.7	17.01	328.6	29.04
288.3	12.55	307.2	15.99	334.5	31.00
290.8	13.19	307.2	20.40	351.0	34.26
291.1	13.17	309.4	22.78	352.6	34.26
293.5	14.48	310.2	22.94	373.4	37.96
293.9	14.50	315.2	25.07	374.8	37.96
297.9	16.07	315.4	23.97	386.8	39.94
298.7	14.01	320.7	27.05	388.8	40.25

The presence of some Pb had no influence on the solubility of Au in Hg.

AUXILIARY INFORMATION

METHOD/APPARATUS/PROCEDURE:	SOURCE AND PURITY OF MATERIALS:
Temperature at which the last crystal of the solid phase disappeared was determined in an evacuated Pyrex glass apparatus. The dissolution of the solid at various temperatures was observed as the liquid was passed over the solid which was retained on top of the capillary section of the apparatus. The amalgams were analyzed gravimetrically.	Gold was 99.98% pure. Mercury was purified with HNO_3 then distilled several times.
	ESTIMATED ERROR: Soly: precision no better than ± 1%. Temp: precision ± 0.1 K.
	REFERENCES:

COMPONENTS:	ORIGINAL MEASUREMENTS:
(1) Gold; Au; [7440-57-5] (2) Mercury; Hg; [7439-97-6]	Sunier, A.A.; Weiner, L.G. J. Am. Chem. Soc. 1931, 53, 1714-21.
VARIABLES: Temperature: 200-300°C	PREPARED BY: C. Guminski; Z. Galus

EXPERIMENTAL VALUES:
Solubility of gold in mercury:

t/°C	Soly/at %	% Ave. Dev.
200.0	2.99	1.8
219.6	3.67	2.9
239.2	5.07	1.7
260.2	6.50	2.4
269.6	7.81	3.6
279.6	9.07	3.0
290.6[a]	10.89	-
292.6	12.58	4.5
299.5	13.95	1.5
299.7[a]	14.27	-

[a] Determined by thermal analysis.

AUXILIARY INFORMATION

METHOD/APPARATUS/PROCEDURE:	SOURCE AND PURITY OF MATERIALS:
An excess of Au was mixed with Hg in glass tubes and the latter were sealed after pressurizing with slightly less than an atmosphere of H_2. The tubes were equilibrated in an air bath, then the amalgams filtered through capillaries and analyzed gravimetrically by evaporating off the Hg. The thermal analyses were made by visual observation of disappearance and reappearance of the Au as the known mixture was heated and cooled.	Gold was 99.99% pure. Mercury was purified with HNO_3 then distilled several times.
	ESTIMATED ERROR: Soly: precision better than ± 4%. Temp: precision ± 0.2 K.
	REFERENCES:

COMPONENTS:	ORIGINAL MEASUREMENTS:
(1) Gold; Au; [7440-57-5] (2) Mercury; Hg; [7439-97-6]	Mees, G. *J. Am. Chem. Soc.* <u>1938</u>, 870-71.

VARIABLES:	PREPARED BY:
Temperature: 190-322°C	C. Guminski; Z. Galus

EXPERIMENTAL VALUES:

Solubility of gold in mercury:

	Experimental Solubility			Smoothed Solubility	
$t/°C$	at %	% Ave. Dev.		$t/°C$	at %
192.5	2.742	1.0		190.0	2.68
207.2	3.40	3.1		200.0	2.92
220.9	3.506	0.68		210.0	3.24
224.8	4.112	0.16		220	3.65
251.7	5.810	0.24		230	4.17
265.9	7.33	3.0		240	4.80
282.3	10.02	0.15		250	5.58
283.6	10.42	1.7		260	6.55
296.1	14.01	2.5		270	7.83
307.8	15.11	0.15		280	9.50
322.6	∼25			290	11.80
				300	15.42

AUXILIARY INFORMATION

METHOD/APPARATUS/PROCEDURE:	SOURCE AND PURITY OF MATERIALS:
Experimental details were identical to those of ref. (1).	Same as in ref. (1).
	ESTIMATED ERROR: Soly: precision better than ± 3%. Temp: ± 0.02 K.
	REFERENCES: 1. Sunier, A.A.; Weiner, L.G. *J. Am. Chem. Soc.* <u>1931</u>, *53*, 1714.

COMPONENTS:	ORIGINAL MEASUREMENTS:
(1) Gold; Au; [7440-57-5] (2) Mercury; Hg; [7439-97-6]	Rolfe, C.; Hume-Rothery, W. *J. Less-Common Metals* 1967, *13*, 1-10.
VARIABLES: Temperature: 129-1051°C	PREPARED BY: C. Guminski; Z. Galus

EXPERIMENTAL VALUES:

Liquidus temperatures of the gold-mercury system:

t/°C	at % Hg	at % Au	t/°C	at % Hg	at % Au	t/°C	at % Hg	at % Au
129	99.1	0.9	351	64.9	35.1	861	19.9	80.1
172	97.8	2.2	375	59.9	40.1	893	18.3	81.7
202	96.7	3.3	418	55.1	44.9	940	14.8	85.2
290	92.5	7.5	469	50.0	50.0	958	13.0	87.0
292	90.0	10.0	514	45.0	55.0	978	11.1	88.9
303	85.1	14.9	567	40.2	59.8	984	9.8	90.2
308	79.7	20.3	629	35.1	64.9	998	8.1	91.9
321	75.1	24.9	680	30.2	69.8	1030	5.2	94.8
328	70.0	30.0	768	25.2	74.8	1051	4.1	95.9

Au_4Hg, Au_3Hg and Au_2Hg were found as solid phases.

AUXILIARY INFORMATION

METHOD/APPARATUS/PROCEDURE:	SOURCE AND PURITY OF MATERIALS:
30 g of Au was heated with the required weight of Hg in evacuated silica capsules. The latter were very slowly heated to temperatures exceeding the freezing point of the alloy, then cooling and heating curves were recorded with calibrated thermocouples. After the experiments, the thermal analysis ingots were analyzed chemically by Johnson—Matthey Co., Ltd.	Spectrographically pure mercury and 99.99% pure gold were obtained from Johnson—Matthey Co., Ltd.
	ESTIMATED ERROR: Temp: precision $\pm$ 2 K. Analysis of amalgam: precision better than $\pm$ 1%.
	REFERENCES:

COMPONENTS:	ORIGINAL MEASUREMENTS:
(1) Gold; Au; [7440-57-5] (2) Mercury; Hg; [7439-97-6]	Kozin, L.F.; Dergacheva, M.B.; Nikushkina, N.L. *Tr. Inst. Khim. Nauk Akad. Nauk Kaz. SSR* 1976, *42*, 82-7.
VARIABLES:	PREPARED BY:
One temperature: 22°C	C. Guminski; Z. Galus

EXPERIMENTAL VALUES:

Solubility of gold in mercury at 22°C was reported to be 0.135 at %.

It was reported that bismuth and lead had no affect on the solubility of gold at this temperature.

AUXILIARY INFORMATION

METHOD/APPARATUS/PROCEDURE:	SOURCE AND PURITY OF MATERIALS:
Gold amalgams were obtained by electrolysis of $HAuCl_4$ solutions. The solubility was determined by a hydrostatic separation method: the samples from various parts of a capillary, standing perpendicularly for a long time, were analyzed by evaporating the Hg under vacuum and treating the residue with nitric acid to determine the gold content.	Pure $HAuCl_4$ was used. Hg purity not specified.
	ESTIMATED ERROR: Soly: nothing specified. Temp: precision ± 0.1 K.
	REFERENCES:

COMPONENTS:	EVALUATOR:
(1) Zinc; Zn; [7440-66-6] (2) Mercury; Hg; [7439-97-6]	C. Guminski; Z. Galus Department of Chemistry University of Warsaw Warsaw, Poland July, 1985

CRITICAL EVALUATION:

Tammann (1) observed that the addition of 0.805 at % of Zn in Hg depressed the melting point of Hg by 1.66 K. Gouy (2) found from a filtration method that saturated zinc amalgam contains 5.3 at % Zn at 288-291 K. Kerp and coworkers (3) determined the solubility in the temperature range of 273-372 K; they found that between 273 and 354.5 K the results were reproducible and that the solubility increased monotonically from 4.72 to 13.57 at %, and there was an abrupt decrease in solubility at temperatures higher than 355 K. However, from comparison with later works only the solubility of 6.17 at % at 298 K is reliable. By thermal analysis, Pushin (4) determined a smooth liquidus curve of the Zn-Hg system over the complete range of compositions. Cohen and Inouye (5) carefully determined the solubility by equilibration and filtration of the amalgam at temperature, as well as some thermal experiments, and showed that Kerp's (3) results are too low in the higher temperature range, and that the abrupt change reported by the latter was not reliable; it was also shown that Pushin's data were too high in the low temperature range. From careful measurements, Crenshaw (6) found that 6.377 at % of Zn is soluble in mercury at 298 K; this result is in good agreement with Cohen and Inouye.

Peshkov (7) investigated the region of the eutectic point by thermal analysis and reported the eutectic at 231.6 K at a zinc concentration of 1.69 at %. However, Hajicek (8) calculated that the eutectic point is at 230 K and 3.26 at % Zn; the latter concentration is nearly twofold too high and is rejected. The eutectic point found by Pushin (4), 2.6 at % at 230.5 K, lies between those of (7) and (8); however, the composition and temperature given by Peshkov seem to be most reliable.

Jangg and Kirchmayr (9), from potentiometric experiments at 288 K, determined a solubility of 5.33 at %. Bennett and Lewis (10, 11) reported solubilities of 6.99 and 8.28 at % at 303 and 313 K, respectively. Schadler and Grace (12) employed a zinc amalgam concentration cell and determined a solubility of 6.75 at % at 303 K, and they also quoted an unpublished solubility of 6.32 at % at 298 K; the latter determination was made at the New Jersey Zinc Co. Dayananda and Grace (13) carried out a precise determination of zinc content in its saturated amalgam and found 9.66 at % at 323.2 K. Very precise solubility determinations also were made by EMF measurements by Benjamin and Strickland-Constable (14) and by Walls and Upthegrove (15). All of the results reported by (10) to (15) agreed with those of Cohen and Inouye (5). However, the results of thermometric titration by Zebreva and coworkers (16-18), 5.6 and 8.2 at % at 298 and 313 K, respectively, are lower than those of the above authors.

Kozin's prediction (19) of the zinc solubility, 5.73 at % at 298 K, is in fair agreement with the experimental results.

The solubility at room temperature reported by Strachan and Harris (20) is too low and is rejected. Kozin (21) determined the solubility potentiometrically at 298 to 353 K, and found that the solubility increased from 5.5 to 13.1 at % in this temperature range; these results are up to 10% too low as compared to the more precise determinations discussed above.

Kozin and Maltsev (22) showed that the solubility of zinc in gallium amalgams may be as much as 40% higher than in mercury.

The Zn-Hg phase diagram (23) is shown in Figure 1.

(Continued next page)

COMPONENTS:	EVALUATOR:
(1) Zinc; Zn; [7440-66-6] (2) Mercury; Hg; [7439-97-6]	C. Guminski; Z. Galus Department of Chemistry University of Warsaw Warsaw, Poland July, 1985

CRITICAL EVALUATION: (continued)

Recommended (r) and tentative values of the solubility of Zn in Hg:

T/K	Soly/at %	Reference
231.6	1.7	[7]
273.2	4.1	[5]
293.2	5.88 (r)[a]	[5,11,22]
298.2	6.32 (r)	[3,6,22,31]
323.2	9.64 (r)[b]	[5,23,33]
373	19[c]	[5,23]
473	45	[4]
573	70	[4]
673	95	[4]

[a] Interpolated value from cited references.

[b] Mean value from cited references.

[c] Extrapolated value from cited references.

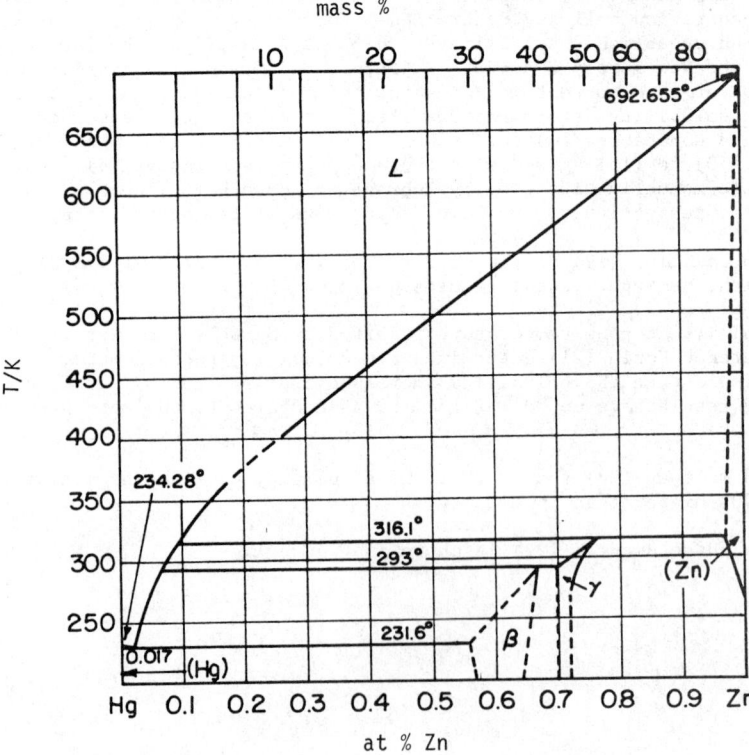

Fig. 1. The Zn-Hg system (23).

COMPONENTS:	EVALUATOR:
(1) Zinc; Zn; [7440-66-6] (2) Mercury; Hg; [7439-97-6]	C. Guminski; Z. Galus Department of Chemistry University of Warsaw Warsaw, Poland July, 1985

CRITICAL EVALUATION:

References

1. Tammann, G. *Z. Phys. Chem.* 1889, *3*, 441.
2. Gouy, M. *J. Phys.* 1895, *4*, 320.
3. Kerp, W.; Böttger, W.; Iggena, H. *Z. Anorg. Chem.* 1900, *25*, 1.
4. Pushin, N. *Z. Anorg. Chem.* 1903, *36*, 201; *Zh. Russ. Fiz. Khim. Obshch., Ser. Khim.* 1902, *34*, 856.
5. Cohen, E.; Inouye, K. *Z. Phys. Chem.* 1910, *71*, 625; 1911, *75*, 437.
6. Crenshaw, J.L. *J. Phys. Chem.* 1910, *14*, 158.
7. Peshkov, W. *Zh. Fiz. Khim.* 1946, *20*, 835; *Acta Physicochim. URSS* 1946, *21*, 109.
8. Hajicek, O. *Hutnicke Listy* 1948, *3*, 265.
9. Jangg, G.; Kirchmayr, H. *Z. Chem.* 1963, *3*, 47.
10. Bennett, J.A.R.; Lewis, J.B. *J. Chim. Phys.* 1958, *55*, 83.
11. Bennett, J.A.R.; Lewis, J.B. *Am. Inst. Chem. Eng. J.* 1958, *4*, 418.
12. Schadler, H.W.; Grace, R.E. *Trans. Met. Soc. AIME* 1959, *215*, 559.
13. Dayananda, M.A.; Grace, R.E. *U.S. At. Ener. Comm. Rep.* TID-11742, 1961.
14. Benjamin, L.; Strickland-Constable, R.F. *Acta Met.* 1960, *8*, 362.
15. Walls, H.A.; Upthegrove, W.R. *J. Chem. Eng. Data* 1964, *9*, 184.
16. Filippova, L.M.; Gayfullin, A.Sh.; Zebreva, A.I. *Prikl. Teor. Khim.*, Alma-Ata 1974, No. 5, 76.
17. Zebreva, A.I.; Filippova, L.M.; Omarova, N.D. *Izv. Vyssh. Ucheb. Zaved., Khim. Khim. Tekhnol.* 1977, *20*, 19.
18. Filippova, L.M.; Zebreva, A.I.; Omarova, N.D.; Korobkina, N.P. *Izv. Vyssh. Ucheb. Zaved., Khim. Khim. Tekhnol.* 1978, *21*, 316.
19. Kozin, L.F. *Fiziko-Khimicheskie Osnovy Amalgamnoi Metallurgii*, Nauka, Alma-Ata, 1964.
20. Strachan, J.F.; Harris, N.L. *J. Inst. Metals* 1956-57, *85*, 17.
21. Kozin, L.F. *Tr. Inst. Khim. Nauk Akad. Nauk Kaz. SSR* 1962, *9*, 71.
22. Kozin, L.F.; Maltsev, Yu.T. *Izv. Akad. Nauk Kaz. SSR, Ser. Khim.* 1969, No. 5, 38.
23. Hultgren, R.; Desai, P.D.; Hawkins, D.T.; Gleiser, M.; Kelley, K.K. *Selected Values of the Thermodynamic Properties of Binary Alloys*, American Soc. Metals, Metals Park, Ohio, 1973, pp. 999-1003.

COMPONENTS:	ORIGINAL MEASUREMENTS:
(1) Zinc; Zn; [7440-66-6] (2) Mercury; Hg; [7439-97-6]	Tammann, G. Z. Phys. Chem. 1889, 3, 441-9.
VARIABLES: Temperature: 232-234 K	PREPARED BY: C. Guminski; Z. Galus

EXPERIMENTAL VALUES:

Freezing point depression, ΔT/K, of mercury as a function of zinc content in the amalgams.

	Zinc Content	
ΔT/K	g Zn/100 g Hg	at %[a]
0.53	0.102	0.306
1.13	0.168	0.507
1.66	0.266	0.805

[a] by compilers

The melting point of mercury is reported to be 244 instead of 234 K, but it is the opinion of the compilers that the former value was a typographical error in the original publication.

AUXILIARY INFORMATION

METHOD/APPARATUS/PROCEDURE:	SOURCE AND PURITY OF MATERIALS:
Melting temperatures of amalgams were determined. No further details were presented.	Nothing specified.
	ESTIMATED ERROR: Soly: nothing specified. Temp: precision ± 0.05 K.
	REFERENCES:

COMPONENTS:	ORIGINAL MEASUREMENTS:
(1) Zinc; Zn; [7440-66-6] (2) Mercury; Hg; [7439-97-6]	Kerp, W.; Böttger, W.; Iggena, H. Z. Anorg. Chem. 1900, 25, 1-71.
VARIABLES: Temperature: 0-99°C	PREPARED BY: C. Guminski; Z. Galus

EXPERIMENTAL VALUES:

Solubility of zinc in mercury:

t/°C	Soly/mass %	Soly/at %[a]
0	1.59±0.10	4.72
25	2.10±0.03	6.17
46.5	2.94±0.02	8.50
56	3.09±0.07	8.91
64.5	3.33±0.13	9.56
81.5	4.87±0.17	13.57
89.5	3.74±0.20	10.65
99	4.52±0.24	12.68

[a] by compilers

The results at 0° and between 46.5 and 99°C are too low, but the value at 25° is reliable.

AUXILIARY INFORMATION

METHOD/APPARATUS/PROCEDURE:	SOURCE AND PURITY OF MATERIALS:
The amalgams were prepared by electrolysis of saturated ZnSO$_4$ with mercury as the cathode. The amalgams were then washed and filtered. The zinc content was determined by treating the filtrate with concentrated HCl, then precipitating the zinc as the carbonate, and subsequently heating to ZnO. The mercury content was determined gravimetrically by washing then drying the residual Hg after the HCl treatment. The procedure for equilibration at various temperatures was not described in detail.	No impurities were found in the recrystallized ZnSO$_4$. Hg purity not specified.
	ESTIMATED ERROR: Soly: precision better than ± 5%. Temp: nothing specified.
	REFERENCES:

COMPONENTS:	ORIGINAL MEASUREMENTS:
(1) Zinc; Zn; [7440-66-6] (2) Mercury; Hg; [7439-97-6]	Pushin, N. *Zh. Russ. Fiz. Khim. Obshch., Ser. Khim.* 1902, *34*, 856-904. *Z. Anorg. Chem.* 1903, *36*, 201-254.
VARIABLES:	PREPARED BY:
Temperature: (-41)-396°C	C. Guminski; Z. Galus

EXPERIMENTAL VALUES:

Temperatures of crystallization of the saturated zinc amalgams:

t/°C	at % Zn	t/°C	at % Zn
396	94.9	209.75	46.4
372	89.4	196.75	43.2
354	84.9	184	40.0
342.5	82.5	172.25	37.1
334	79.6	155	33.4
325.75	77.2	134.75	28.6
317	75	120	25.1
300	70.5	103.5	21.5
285	66.7	88.25	18
274.5	63.2	72	14.2
262.25	60	51.5	10.6
246.75	56.1	∼36	8.4
233.5	52.7	∼13	5.7
223.75	50	∼-41.5	2.6

AUXILIARY INFORMATION

METHOD/APPARATUS/PROCEDURE:	SOURCE AND PURITY OF MATERIALS:
The amalgams were prepared by mixing weighed portions of the metals. The crystallization temperatures were determined from cooling curves. The amalgams were protected from oxidation with a layer of paraffin or vaseline on the surface.	Nothing specified.
	ESTIMATED ERROR: Soly: nothing specified. Temp: precision ± 0.5 K.
	REFERENCES:

COMPONENTS:	ORIGINAL MEASUREMENTS:
(1) Zinc; Zn; [7440-66-6] (2) Mercury; Hg; [7439-97-6]	Cohen, E.; Inouye, K. Z. Phys. Chem. 1910, 71, 625-35.
VARIABLES: Temperature: 0-100°C	PREPARED BY: C. Guminski; Z. Galus

EXPERIMENTAL VALUES:

Solubility of zinc in mercury at various temperatures:

$t/°C$	Soly/mass %	Soly/at %[a]
0.3	1.37±0.02	4.09
19.9	1.99±0.01	5.86
30.0	2.39±0.01	6.99
39.95	2.86±0.01	8.28
50.0	3.37±0.04	9.66
64.75	4.28±0.14	12.06
80.1	5.36±0.10	14.80
89.5	6.10±0.08	16.62
94.8	6.59±0.10	17.79
99.6	7.04±0.13	18.85

[a] by compilers

AUXILIARY INFORMATION

METHOD/APPARATUS/PROCEDURE:	SOURCE AND PURITY OF MATERIALS:
The amalgams were prepared by the dissolution of zinc in mercury at temperatures higher than those of the experimental measurements. The tubes with the amalgams were then shaken for one to a few days in a thermostat. The amalgams were then filtered and the filtrates were treated with HCl and the residual mercury was determined gravimetrically after being washed and dried.	Very pure zinc was obtained from Kahlbaum; mercury was purified chemically then double distilled before use.
	ESTIMATED ERROR: Soly: precision as high as ± 3%, but the mean was ± 1%. Temp: precision ± 0.2 K.
	REFERENCES:

COMPONENTS:	ORIGINAL MEASUREMENTS:
(1) Zinc; Zn; [7440-66-6] (2) Mercury; Hg; [7439-97-6]	Crenshaw, J. L. *J. Phys. Chem.* 1910, *14*, 158-170.
VARIABLES: Temperature: 25°C	PREPARED BY: C. Guminski; Z. Galus

EXPERIMENTAL VALUES:

Solubility of zinc in 100 g of mercury at 25.0°C was determined to be 2.220 ± 0.007 g. The solubility in atomic % calculated by the compilers is 6.377 at %.

AUXILIARY INFORMATION

METHOD/APPARATUS/PROCEDURE:	SOURCE AND PURITY OF MATERIALS:
The amalgams were prepared by mixing precisely weighed quantities of the metals in a special apparatus with mercury under distilled water and polarized up to 12 V. This procedure protected the zinc from oxidation. The saturated amalgams were equilibrated for several weeks in a rotated tube immersed in a thermostat. Determination of the zinc concentration was made by two methods: (I) Densities of the saturated and diluted amalgams of exact composition were determined pycnometrically, and the saturated amalgam concentration was obtained from a calibration curve. (II) The saturated zinc amalgam was filtered and a known quantity of the filtrate was treated with concentrated HCl; the mercury was washed and dried and its concentration determined gravimetrically.	Mercury was chemically purified and then distilled. $ZnSO_4$ was purified by precipitation of all other heavy metals with H_2S, then recrystallized 3 times. Metallic zinc was obtained by electrolysis of the purified $ZnSO_4$ solution, and the metal was vacuum distilled.
	ESTIMATED ERROR: Soly: accuracy ± 0.3%. Temp: precision ± 0.02 K.
	REFERENCES:

COMPONENTS:	ORIGINAL MEASUREMENTS:
(1) Zinc; Zn; [7440-66-6] (2) Mercury; Hg; [7439-97-6]	Peshkov, V. *Zh. Fiz. Khim.* 1946, *20*, 835-51.
VARIABLES:	PREPARED BY:
Temperature: (-39)-(-20)°C	C. Guminski; Z. Galus

EXPERIMENTAL VALUES:

Crystallization temperatures of dilute zinc amalgams:

$t/°C$	Soly/mass %	Soly/at %[a]
-39.33	0.100	0.306
-40.38	0.300	0.915
-41.63[b]	0.534	1.62
-41.75	0.569	1.72
-34.7	0.717	2.17
-19.9 (-17.6)	1.046	3.14

[a] by compilers
[b] eutectic point

In another paper by the same author somewhat different values of temperatures are given (1).

AUXILIARY INFORMATION

METHOD/APPARATUS/PROCEDURE:	SOURCE AND PURITY OF MATERIALS:
The amalgams were obtained by mixing the two metals. Temperatures at the start of crystallization and the end of melting were determined. Microscopic examinations also were made of the amalgams.	Mercury purity: 99.999% Zinc purity: 99.97%
	ESTIMATED ERROR:
	Soly: nothing specified. Temp: precision ± 0.01 K.
	REFERENCES:
	1. Peshkov, V. *Acta Physicochem. URSS* 1946, *21*, 109.

COMPONENTS:	ORIGINAL MEASUREMENTS:

(1) Zinc; Zn; [7440-66-6]
(2) Mercury; Hg; [7439-97-6]
	1. Bennett, J.A.R.; Lewis, J.B.
	   J. Chim. Phys. 1958, *55*, 83-7.
	2. Bennett, J.A.R.; Lewis, J.B.
	   Am. Inst. Chem. Eng. J. 1958, 418-22.

VARIABLES: Temperature: 30-40°C	PREPARED BY: C. Guminski; Z. Galus

EXPERIMENTAL VALUES:

Solubility of zinc in mercury at 30 and 40°C was reported to be 2.39 and 2.86 mass %; 6.99 and 8.28 at %, respectively.

AUXILIARY INFORMATION

METHOD/APPARATUS/PROCEDURE:	SOURCE AND PURITY OF MATERIALS:
The amalgams were prepared by dissolution of a rotating zinc cylinder in Hg. The dissolution vessel was mounted inside a glove box filled with pure argon. After equilibration the amalgams were analyzed by distilling out mercury at 300°C in nitrogen atmosphere. The residue was dissolved in aqua regia and then analyzed by polarography.	99.99% pure metals were used.
	ESTIMATED ERROR: Soly: nothing specified; no better than ± 3% (compilers). Temp: precision ± 0.2 K.
	REFERENCES:

COMPONENTS:	ORIGINAL MEASUREMENTS:
(1) Zinc; Zn; [7440-66-6] (2) Mercury; Hg [7439-97-6]	Schadler, H.W.; Grace, R.E. *AIME Trans.* 1959, *215*, 559-66.
VARIABLES: Temperature: 30°C	PREPARED BY: C. Guminski; Z. Galus

EXPERIMENTAL VALUES:

The solubility of zinc in mercury at 30°C was determined to be 6.75 at %.

The authors also quote unpublished solubility data, determined at the New Jersey Zinc Co., of 2.147 ± 0.01 and 2.157 ± 0.01 mass % at 25°C.

AUXILIARY INFORMATION

METHOD/APPARATUS/PROCEDURE:	SOURCE AND PURITY OF MATERIALS:
The amalgams were prepared either by electrolysis or by dissolving solid zinc in Hg. EMF of the cell, $Hg(Zn)_{sat} | Zn^{++}(0.1 \text{ mol dm}^{-3}) | Hg(Zn)_x$ were measured for a series of amalgams, including the saturated amalgam. Although not described, the solubility was probably determined from the breakpoint in the plot of EMF vs. amalgam concentration.	Hg: ACS Reagent Grade from Goldsmith Bros. Zn: Cast rod from New Jersey Zinc Co. with Pb <0.002%, Cd <0.00005%, and Fe <0.0003%.
	ESTIMATED ERROR: Soly: not specified; accuracy probably better than ± 1% (compilers). Temp: precision ± 0.03 K.
	REFERENCES:

COMPONENTS:	ORIGINAL MEASUREMENTS:
(1) Zinc; Zn; [7440-66-6] (2) Mercury; Hg; [7439-97-6]	Benjamin, L.; Strickland-Constable, R.F. *Acta Met.* 1960, *8*, 362-72.
VARIABLES:	PREPARED BY:
Temperature: 23-41°C	C. Guminski; Z. Galus

EXPERIMENTAL VALUES:

Solubility of zinc in mercury at three temperatures was reported.

$t/°C$	Soly/mass %	Soly/at %[a]
23.21	2.08	6.12
36.87	2.65	7.71
40.90	2.90	8.39

[a] by compilers

Kinetics of nucleation and crystal growth from zinc amalgam also were studied.

AUXILIARY INFORMATION

METHOD/APPARATUS/PROCEDURE:	SOURCE AND PURITY OF MATERIALS:
The amalgams were prepared by dissolution of Zn in Hg, and concentration cells of the type, $Zn(Hg)_x | 2 \text{ mol dm}^{-3} \text{ ZnSO}_4 | Zn(Hg)_y$ were constructed. Nitrogen was bubbled through the ZnSO$_4$ solution after it had been allowed to boil. The cells were equilibrated for a day or two before EMF measurements were made.	High purity zinc from U.K.A.E.A., Harwell. Purity of other substances not specified.
	ESTIMATED ERROR:
	Soly: nothing specified; precision better than $\pm$ 1% (compilers). Temp: precision $\pm$ 0.02 K.
	REFERENCES:

COMPONENTS:	ORIGINAL MEASUREMENTS:
(1) Zinc; Zn; [7440-66-6] (2) Mercury; Hg; [7439-97-6]	Dayananda, M.A.; Grace, R.E. U.S. At. Ener. Comm. Rep., TID-11742, <u>1961</u>.
VARIABLES:	PREPARED BY:
One temperature: 50°C	C. Guminski; Z. Galus

EXPERIMENTAL VALUES:

At 50.0°C the solubility of Zn in Hg was determined to be 3.37 mass %. The solubility in atomic % calculated by the compilers is 9.66 at %.

AUXILIARY INFORMATION

METHOD/APPARATUS/PROCEDURE:	SOURCE AND PURITY OF MATERIALS:
Single crystals of zinc were first immersed in H_2O_2 for a day, then briefly dipped in 3:1 HNO_3 and rinsed with water. The dissolution of the zinc in its unsaturated amalgam was followed by determination of its activity in the amalgam as a function of time. For this measurement, a sample of amalgam removed from the dissolution flask was used in the cell, $Zn(Hg)_{sat} | 0.1$ mol dm^{-3} $ZnSO_4	Zn(Hg)_x$, and the activity determined from the EMF. To prevent oxidation of the amalgam, 18 V was applied between the solution (anode) and the amalgam (cathode).	Zn purity was 99.999%; impurities were Pb, Cd, and Fe at 2×10^{-4}, 5×10^{-5}, and 3×10^{-4} %, respectively. Hg contained 5×10^{-4} % Ag + Au and less than 1×10^{-4} % base metal.
	ESTIMATED ERROR:
	Temperature: precision $\pm$ 0.1 K. Stability of EMF was $\pm 3 \times 10^{-6}$ V.
	REFERENCES:

COMPONENTS:	ORIGINAL MEASUREMENTS:
(1) Zinc; Zn; [7440-66-6] (2) Mercury; Hg; [7439-97-6]	Kozin, L.F. *Tr. Inst. Khim. Nauk Akad. Nauk Kaz. SSR* 1962, *9*, 71-80.

VARIABLES:	PREPARED BY:
Temperature: 25-80°C	C. Guminski; Z. Galus

EXPERIMENTAL VALUES:

Solubility of zinc in mercury:

t/°C	Soly/at %
25	5.5
40	7.9
60	10.6
80	13.1

AUXILIARY INFORMATION

METHOD/APPARATUS/PROCEDURE:	SOURCE AND PURITY OF MATERIALS:
The amalgams were prepared by dissolution of zinc in mercury, and EMF's were measured of the cell, Zn(Hg)$_x$ | 0.1 mol dm^{-3} Zn(ClO$_4$)$_2$, 0.9 mol dm^{-3} NaClO$_4$ | NaCl, Hg$_2$Cl$_2$, Hg. The solutions were protected against oxygen with a stream of pure nitrogen. The breakpoint in the curve relating EMF to logarithm of zinc concentration corresponded to the saturation in the amalgam.	The salts were recrystallized twice. Mercury was purified chemically then doubly distilled. Zinc was 99.999% pure.
	ESTIMATED ERROR: Soly: nothing specified. Temp: precision ± 0.2 K.
	REFERENCES:

COMPONENTS:	ORIGINAL MEASUREMENTS:
(1) Zinc; Zn; [7440-66-6] (2) Mercury; Hg; [7439-97-6]	Jangg, G.; Kirchmayr, H. Z. Chem. 1963, 3, 47-56.
VARIABLES:	PREPARED BY:
Temperature: 15°C	C. Guminski; Z. Galus

EXPERIMENTAL VALUES:

Solubility of zinc in mercury at 15°C was reported to be 3.68 ± 0.10 mol dm^{-3}.

The solubility in atomic % calculated by the compilers is 5.33 at %.

AUXILIARY INFORMATION

METHOD/APPARATUS/PROCEDURE:	SOURCE AND PURITY OF MATERIALS:
Amalgams were prepared by electrolysis, and a cell was constructed as follows: $Zn(Hg)_x	ZnSO_4	KCl, Hg_2Cl_2, Hg$. The concentration of the saturated amalgam was determined from the breakpoint in the curve of EMF vs. log $C_{Zn(Hg)}$, where $C_{Zn(Hg)}$ is the amalgam concentration. The experiments were conducted in an inert gas atmosphere.	Nothing specified.
	ESTIMATED ERROR: Soly: precision better than $\pm$ 3%. Temp: nothing specified.
	REFERENCES:

COMPONENTS:	ORIGINAL MEASUREMENTS:
(1) Zinc; Zn; [7440-66-6] (2) Mercury; Hg; [7439-97-6]	Walls, H.A.; Upthegrove, W.R. *J. Chem. Eng. Data* 1964, *9*, 184-187.
VARIABLES:	PREPARED BY:
Temperature: 323-366 K	C. Guminski; Z. Galus

EXPERIMENTAL VALUES:

Solubility of zinc in mercury:

T/K	Soly/mass %	Soly/at %
323.2	3.348	9.608
343.4	4.645	13.003
366.4	6.540	17.676

AUXILIARY INFORMATION

METHOD/APPARATUS/PROCEDURE:	SOURCE AND PURITY OF MATERIALS:
The amalgams were prepared by directly dissolving zinc into mercury, and the concentration ascertained from the known weights of each component. The solubilities were determined by extrapolating the concentration versus EMF curve to zero potential for the cell, $Zn(Hg)_{sat} \mid 0.1 \text{ mol dm}^{-3} \text{ ZnSO}_4 \mid Zn(Hg)_x$. The saturated amalgams were prepared at temperatures slightly above experimental and slowly cooled to ascertain equilibration. The amalgams and electrolyte were handled under a blanket of argon to exclude air. Precision potentiometer and galvanometer were used; calibrated thermocouples used for temperature measurement.	All materials were ACS Reagent Grade or better.
	ESTIMATED ERROR: Temp: precision ± 0.01 K. EMF measurement: precision better than ± 0.05 mV. Concentration: accuracy ± 0.001%.
	REFERENCES:

COMPONENTS:	ORIGINAL MEASUREMENTS:
(1) Zinc; Zn [7440-66-6] (2) Mercury; Hg; [7439-97-6]	Filippova, L.M.; Zebreva, A.I.; Omarova, N.D.; Korobkina, N.P. *Izv. Vyssh. Ucheb. Zaved., Khim. Khim. Tekhnol.* 1978, *21*, 316-9.
VARIABLES:	PREPARED BY:
Temperature: 25-40°C	C. Guminski; Z. Galus

EXPERIMENTAL VALUES:

Solubility of zinc in mercury at 25 and 40°C were reported to be 5.6 $\pm$ 0.5 and 8.2 $\pm$ 0.1 at %, respectively.

The same solubility at 25°C is reported also in (2) and (3).

AUXILIARY INFORMATION

METHOD/APPARATUS/PROCEDURE:	SOURCE AND PURITY OF MATERIALS:
The heterogeneous and homogeneous amalgams were prepared by direct dissolution of zinc in mercury. The amalgams were thermometrically titrated by the addition of mercury in a specially constructed apparatus (1). The zinc solubility was determined from the breakpoint of the curve relating composition to the thermal effect. All operations were performed in an argon atmosphere.	Zinc was specified as "for analysis". Mercury purity not specified.
	ESTIMATED ERROR:
	Soly: accuracy no better than $\pm$ 10%. Temp: precision $\pm$ 0.5 K.
	REFERENCES: 1. Zebreva, A.I.; Filippova, L.M.; Omarova, N.D.; Gayfullin, A.Sh. *Izv. Vyssh. Ucheb. Zaved., Khim. Khim. Teknol.* 1976, *19*, 1043-6. 2. Filippova, L.M.; Gayfullin, A.Sh.; Zebreva, A.I. *Prikl. Teor. Khim.*, Alma-Ata, 1974, *5*, 76-82. 3. Zebreva, A.I.; Filippova, L.M.; Omarova, N.D. *Izv. Vyssh. Ucheb. Zaved, Khim. Khim. Tekhnol.* 1977, *20*, 19-22.

COMPONENTS:	EVALUATOR:
(1) Cadmium; Cd; [7440-43-9] (2) Mercury; Hg; [7439-97-6]	C. Guminski; Z. Galus Department of Chemistry University of Warsaw Warsaw, Poland July, 1985

CRITICAL EVALUATION:

In the earliest report on this system, Tammann (1) observed that the melting point of mercury was elevated by 1.8 K when 0.55 at % of cadmium was dissolved into the mercury. Heycock and Neville (2) conducted similar thermal analyses in the Cd-rich region and observed that the M.P. was depressed up to 15 K by the dissolution of up to 5.19 at % of Hg. Later measurements by Honda and Ishigaki (3) confirmed the results of ref. (2).

Gouy (4) reported a solubility of 6.8 at % at 288-291 K, but this result is rejected because it is 10% lower than the most precise measurements. From the potentiometric measurements of Jaeger (5) at 288 K a solubility of approximately 8.6 at % may be estimated; this result is rejected.

Hulett and DeLury (6) determined the solubility of cadmium at 298 K by equilibration of the two metals, followed by careful analysis of the saturated liquid. The solubility reported by these authors was 9.529 at.%. The solubility of 9.6 at % at 298 K, reported by Zebreva and coworkers (7,8) from thermometric titrations of homo- and heterogeneous amalgams, is in good agreement with the above value. Strachan and Harris (9) reported a solubility of 9.41 at % at room temperature.

Moesveld and De Meester (10) determined the solubility of Cd between 273 and 314 K from careful potentiometric measurements, and they found that the solubility increased from 4.82 to 13.76 at % in this temperature range. These authors fitted their solubility to a parabolic function of the temperature. Walls and Upthegrove (11), from careful EMF measurements, reported solubilities of 16.10 to 29.03 at % at 323.2 to 366.4 K. Kerp and coworkers (12) determined the solubility at 273-372 K by an analytical method, and their results near room temperature are in good agreement with other precise measurements; however, the results at the higher temperatures are too low, while the solubilities near 273 K are too high. Smith (13) investigated the Weston normal cell over a temperature range of 273-338 K; from the data presented in this work the solubility was estimated to increase from 5.2 to 20.3 at % over the given temperature range.

Bijl (14) and Pushin (15) reported the liquidus curve for the complete Cd-Hg system and the results from thermal analyses were in good agreement with other reported determinations; however, Bijl also determined some of the solubilities by potentiometric measurements and these results were slightly lower than those determined by thermal analysis. The liquidus determined by Jänecke (16), at 20-80 at % Cd, was in good agreement with those of refs. (14) and (15); similar agreement with the latter works was reported by Teeter (17) and by Mehl and Barrett (18). Schulze (19) determined the crystallization temperatures for compositions ranging from 13.76 to 23.74 at %, but his liquidus temperatures are slightly too low. The complete phase diagram was redetermined by Semibratova and coworkers (20), but these authors found lower liquidus temperatures in the Hg-rich and higher temperatures in the Cd-rich regions as compared to those of refs. (14),(15), and (18); the results for the remainder of the liquidus agreed well with the earlier measurements. Campbell and Kartzmark (21) conducted thermoanalytical measurements and confirmed the results of Bijl; however, the former authors did not observe the peritectic at about 463 K. Bukhman and coworkers (22) determined the Cd content in the saturated amalgams at 290-296 K and obtained solubilities of 6.59 to 9.69 at % in this temperature range; however, the temperature dependence of the solubility is too high, and only the result at 295 K is acceptable from comparison with other works.

Kozin's (23) prediction of 5.16 at % at 298 K is too low, and an estimate from Spencer's (24) EMF measurement is too imprecise.

The saturated cadmium amalgams are in equilibrium with the rather unstable ω-phase or with pure cadmium; see the most recent phase diagram (25) in Fig. 1. However, Bukhman and coworkers (22) demonstrated that $CdHg_3$ is in equilibrium with the saturated amalgams at room temperature.

The solubility of Cd in the amalgams of Bi, Pb, Sn, Tl, and Zn was reported by (26). It also was reported that the presence of Mn in the amalgam decreased the solubility of Cd only slightly (27).

(continued next page)

COMPONENTS:	EVALUATOR:
(1) Cadmium; Cd; [7440-43-9] (2) Mercury; Hg; [7439-97-6]	C. Guminski; Z. Galus Department of Chemistry University of Warsaw Warsaw, Poland July, 1985

CRITICAL EVALUATION: (continued)

The recommended (r) and tentative values of the solubility of Cd in Hg:

T/K	Soly/at %	Source
239	1.3 peritectic	[14]
273.2	4.9 (r)	[7,13]
293.2	8.6 (r)[a]	[5,13,14]
298.2	9.53 (r)	[5,10,6,25,27]
323.2	16.1 (r)	[6,14,24]
373	32 (r)[a]	[9,24,25]
473	67 (r)	[7,9,25]
573	91 (r)	[7,25]

a. Interpolated from data of cited references.

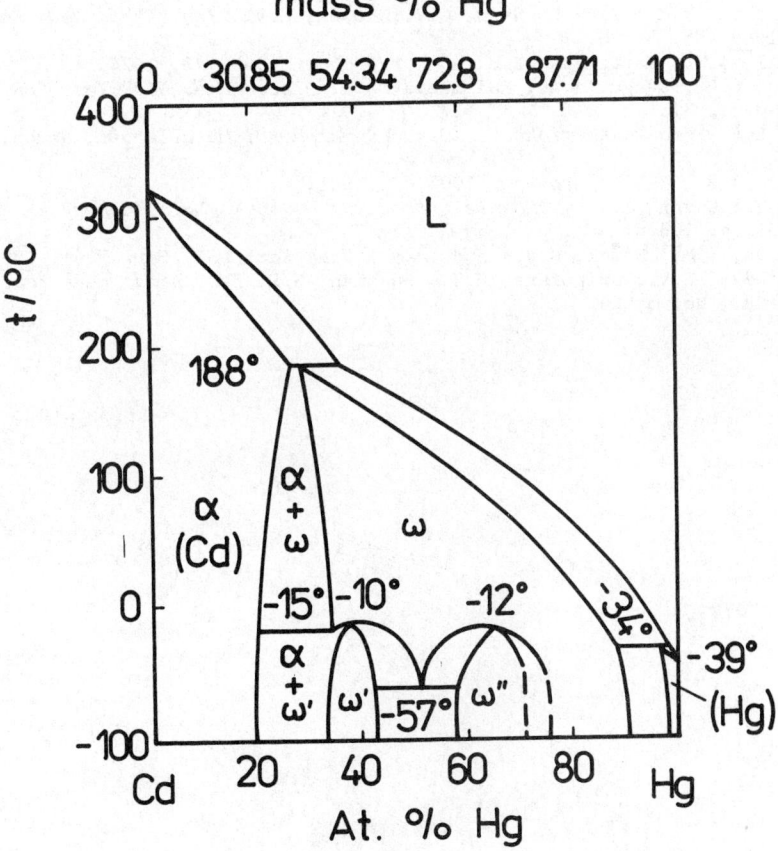

Fig. 1. The Cd-Hg system (25).

Cadmium

COMPONENTS:	EVALUATOR:
(1) Cadmium; Cd; [7440-43-9] (2) Mercury; Hg; [7439-97-6]	C. Guminski; Z. Galus Department of Chemistry University of Warsaw Warsaw, Poland July, 1985

CRITICAL EVALUATION:

REFERENCES

1. Tammann, G. *Z. Phys. Chem.* 1889, *3*, 441.
2. Heycock, C.T.; Neville, F.H. *J. Chem. Soc.* 1892, 888.
3. Honda, K.; Ishigaki, T. *Sci. Rep. Tohoku Univer.* 1925, *14*, 219.
4. Gouy, M. *J. Phys.* 1895, *4*, 320.
5. Jaeger, W. *Wied. Ann.* 1898, *65*, 106.
6. Hulett, G.A.; De Lury, R.H. *J. Am. Chem. Soc.* 1908, *30*, 1805.
7. Zebreva, A.I.; Filippova, L.M.; Omarova, N.D.; Gayfullin, A.Sh. *Izv. Vyssh. Ucheb. Zaved., Khim. Khim. Tekhnol.* 1976, *19*, 1043.
8. Filippova, L.M.; Zebreva, A.I.; Espenbetov, A.A. *Izv. Vyssh. Ucheb. Zaved., Khim. Khim. Tekhnol.* 1977, *20*, 1468.
9. Strachan, J.F.; Harris, N.L. *J. Inst. Metals* 1956-57, *85*, 17.
10. Moesveld, A.L.T.; De Meester, W.A.T. *Z. Phys. Chem.* 1927, *130*, 146.
11. Walls, H.A.; Upthegrove, W.R. *J. Chem. Eng. Data* 1964, *9*, 184.
12. Kerp, W.; Böttger, W.; Iggena, H. *Z. Anorg. Chem.* 1900, *25*, 1.
13. Smith, F.E. *Phil. Mag., Ser. 6* 1910, *19*, 250; *Z. Phys. Chem.* 1920, *95*, 293.
14. Bijl, H.C. *Z. Phys. Chem.* 1902, *41*, 641.
15. Pushin, N. *Z. Anorg Chem.* 1903, *36*, 201; *Zh. Russ. Fiz. Khim. Obshch., Ser. Khim.* 1902, *34*, 856.
16. Jänecke, E. *Z. Phys. Chem.* 1907, *60*, 399.
17. Teeter, C.E. *J. Am. Chem. Soc.* 1931, *53*, 3927.
18. Mehl, R.F.; Barrett, C.S. *Trans. AIME* 1930, *89*, 575; *Met. Erz.* 1930, *27*, 624.
19. Schulze, A. *Z. Phys. Chem.* 1923, *105*, 177.
20. Semibratova, N.M.; Yan-Sho-Syan, G.V.; Nosek, M.V. *Izv. Akad. Nauk Kaz. SSR, Ser. Khim.* 1969, No. 5, 30.
21. Campbell, A.N.; Kartzmark, E.M. *Can. J. Chem.* 1965, *43*, 1924.
22. Bukhman, S.P.; Lange, A.A.; Kairbaeva, A.A. *Izv. Akad. Nauk Kaz. SSR, Ser. Khim.* 1984, No. 1, 31.
23. Kozin, L.F. *Fiziko-Khimicheskie Osnovy Amalgamnoi Metallurgii*, Nauka, Alma-Ata, 1964.
24. Spencer, J.F. *Z. Elektrochem.* 1905, *11*, 681.
25. Vol, A.E.; Kagan, I.K. *Stroenie i Svoistva Dvoinykh Metallicheskikh Sistem*, Moskva, 1979, *IV*, p. 168.
26. Atamanova, N.M.; Nosek, M.V. *Izv. Akad. Nauk Kaz. SSR, Ser. Khim.* 1983, No. 3, 51.
27. Shirinskikh, A.V.; Grigoreva, M.J.; Bukhman, S.P. *Izv. Akad. Nauk Kaz. SSR, Ser. Khim.* 1983, No. 5, 20.

COMPONENTS:	ORIGINAL MEASUREMENTS:
(1) Cadmium; Cd; [7440-43-9] (2) Mercury; Hg; [7439-97-6]	Tammann, G. Z. Phys. Chem. 1889, 3, 441-9.
VARIABLES:	PREPARED BY:
Temperature: 234-236 K	C. Guminski; Z. Galus

EXPERIMENTAL VALUES:

Changes in freezing point of mercury, ΔT, upon addition of small amounts of cadmium.

ΔT/K	g Cd/100 g Hg	at % Cd[a]
0.4	0.073	0.13
0.85	0.143	0.255
1.5	0.270	0.479
1.8	0.310	0.550

[a] by compilers

The melting point of Hg was reported to be 244 K instead of 234 K; in the opinion of the compilers there was a typographical error in the reported melting point of Hg.

AUXILIARY INFORMATION

METHOD/APPARATUS/PROCEDURE:	SOURCE AND PURITY OF MATERIALS:
Melting points of amalgams were determined. Details of experimental procedure not presented.	Nothing specified.
	ESTIMATED ERROR: Soly: nothing specified. Temp: better than $\pm$ 0.05 K.
	REFERENCES:

COMPONENTS:	ORIGINAL MEASUREMENTS:
(1) Cadmium; Cd; [7440-43-9] (2) Mercury; Hg; [7439-97-6]	Heycock, C.T.; Neville, F.H. J. Chem. Soc. 1892, 888-914.
VARIABLES: Temperature: 305-321°C	PREPARED BY: C. Guminski; Z. Galus

EXPERIMENTAL VALUES:

Temperatures of crystallization of saturated cadmium amalgams:

t/°C	atoms Hg/100 atoms Cd	at % Hg[a]
320.52	0.0285	0.0285
320.3	0.118	0.118
319.92	0.259	0.258
319.05	0.584	0.581
317.59	1.106	1.094
314.93	2.063	2.021
311.59	3.288	3.183
305.5	5.477	5.193

[a] by compilers

AUXILIARY INFORMATION

METHOD/APPARATUS/PROCEDURE:	SOURCE AND PURITY OF MATERIALS:
Weighed quantities of the metals were placed in a hard glass tube then evacuated prior to sealing. The tube was heated to red heat and well shaken. The melting temperatures were determined with carefully calibrated thermometers.	Nothing specified.
	ESTIMATED ERROR: Soly: nothing specified. Temp: precision ± 0.05 K (compilers)
	REFERENCES:

COMPONENTS:	ORIGINAL MEASUREMENTS:
(1) Cadmium; Cd; [7440-43-9] (2) Mercury; Hg; [7439-97-6]	Kerp, W.; Böttger, W.; Iggena, H. Z. Anorg. Chem. 1900, 25, 1-71.
VARIABLES: Temperature: 0-99°C	PREPARED BY: C. Guminski; Z. Galus

EXPERIMENTAL VALUES:

Solubility of cadmium in mercury at 0 to 99°C:

	Soly	
$t/°C$	at %[a]	mass %
0	5.52	3.17 ± 0.12
18	8.25	4.80 ± 0.05
25	9.54	5.58 ± 0.13
30	10.65	6.26 ± 0.02
35	11.83	6.99 ± 0.03
38	12.61	7.48 ± 0.03
40.5	13.09	7.78 ± 0.06
44	14.05	8.39 ± 0.10
56.8	17.07	10.34 ± 0.12
63	18.66	11.39 ± 0.10
73	22.09	13.71 ± 0.17
82	25.09	15.80 ± 0.20
89	27.39	17.45 ± 0.15
99	30.36	19.63 ± 0.03

[a] by compilers

The most reliable solubilities were obtained near room temperature. The solubilities at higher temperatures are slightly lower than those from the most reliable determinations, whereas at 0°C the solubility is too high.

AUXILIARY INFORMATION

METHOD/APPARATUS/PROCEDURE:	SOURCE AND PURITY OF MATERIALS:
Amalgams were prepared electrolytically from saturated $CdSO_4$ solution with Hg as the cathode. The heterogeneous amalgams were filtered, and the filtrates were treated with HCl. The cadmium concentrations were determined from the difference in weight between the original amalgam and the residual mercury after the acid treatment.	Nothing specified.
	ESTIMATED ERROR: Soly: precision better than ± 3%. Temp: nothing specified.
	REFERENCES:

COMPONENTS:	ORIGINAL MEASUREMENTS:
(1) Cadmium; Cd; [7440-43-9] (2) Mercury; Hg; [7439-97-6]	Bijl, H.C. Z. Phys. Chem. 1902, 41, 641-71.
VARIABLES: Temperature: (-36)-273°C	PREPARED BY: C. Guminski; Z. Galus

EXPERIMENTAL VALUES:

Temperatures of crystallization of cadmium amalgams were determined from cooling curves, A, and from potentiometric measurements, B.

A		B	
$t/°C$	at % Cd	$t/°C$	at % Cd
-36.4	0.47	25	9
-34.6	0.94	50	16
-1.6	5.52	75	23
34.0	12.44		
54.4	18.39		
68.8	22.21		
84.6	27.22		
121.8	40.04		
149.6	50.28		
163.6	55.10		
190.8	64.33		
214.6	70.90		
237.3	74.58		
273.4	84.96		

AUXILIARY INFORMATION

METHOD/APPARATUS/PROCEDURE:	SOURCE AND PURITY OF MATERIALS:
The amalgams were prepared by mixing the metals and heating in CO_2 atmosphere. Cadmium was previously cleaned with HCl then dried. The cooling curves of the amalgams were recorded. Also, potentials of the following cell were determined: $Cd(Hg)_x \mid CdSO_4(aq.) \mid Cd(Hg)_a$ where a = 12.04%.	Cadmium supplied by Merck. Mercury was purified; method not specified.
	ESTIMATED ERROR: Soly: nothing specified. Temp: precision ± 0.2 K.
	REFERENCES:

COMPONENTS:	ORIGINAL MEASUREMENTS:
(1) Cadmium; Cd; [7440-43-9] (2) Mercury; Hg; [7439-97-6]	Pushin, N.A. *Zh. Russ. Fiz. Khim. Obshch., Ser. Khim.* 1902, *34*, 856-904; *Z. Anorg. Chem.* 1903, *36*, 201-254.
VARIABLES: Temperature: (-11)-316°C	PREPARED BY: C. Guminski; Z. Galus

EXPERIMENTAL VALUES:

Crystallization temperatures of amalgams as a function of mercury concentration.

t/°C	at % Hg	t/°C	at % Hg
316.25	1.7	176.5	40.0
310.0	3.8	166.5	43.4
297.0	7.8	154.5	47.7
281.0	12.5	143.75	51.8
261.75	17.7	129.5	56.8
243.5	22.6	114.5	62.3
222.0	28.0	102.5	66.6
212.75	30.4	89.25	71.2
207.5	31.6	78.25	75.0
200	33.3	70.50	77.5
199.25	33.5	62.5	80.0
196.0	34.3	51.25	83.4
192.0	35.3	40.5	86.4
187	36.4	31.0	88.9
183.75	37.5	12.5	92.8
181	38.4	-6.0	95.8
179	38.9	-11.0	96.6
178	39.4		

AUXILIARY INFORMATION

METHOD/APPARATUS/PROCEDURE:	SOURCE AND PURITY OF MATERIALS:
The amalgams were obtained by mixing the metals, followed by heating. Cooling curves were recorded on the amalgams; the amalgams were protected from oxidation by a film of paraffin or vaseline during the measurements.	Nothing specified.
	ESTIMATED ERROR: Soly: nothing specified. Temp: precision ± 0.5 K.
	REFERENCES:

COMPONENTS:	ORIGINAL MEASUREMENTS:
(1) Cadmium; Cd; [7440-43-9] (2) Mercury; Hg; [7439-97-6]	Jänecke, E. Z. Phys. Chem. 1907, 60, 399-412.
VARIABLES: Temperature: 67-248°C	PREPARED BY: C. Guminski; Z. Galus

EXPERIMENTAL VALUES:

Temperatures of crystallization were determined for saturated cadmium amalgams of various compositions.

$t/°C$	at % Cd
248	80
199	66.5
147	50
102	33.5
67	20

AUXILIARY INFORMATION

METHOD/APPARATUS/PROCEDURE:	SOURCE AND PURITY OF MATERIALS:
The melting points were determined from cooling curves with the temperatures observed either with a mercury thermometer or with a thermoelement. Microscopic observations also were carried out in parallel.	Nothing specified.
	ESTIMATED ERROR: Soly: nothing specified. Temp: $\pm$ 1 K (compilers).
	REFERENCES:

COMPONENTS:	ORIGINAL MEASUREMENTS:
(1) Cadmium; Cd; [7440-43-9] (2) Mercury; Hg; [7439-97-6]	Hulett, G.A.; DeLury, R.H. *J. Am. Chem. Soc.* 1908, *30*, 1805-27.
VARIABLES: Temperature: 25°C	PREPARED BY: C. Guminski; Z. Galus

EXPERIMENTAL VALUES:

Solubility of cadmium in mercury at 25.00°C was reported to be 5.573 $\pm$ 0.002 mass %. The solubility in atomic % calculated by the compilers is 9.529 at %.

AUXILIARY INFORMATION

METHOD/APPARATUS/PROCEDURE:	SOURCE AND PURITY OF MATERIALS:
The amalgams were made by mixing the exactly weighed metals. The cadmium dissolution was carried out in special apparatus where Hg was polarized at 10 V under distilled water. This prevented the oxidation of Cd. Saturation of the amalgam was carried out in a tube which was rotated for several days in a thermostat. The Cd concentration was determined by two methods: I. Densities of the saturated and diluted amalgams were determined pycnometrically, and the concentration of the saturated amalgam was obtained from a calibration curve. II. The saturated amalgam was filtered and the weighed filtrate treated with HCl to dissolve the Cd. The mass difference between amalgam and residual Hg gave the Cd content. Correction applied for dissolution of traces of Hg.	Mercury was chemically purified and then distilled. $CdSO_4$ was purified by first precipitating CdS with H_2S, then the CdS was dissolved in H_2SO_4 to form $CdSO_4$; the latter was recrystallized. Metallic Cd was obtained by electrolysis and the Cd was double distilled.
	ESTIMATED ERROR: Soly: precision $\pm$ 0.03%. Temp: precision $\pm$ 0.01 K.
	REFERENCES:

COMPONENTS:	ORIGINAL MEASUREMENTS:
(1) Cadmium; Cd; [7440-43-9] (2) Mercury; Hg; [7439-97-6]	Smith, F.E. *Phil. Mag., Ser. 6* 1910, *19*, 250-276.
VARIABLES: Temperature: 0-65°C	PREPARED BY: C. Guminski; Z. Galus

EXPERIMENTAL VALUES:

The author investigated the Weston normal cell over a range of cadmium amalgam concentrations and temperatures. From the extensive data, the following cadmium solubilities have been derived by the compilers.

		Soly	
$t/°C$		mass %	at %
0		3	5.2
5		3.5	6.0
10		4	6.9
15		5	8.6
20		5.5	9.2
25		6	10.2
30		7	11.9
35		8	13.4
40		9	15.0
45		10	16.5
50		10.5	17.2
55		11	18.0
60		12	19.5
65		12.5	20.3

The measurements were not concerned with the solubility determinations so that precise results were not obtained. The same results are also presented in a later paper (1).

AUXILIARY INFORMATION

METHOD/APPARATUS/PROCEDURE:	SOURCE AND PURITY OF MATERIALS:
EMF measurements were made on cells of the type, $Cd(Hg)_x	CdSO_4(sat)	Hg_2SO_4, Hg$. The EMF attained a constant value when saturation was reached.	Cadmium was obtained from Kahlbaum, Merck, Baird and Tallock, and from Harrington. The Hg, $CdSO_4$, and Hg_2SO_4 were purified by prior methods (2).
	ESTIMATED ERROR: Soly: nothing specified. Temp: nothing specified.

	REFERENCES:
	1. Smith, F.E.
	   Z. Phys. Chem. 1920, *95*, 293.
	2. Smith, F.E.
	   Phil. Trans. Roy. Soc. 1908, *207*, 393.

COMPONENTS:	ORIGINAL MEASUREMENTS:
(1) Cadmium; Cd; [7440-43-9] (2) Mercury; Hg; [7439-97-6]	Schulze, A. Z. Phys. Chem. 1923, 105, 177-203.
VARIABLES: Temperature: 20-72°C	PREPARED BY: C. Guminski; Z. Galus

EXPERIMENTAL VALUES:

Crystallization temperatures of the saturated cadmium amalgams were reported as a function of cadmium concentration.

$t/°C$	at %[a]	mass %
20.0	9.2	5.4
38.9	13.76	8.21
45.4	15.51	9.33
57.4	19.06	11.66
61.6	20.46	12.60
72.3	23.74	14.85

[a] by compilers

AUXILIARY INFORMATION

METHOD/APPARATUS/PROCEDURE:	SOURCE AND PURITY OF MATERIALS:
The amalgams were prepared by mixing the metals, and cooling curves were obtained. The saturated liquid phase was also analyzed by dissolution of the amalgam with HNO_3, followed by precipitation of CdS with H_2S. The CdS was subsequently dissolved in HNO_3. Details of experimental method not given.	Cadmium supplied by Kahlbaum. Mercury was "purest" grade which was further vacuum distilled a few times.
	ESTIMATED ERROR: Soly: nothing specified. Temp: precision ± 0.1 K (compilers)
	REFERENCES:

COMPONENTS:	ORIGINAL MEASUREMENTS:
(1) Cadmium; Cd; [7440-43-9] (2) Mercury; Hg; [7439-97-6]	Honda, K.; Ishigaki, T. *Sci. Rep. Tohoku Univ.* 1925, *14*, 219-33.
VARIABLES:	PREPARED BY:
Temperature: 590-594 K	C. Guminski; Z. Galus

EXPERIMENTAL VALUES:

Depression of freezing point of cadmium was reported to be 1.29 and 3.75 K for 99.5 and 98.5 at % Cd amalgam, respectively. The melting point of Cd was assumed to be 594.1 K.

AUXILIARY INFORMATION

METHOD/APPARATUS/PROCEDURE:	SOURCE AND PURITY OF MATERIALS:
The usual method of thermal analysis was used. The alloys were melted in an alundum tube, and the melts were protected from oxidation with a thick layer of asbestos wool, over which was poured fluid paraffin or vaseline. Temperatures were measured with a copper-constantan thermocouple.	Extra pure metals from Merck were probably used.
	ESTIMATED ERROR: Soly: nothing specified. Temp: precision better than ± 0.5 K.
	REFERENCES:

COMPONENTS:	ORIGINAL MEASUREMENTS:
(1) Cadmium; Cd; [7440-43-9] (2) Mercury; Hg; [7439-97-6]	Moesveld, A.L.T.; De Meester, W.A.T. Z. Phys. Chem. 1927, 130, 146-53.
VARIABLES: Temperature: 0-41°C	PREPARED BY: C. Guminski; Z. Galus

EXPERIMENTAL VALUES:

Solubility of cadmium in mercury:

t/°C	Soly/mass %	Soly/at %[a]
0.00	2.76	4.82
9.00	3.70	6.42
17.00	4.63	7.97
25.00	5.70	9.74
33.00	6.86	11.62
41.00	8.21	13.76

[a] by compilers

AUXILIARY INFORMATION

METHOD/APPARATUS/PROCEDURE:	SOURCE AND PURITY OF MATERIALS:
Potential difference between saturated amalgam electrode and amalgam electrodes of various concentrations were measured; a saturated CdSO$_4$ solution was used as the electrolyte. The potential difference was equal to zero when both half-cells contained the saturated amalgam. The increase in the cadmium concentration in the second half-cell was obtained by electrolysis of the CdSO$_4$ solution.	Cadmium from Kahlbaum and pure mercury were used.
	ESTIMATED ERROR: Soly: accuracy ± 1% (compilers). Temp: nothing specified.
	REFERENCES:

COMPONENTS:	ORIGINAL MEASUREMENTS:
(1) Cadmium; Cd; [7440-43-9] (2) Mercury; Hg; [7439-97-6]	Mehl, R.F.; Barrett, Ch.S. *Trans. AIME* 1930, *89*, 575-88.
VARIABLES: Temperature: (-35)-88°C	PREPARED BY: C. Guminski; Z. Galus

EXPERIMENTAL VALUES:

The authors present their data in graphical form. The compilers read off the following liquidus data points from the curve.

$t/°C$	Soly/at %
-35	0.8
-34	1.3
-25	2
-19	2.5
-13	3
-10	3.5
-2	5
+6.5	6.5
12.5	7.5
17.5	8.5
28.5	10.5
48	15.5
57	18.5
65.5	20
74	22
76	24
85.5	27.5
88	28

AUXILIARY INFORMATION

METHOD/APPARATUS/PROCEDURE:	SOURCE AND PURITY OF MATERIALS:
The amalgams were prepared by mixing the metals in a Pyrex tube. Heating and cooling curves were recorded with calibrated iron-constantan thermocouples; thermo-potentials were measured with a precision potentiometer.	Mercury was purified with nitric acid then twice distilled at a low pressure. Cadmium was 99.90% pure with traces of Zn, Pb, and Fe.
	ESTIMATED ERROR: Soly: nothing specified. Temp: precision ± 0.1 K in original measurements; accuracy ± 1 K at best for values read from graph.
	REFERENCES:

COMPONENTS:	ORIGINAL MEASUREMENTS:
(1) Cadmium; Cd; [7440-43-9] (2) Mercury; Hg; [7439-97-6]	Walls, H.A.; Upthegrove, W.R. *J. Chem. Eng. Data* 1964, *9*, 184-7.
VARIABLES: Temperature: 323-366 K	PREPARED BY: C. Guminski; Z. Galus

EXPERIMENTAL VALUES:

Solubility of cadmium in mercury:

T/K	Soly/mass %	Soly/at %
323.2	9.710	16.102
343.4	13.758	22.161
366.4	19.310	29.927

AUXILIARY INFORMATION

METHOD/APPARATUS/PROCEDURE:	SOURCE AND PURITY OF MATERIALS:
Amalgams were prepared by directly dissolving Cd in Hg, and the concentration ascertained from the known weights of each component. The solubilities were determined by extrapolating the concentration versus EMF curve to zero potential for the cell, $Cd(Hg)_{sat} \vert CdSO_4(aq) \vert Cd(Hg)_x$. The saturated amalgams were prepared at temperatures slightly above experimental and slowly cooled to assure equilibrium. The amalgams and electrolyte were handled under a blanket of argon to exclude air. Precision potentiometer and galvanometer were used; calibrated thermocouples were used for temperature measurements.	All materials were ACS Reagent Grade or better. ESTIMATED ERROR: Soly: accuracy better than ± 0.001%. Temp: precision ± 0.01 K. REFERENCES:

COMPONENTS:	ORIGINAL MEASUREMENTS:
(1) Cadmium; Cd; [7440-43-9] (2) Mercury; Hg; [7439-97-6]	Semibratova, N.M.; Yan-Sho-Syan, G.V.; Nosek, M.V. *Izv. Akad. Nauk Kaz. SSR, Ser. Khim* 1969, No. 5, 30-8.
VARIABLES: Temperature: (-38)-296°C	PREPARED BY: C. Guminski; Z. Galus

EXPERIMENTAL VALUES:

Liquidus temperatures for the Cd-Hg system were reported.

t/°C	Soly/at %
-38	1.0
-25	2.5
-10	5.0
+27	10.0
44	15.0
61	20.0
69	22.5
94	30.0
107	35.0
116	38.0
122	40.0
128	42.5
149	50.0
177	60.0
184	62.5
202	65.0
221	70.0
237	72.5
239	75.0
257	78.0
261	80.0
273	82.5
287	85.0
296	90.0

AUXILIARY INFORMATION

METHOD/APPARATUS/PROCEDURE:	SOURCE AND PURITY OF MATERIALS:
The amalgams were prepared by dissolution of solid cadmium in mercury. Samples of the alloys were encapsulated in tubes with dry CO_2. The tubes were heated up to 350°C and the cooling curves were recorded with the use of a copper-constantan thermocouple.	Cadmium of purity "0". Mercury was purified chemically, electrochemically and doubly distilled under vacuum.
	ESTIMATED ERROR: Soly: not specified. Temp: accuracy $\pm$ 2 K.
	REFERENCES:

COMPONENTS:	ORIGINAL MEASUREMENTS:
(1) Cadmium; Hg; [7440-43-9] (2) Mercury; Hg; [7439-97-6]	Zebreva, A.I.; Filippova, L.M.; Omarova, N.D. *Izv. Vysch. Uchebn. Zaved., Khim. Khim. Tekhnol.* 1976, *19*, 1043-6.
VARIABLES: Temperature: 25°C	PREPARED BY: C. Guminski; Z. Galus

EXPERIMENTAL VALUES:

Solubility of cadmium in mercury at 25°C was reported to be 9.6 at %.

AUXILIARY INFORMATION

METHOD/APPARATUS/PROCEDURE:	SOURCE AND PURITY OF MATERIALS:
The heterogeneous and homogeneous amalgams were prepared by direct dissolution of Cd in Hg. The amalgams were thermometrically titrated by the addition of Hg in a specially constructed apparatus. The Cd solubility was determined from the break-point of the curve relating composition to the thermal effect. All operations were performed in an argon atmosphere.	Nothing specified.
	ESTIMATED ERROR: Soly: precision ± 5%. Temp: precision ± 0.5 K.
	REFERENCES:

COMPONENTS:	ORIGINAL MEASUREMENTS:
(1) Cadmium; Cd; [7440-43-9] (2) Mercury; Hg; [7439-97-6]	Bukhman, S.P.; Lange, A.A.; Kairbaeva, A.A. *Izv. Akad. Nauk Kaz. SSR, Ser. Khim.* <u>1984</u>, No. 1, 31-4.
VARIABLES:	PREPARED BY:
Temperature: 17-23°C	C. Guminski; Z. Galus

EXPERIMENTAL VALUES:

The solubilities of Cd in Hg:

$t/°C$	Soly/mass %	Soly/at %[a]
17	3.80	6.59
20	4.61	7.94
	4.72	8.12
	4.57	7.87
21	4.86	8.35
22	5.20	9.25
	5.27	9.37
	5.27	9.37
23	5.67	9.69

[a] by compilers

The results at lower temperatures are understated. The $CdHg_3$ solid phase was identified to be in equilibrium with the saturated amalgam.

AUXILIARY INFORMATION

METHOD/APPARATUS/PROCEDURE:	SOURCE AND PURITY OF MATERIALS:
The Cd amalgam was obtained by electrolysis of solution of $CdSO_4$ in 1 mol dm^{-3} H_2SO_4. The amalgam was conditioned 24-30 h at cathodic polarization and then filtered. The filtrate was dissolved completely in HNO_3. Hg(II) was reduced with formic acid and Cd(II) was analyzed by atomic absorption spectroscopy or by titration with EDTA.	$CdSO_4$ was analytically pure. Mercury purity not specified.
	ESTIMATED ERROR: Soly: precision better than ± 2% (compilers). Temp: nothing specified.
	REFERENCES:

COMPONENTS:	EVALUATOR:
(1) Radioactive Elements (2) Mercury; Hg; [7439-97-6]	C. Guminski; Z. Galus Department of Chemistry University of Warsaw Warsaw, Poland July, 1985

CRITICAL EVALUATION:

No experimental determinations have been reported for the solubility of technetium, promethium, polonium, francium, radium, actinium, and protactinium in mercury. On the other hand, experimental data have been reported for the actinides which are of importance to the nuclear energy programs, and the solubility of these elements have been reported separately. The only data reported for the former seven elements are the predicted solubilities of Kozin (1,2) at 298 K; these are summarized in Table I. It is the opinion of the evaluators that the data from (2) are nearer to the correct value, although some of these data also are clearly incorrect. The value predicted for promethium (2) appears to be of the correct magnitude by comparison with the solubility of the other lanthanides in mercury at 298 K. However, by comparison with the solubility of elements in the same groups, those predicted for polonium, francium, and radium appear too high to the evaluators. In the case of francium, the predicted value of 99.9 at % would be of the correct magnitude for the Fr-rich region, similar to that for the Cs-Hg system.

The saturated polonium amalgam should be in equilibrium with solid PoHg (3).

TABLE I

Kozin's Predicted Solubility of Radioactive
Elements in Mercury at 298 K

Element	Soly/at %	Reference
Technetium; Tc; [7440-26-8]	3.0×10^{-13}	1
	1.1×10^{-9}	2
Promethium, Pm; [7440-12-2]	6.2×10^{-3}	1
	1.1×10^{-2}	2
Polonium; Po; [7440-08-6]	1.6	2
Francium; Fr; [7440-73-5]	99.9	2
Radium; Ra; [7440-14-4]	1.1	1
Actinium; Ac; [7440-34-8]	3.6×10^{-4}	1
	1.2×10^{-3}	2
Protactinium; Pa; [7440-13-3]	2.4×10^{-4}	1
	6.9×10^{-4}	2

References

1. Kozin, L.F. *Tr. Inst. Khim. Nauk Akad. Nauk Kaz. SSR* 1962, *9*, 101.
2. Kozin, L.F. *Fiziko Khimicheskie Osnovy Amalgamnoi Metallurgii*, Nauka, Alma-Ata, 1964.
3. Witteman, G.W.; Giorgi, A.L.; Vier, D.T. *J. Phys. Chem.* 1960, *64*, 434.

COMPONENTS:	EVALUATOR:
(1) Thorium; Th; [7440-29-1] (2) Mercury; Hg; [7439-97-6]	C. Guminski; Z. Galus Department of Chemistry University of Warsaw Warsaw, Poland July, 1985

CRITICAL EVALUATION:

Messing and Dean (1) found that the solubility of thorium in mercury increased from 1.82×10^{-3} to 2.55×10^{-2} at % in the temperature range of 313 to 629 K. Jangg and Palman (2) determined thorium solubilities ranging from 1.3×10^{-3} to 3.5×10^{-2} at % at 293 to 673 K. The solubilities reported by (1) and (2) are similar, and in the opinion of the evaluators these are the most accurate data; both groups of workers employed equilibration, filtration, and chemical analyses of the amalgams for their solubility determinations. Room temperature determinations reported by other workers, 7×10^{-3} (3) and 1.36×10^{-2} at % (4) at 298 K, are rejected because they are much higher than those determined by (1) and (2). Much higher solubilities were obtained by Domagala and coworkers (5) who reported 0.53 to 4.8 at % in the temperature range of 337 to 571 K. Kozin's (6) predicted value of 7.3×10^{-5} at % at 298 K is much too low.

The saturated thorium amalgams are in equilibrium with the compounds $ThHg_3$, $ThHg_2$ and ThHg which are stable up to 773, 860 and 920 K, respectively (5,7).

The solubility of thorium in saturated uranium amalgam has been reported to be approximately one-half that in mercury (1).

Tentative values of the solubility of Th in Hg:

T/K	Soly/at %	Reference
293	1.3×10^{-3}	[2]
298	1.5×10^{-3} [a]	[1,2]
323	2.3×10^{-3} [a]	[1,2]
373	4.6×10^{-3} [a]	[1,2]
473	1.2×10^{-2} [a]	[1,2]
573	2.1×10^{-2} [a]	[1,2]
673	3.2×10^{-2} [a]	[1,2]

[a] Interpolated value from data of (1) and (2).

References

1. Messing, A.F.; Dean, O.C. *U.S. At. Ener. Comm. Rep.*, ORNL-2871, 1960.
2. Jangg, G.; Palman, H. *Z. Metallk.* 1963, 54, 364.
3. Strachan, J.F.; Harris, N.L. *J. Inst. Metals* 1956-57, 85, 17.
4. Parks, W.G.; Prime, G.E. *J. Am. Chem. Soc.* 1936, 58, 1413.
5. Domagala, R.F.; Elliott, R.P.; Rostocker, W. *Trans. AIME* 1958, 212, 393.
6. Kozin, L.F. *Fiziko-Khimicheskie Osnovy Amalgamnoi Metallurgii*, Nauka, Alma-Ata, 1964.
7. Jangg, G.; Steppan, F. *Z. Metallk.* 1965, 56, 172.

COMPONENTS:	ORIGINAL MEASUREMENTS:
(1) Thorium; Th; [7440-29-1] (2) Mercury; Hg; [7439-97-6]	Messing, A.F.; Dean, O.C. *U.S. At. Ener. Comm. Rep.*, ORNL-2871, 1960.
VARIABLES: Temperature: 40-356°C	PREPARED BY: C. Guminski; Z. Galus

EXPERIMENTAL VALUES:

The solubility of thorium in mercury.

t/°C	Soly/mass %	Soly/at %
40	0.00211	0.00182
60	0.00313	0.00270
120	0.00675	0.00583
160	0.00921	0.00790
200	0.0120	0.0104
220	0.0151	0.0130
280	0.0203	0.0175
300	0.0235	0.0203
356	0.0295	0.0255

The authors observed that the solubility of thorium in saturated uranium amalgam is approximately one-half that in pure mercury.

AUXILIARY INFORMATION

METHOD/APPARATUS/PROCEDURE:	SOURCE AND PURITY OF MATERIALS:
Mercury and thorium, after drying and outgassing in the stainless steel dissolver, were kept for several days at the desired temperature. After equilibration, a sample of liquid amalgam was forced through the filter. The sample was collected, dissolved in nitric acid, and analyzed for thorium and mercury.	Nothing specified.
	ESTIMATED ERROR: Soly: standard deviation in fitted equation is 0.02046. Temp: nothing specified.
	REFERENCES:

COMPONENTS:	ORIGINAL MEASUREMENTS:
(1) Thorium; Th; [7440-29-1] (2) Mercury; Hg; [7439-97-6]	Jangg, G.; Palman, H. Z. Metallk. <u>1963</u>, 54, 364-9.
VARIABLES: Temperature: 20-400°C	PREPARED BY: C. Guminski; Z. Galus

EXPERIMENTAL VALUES:

The solubility of thorium is presented graphically as a function of temperature. The data points on the curve were read off by the compilers:

$t/°C$	$Soly/10^3$ at %
20	1.3
50	2.8
100	4.5
150	8.1
200	10
250	17
300	20
350	28
400	35

AUXILIARY INFORMATION

METHOD/APPARATUS/PROCEDURE:	SOURCE AND PURITY OF MATERIALS:
The heterogeneous amalgam was introduced into a specially constructed apparatus made of refractory chromium steel. Such steel apparatus could be used because the solubility of iron in mercury is very low and the chromium (III) oxide film inhibits the wetting of the steel by mercury. After twelve hours of equilibration at the temperature of the experiment, the amalgam was filtered through the sintered iron-frit under the pressure of purified nitrogen. Usually, 3- to 4-fold filtration was necessary. The metal content was then analytically determined in the filtered saturated amalgam. For experiments carried out below 320°C, amalgam was equilibrated in a glass vessel. The analytical procedure is not described in the paper.	Nothing specified.
	ESTIMATED ERROR: Soly: precision ± 5%. Temp: precision ± 2 K.
	REFERENCES:

COMPONENTS:	EVALUATOR:
(1) Uranium; U; [7440-61-1] (2) Mercury; Hg; [7439-97-6]	C. Guminski; Z. Galus Department of Chemistry University of Warsaw Warsaw, Poland July, 1985

CRITICAL EVALUATION:

There have been numerous reports on the solubility determination of uranium in mercury, many of which in more recent years have been from laboratories associated with the processing of nuclear fuels. However, a number of the determinations, especially near room temperature, are either too low or too high. Tammann and Hinnüber (1) reported a solubility of 1.1×10^{-4} at % at 291 K, while several workers only reported the upper limit of 1×10^{-3} at % at room temperature (2,17,22,23,24); these results are all too low and are rejected. Chang and coworkers (12,21) reported solubilities as high as 6.2×10^{-2} at % at room temperature, and these high values are rejected. At higher temperatures, Magel and Dallas (9) obtained a solubility of 0.1 at % at 348 and 536 K; these results also are rejected because of the lack of experimental details. Kozin's (6) predicted solubility of 3.5×10^{-4} at % at 298 K is too low because his equation neglected the U-Hg interactions.

Ahmann and coworkers (3) reported solubilities of uranium at five temperatures between 298 and 623 K, of which the values at 373, 573 and 623 K are acceptable. Jangg and Palman (4) and Messing and Dean (5) employed similar methods for the equilibration and chemical analysis of the amalgams to determine the uranium solubilities over a wide temperature range. The solubilities determined by (4) at 293 to 540 K increased from 4.2×10^{-3} to 0.33 at %, while those determined by (5) at 313 to 629 K increased from 5.6×10^{-3} to 1.02 at % at increasing temperatures. The results of (4) and (5) are in good agreement and are considered by the evaluators to be the most accurate.

Kobayashi and coworkers (8,20) reported an acceptable solubility of 3.7×10^{-3} at % at room temperature, but an earlier determination (19) of 1.8×10^{-2} at % was too high and is rejected. Schweitzer (28) determined the solubilities at 296 to 526 K, with end values of 4.2×10^{-3} and 3.2×10^{-2} at % in this temperature range. Although the solubilities at both ends of the temperature range are acceptable, those at intermediate temperatures are up to 30% too low, and no experimental details are known to the evaluators. Ettmayer and Jangg (27) reported a solubility of 0.6 at % at 573 K. Forsberg (15), from vapor pressure measurements, reported an upper limit of 1.1 at % for the solubility of uranium at 630 K. Wymer (16) estimated a solubility of 0.94 at % at 630 K, while Morrison and Blanco (17,25), without giving details, reported 0.85 at % at the same temperature. Dean and coworkers (23,24) estimated that the saturated amalgam contains 0.95 at % at 630 K and at least 19 at % at 873 K and 23 atm.

Frost (7) presented a complete phase diagram, but the most recent work of Lee and coworkers (11) has shown that the phase diagram presented by Frost is incorrect. Also, the solubilities taken from the liquidus of Frost's phase diagram are of an order of magnitude too high at 373 and 628 K. The error in the work of (7) may be attributed to an incomplete dehydrogenation of the uranium which was used, and to the possible reaction of the amalgam with nitrogen and the quartz container. Moreover, the investigation of Forsberg (15) and Lee et al. (11) showed a strong influence of pressure on the decomposition temperature of the U-Hg solid phases. Based on a thermodynamic analysis of this system, Lee (26) predicted another version of the U-Hg phase diagram with congruent melting of UHg_2 at 913 K and eutectic point at 748 K for 65 at % U. However, thermal analysis experiments of Lee et al. (11) did not confirm the prediction. The determined points on the liquidus reach a value of 10.0 at % U at 1118 K.

Although there have been several empirical equations fitted to the solubilities as a function of the temperature (5,11,14,28), there appears to be relatively poor agreement among these equations. This system needs further work in the composition range of 33 to 100 at % U. The saturated uranium amalgams are in equilibrium with U-Hg solid intermetallic compounds (3,7,10,11,13,26), as shown in the phase diagram, Fig. 1, reported by (11).

Kinetics of U dissolution in Hg and the saturated amalgam was investigated in (17,18,23,24).

Addition of Mg or Bi increases, and addition of Na or Th decreases, the uranium solubility in mercury (5,23,24).

(Continued next page)

COMPONENTS:	EVALUATOR:
(1) Uranium; U; [7440-61-1] (2) Mercury; Hg; [7439-97-6]	C. Guminski; Z. Galus Department of Chemistry University of Warsaw Warsaw, Poland July, 1985

CRITICAL EVALUATION: (Continued)

Tentative values of the solubility of U in Hg:

T/K	Soly/at %	Reference
293	4.0×10^{-3}	[4,8,20,28]
323	9×10^{-3}	[4,5]
373	2.5×10^{-2}	[3,5]
473	0.17	[4]
573	0.5	[3,5,27]
673	1.5^a	[5,11,15]
773	2.5^a	[11]
873	4^a	[11]
973	6^a	[11]
1073	8^a	[11]

[a] Solubility obtained by interpolation of data in cited references.

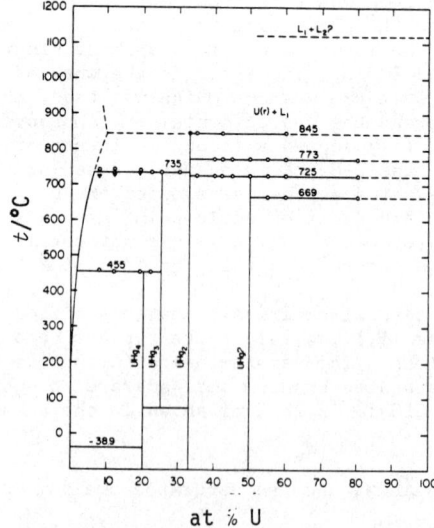

Fig. 1. U-Hg phase diagram under constrained vapor (11).

COMPONENTS:	EVALUATOR:
(1) Uranium; U; [7440-61-1] (2) Mercury; Hg; [7439-97-6]	C. Guminski; Z. Galus Department of Chemistry University of Warsaw Warsaw, Poland July, 1985

CRITICAL EVALUATION:

References

1. Tammann, G.; Hinnüber, J. *Z. Anorg. Chem.* 1927, *160*, 260.
2. Irvin, N.M.; Russell, A.S. *J. Chem. Soc.* 1932, 891.
3. Ahmann, D.H.; Baldwin, R.R.; Wilson, A.S. *U.S. At. Ener. Comm. Rep. CT-2960*, 1945.
4. Jangg, G.; Palman, H. *Z. Metallk.* 1963, *54*, 364.
5. Messing, A.F.; Dean. O.C. *U.S. At. Ener. Comm. Rep. ORNL-2871*, 1960.
6. Kozin, L.F. *Fiziko-Khimicheskie Osnovy Amalgamnoi Metallurgi*, Nauka, Alma-Ata, 1964.
7. Frost, B.R.T. *J. Inst. Metals* 1953-54, *82*, 456; *At. Ener. Res. Establ., M/R-1208*, 1953.
8. Kobayashi, Y.; Saito, A. *J. Nucl. Sci. Technol.* 1975, *12*, 48.
9. Magel, T.T.; Dallas, H.S. *U.S. At. Ener. Comm. Rep. CK-591*, 1943, as cited by 3.
10. Jangg, G.; Steppan, F. *Z. Metallk.* 1965, *56*, 172.
11. Lee, T.S.; Chiotti, P.; Mason, J.T. *J. Less-Common Metals* 1979, *66*, 33.
12. Yu, T.L.; Lee, Y.S.; Chuang, Y.D.; Chang, C.T. *J. Nucl. Sci. Technol.* 1979, *16*, 508.
13. Merlo, F.; Fornasini, M.L. *J. Less-Common Metals* 1979, *64*, 221.
14. Walker, R.A.; Pratt, J.N. *Rep. Dep. Phys. Metall. Sci. Mater., Univ. of Birmingham*, 1971.
15. Forsberg, H.C. *U.S. At. Ener. Comm. Rep., ORNL-2885*, 1960.
16. Wymer, R.G. cited by B. H. Morrison, R. E. Blanco in ref. 17.
17. Morrison, B.H.; Blanco, R.E. *U.S. At. Ener. Comm. Rep., CF-56-1-151*, 1956.
18. Segre, G.J. *Ital. At. Ener. Comm. Rep., CNI-16*, 1959.
19. Kobayashi, Y.; Ishimori, T. *J. Inorg. Nucl. Chem.* 1969, *31*, 981.
20. Malan, H.P.; Kobayashi, Y.; Ishimori, T. *J. Inorg. Nucl. Chem.* 1971, *33*, 3097.
21. Lee, H.C.; Wang, L.C.; Hung, H.H.; Chang, C.T. *J. Chem. Soc., Chem. Commun.* 1975, 124.
22. Jangg, G. *Atompraxis* 1962, *8*, 87.
23. Dean, O.C.; Sturch, E.; Morrison, B.H.; Blanco, R.E. *U.S. At. Ener. Comm. Rep., ORNL-2242*, 1957.
24. Dean, O.C. *Progr. Nucl. Ener., Ser. 3* 1958, *2*, 412.
25. Blanco, R.E. *Nucl. Sci. Eng.* 1956, *1*, 409.
26. Lee, T.S. *U.S. At. Ener. Comm. Rep., IS-T-824*, 1979.
27. Ettmayer, P.; Jangg, G. *Monatsh. Chem.* 1973, *104*, 1120.
28. Schweitzer, D.G. *Brookhaven National Laboratory* 1958; unpublished results communicated to the evaluators by J. R. Weeks.

COMPONENTS:	ORIGINAL MEASUREMENTS:
(1) Uranium; U; [7440-61-1] (2) Mercury; Hg; [7439-97-6]	Ahmann, D.H.; Baldwin, R.R.; Wilson, A.S. U.S. At. Ener. Comm. Rep. CT-2960, 1945.

VARIABLES:	PREPARED BY:
Temperature: 25-350°C	C. Guminski; Z. Galus

EXPERIMENTAL VALUES:

The solubility of uranium in mercury:

$t/°C$	Soly/mass %	Soly/at %[a]
25	0.001-0.01	0.0008-0.008
100	0.03	0.025
200	0.05	0.042
300	0.50	0.42
350	1.06	0.89

[a] by compilers

The solid phases in equilibrium with the homogeneous amalgam are UHg_4, UHg_3 and UHg_2.

AUXILIARY INFORMATION

METHOD/APPARATUS/PROCEDURE:	SOURCE AND PURITY OF MATERIALS:
1 to 2% uranium amalgams were placed on a fine-porosity sintered-glass filter in a special apparatus. The amalgams were covered with Na_2CO_3 to protect them to some extent from air. The apparatus was then heated to desired temperature and centrifuged immediately for 20 to 30 sec. Control runs with the asbestos packed centrifuge cup indicated that the temperature dropped only about 10° at 300°C during the centrifugation. The filtrate after a given run was then analyzed for uranium and mercury.	Uranium purity was better than 99.9%. Mercury was washed with nitric acid, then triple distilled in glass.
	ESTIMATED ERROR: Soly: not specified; error probably quite high (compilers). Temp: nothing specified.
	REFERENCES:

COMPONENTS:	ORIGINAL MEASUREMENTS:
(1) Uranium; U; [7440-61-1] (2) Mercury; Hg; [7439-97-6]	Dean, O.C.; Sturch, E.S.; Morrison, B.V.; Blanco, R.E. *U.S. At. Ener. Comm. Rep.*, ORNL-2242, 1957.
VARIABLES:	PREPARED BY:
Temperature: 298-873 K	C. Guminski; Z. Galus

EXPERIMENTAL VALUES:

The solubility of uranium in mercury:

T/K	Soly/mass %	Soly/at %[a]	Pressure/atm
298	<1 x 10^{-3}	--	--
629	1.12	0.95	--
873		≥19	23

These results were also presented in (1); kinetics of dissolution and the solubilities of U in Hg and in Bi, Mg and Na amalgams were investigated in this work.

AUXILIARY INFORMATION

METHOD/APPARATUS/PROCEDURE:	SOURCE AND PURITY OF MATERIALS:
A sample of U was heated in boiling Hg for 30 min. under an argon atmosphere. The amalgam was filtered at 629 and at 298 K, and the filtrates were analyzed after dissolution in nitric acid. The method of estimation at 873 K is not specified.	Nothing specified but probably the same as in (2): i.e., U of highest purity available. Hg purified by filtering, washing with HNO_3 and double distillation under vacuum
	ESTIMATED ERROR: Soly: nothing specified; precision better than ± 5% (compilers). Temp: nothing specified.
	REFERENCES: 1. Dean, O.C. *Progr. Nucl. Ener.*, Ser. 3 1958, 2, 412-9. 2. Forsberg, H.C. *U.S. At. Ener. Comm. Rep.*, ORNL-2885, 1960.

COMPONENTS:	ORIGINAL MEASUREMENTS:
(1) Uranium; U; [7440-61-1] (2) Mercury; Hg; [7439-97-6]	Messing, A.F.; Dean, O.C. *U.S. At. Ener. Comm. Rep.*, ORNL-2871, <u>1960</u>.
VARIABLES:	PREPARED BY:
Temperature: 40-356°C	C. Guminski; Z. Galus

EXPERIMENTAL VALUES:

The solubility of uranium in mercury:

t/°C	Soly/mass %	Soly/at %
40	0.0067	0.0056
50	0.0093	0.0078
70	0.0155	0.0131
100	0.0340	0.0286
145	0.0826±0.0007	0.0696
150	0.0930	0.0783
205	0.234±0.004	0.197
250	0.436±0.017	0.368
300	0.727±0.002	0.613
356	1.21±0.03	1.02

The authors also reported that the solubility of uranium in 0.1 mass % magnesium amalgam was higher than in pure mercury, and that the solubility in saturated thorium amalgam was lower than in mercury.

AUXILIARY INFORMATION

METHOD/APPARATUS/PROCEDURE:	SOURCE AND PURITY OF MATERIALS:
Mercury and uranium, after drying and outgassing in the stainless steel dissolver, were kept for several days at the desired temperature. After equilibration, a sample of liquid amalgam was forced through the filter and filtrate was collected, dissolved in nitric acid, and submitted for analysis for uranium and mercury.	Nothing specified.
	ESTIMATED ERROR: Soly: standard deviation of fitted equation was 0.05136. Temp: nothing specified.
	REFERENCES:

COMPONENTS:	ORIGINAL MEASUREMENTS:
(1) Uranium; U; [7440-61-1] (2) Mercury; Hg; [7439-97-6]	Jangg, G.; Palman, H. Z. Metallk. 1963, 54, 364-9.
VARIABLES: Temperature: 20-267 K	PREPARED BY: C. Guminski; Z. Galus

EXPERIMENTAL VALUES:

The solubility of uranium in mercury was presented graphically as a function of temperature. Numerical values of the data points were read from the curve by the compilers.

$t/°C$	Soly/at %
20	4.2×10^{-3}
50	9.8×10^{-3}
100	3.2×10^{-2}
150	7.4×10^{-2}
162	9.6×10^{-2}
200	0.17
243	0.25
250	0.28
267	0.33

AUXILIARY INFORMATION

METHOD/APPARATUS/PROCEDURE:	SOURCE AND PURITY OF MATERIALS:
The heterogeneous amalgam was introduced into a specially constructed apparatus made of glass. After twelve hours of equilibration at the temperature of the experiment, the amalgam was filtered through the sintered-glass frit under the pressure of purified nitrogen. The metal content was then analytically determined in the filtered saturated amalgam by an unspecified method.	Nothing specified.
	ESTIMATED ERROR: Soly: accuracy ± 5%. Temp: precision ± 2 K.
	REFERENCES:

COMPONENTS:	ORIGINAL MEASUREMENTS:
(1) Uranium; U; [7440-61-1] (2) Mercury; Hg; [7439-97-6]	Ettmayer, P.; Jangg, G. *Monatsh. Chem.* 1973, *104*, 1120-30.
VARIABLES: One temperature: 573 K	PREPARED BY: C. Guminski; Z. Galus

EXPERIMENTAL VALUES:

The solubility of U in Hg at 573 K was reported to be 0.6 mass %. The atomic % solubility calculated by the compilers is 0.5 at %.

AUXILIARY INFORMATION

METHOD/APPARATUS/PROCEDURE:	SOURCE AND PURITY OF MATERIALS:
Uranium amalgam was obtained by dissolution of U turnings in Hg. The materials were placed in a bomb and heated to 723-773 K. The amalgam was filtered and analyzed by an unspecified method.	Nothing specified.
	ESTIMATED ERROR: Soly: nothing specified; about ± 5% (compilers). Temp: precision ± 2 K (compilers).
	REFERENCES:

COMPONENTS:	ORIGINAL MEASUREMENTS:
(1) Uranium; U; [7440-61-1] (2) Mercury; Hg; [7439-97-6]	Lee, T.S.; Chiotti, P.; Mason, J.T. *J. Less-Common Metals* 1979, *66*, 33-40.
VARIABLES: Temperature: 455-845°C Pressure	PREPARED BY: C. Guminski; Z. Galus

EXPERIMENTAL VALUES:

The points on the U-Hg liquidus line were determined under constrained pressure:

t/°C	Soly/at %	Pressure/atm
455	2.0	--
735	<6.5	--
845	9.5-10.0	90

AUXILIARY INFORMATION

METHOD/APPARATUS/PROCEDURE:	SOURCE AND PURITY OF MATERIALS:
Preequilibrated alloy or the separate metals were sealed in tantalum capsules in a He atmosphere; a thermocouple well was sealed to the bottom of the capsule. Differential thermal analysis was made in a He atmosphere by inserting the filled and an empty capsule in a nickel block. Chromel-Alumel thermocouples were used for the DTA; the samples were heated in a split tube furnace.	Reactor grade U and high purity, triply distilled Hg were used. Chemical analysis of U showed $1-5 \times 10^{-2}$ and $0.5-10 \times 10^{-2}$ mass % of oxygen and carbon, respectively. The alloys contained less than 6×10^{-2} mass % of Ta.
	ESTIMATED ERROR: Soly: nothing specified. Temp: nothing specified; ± 5 K (by compilers).
	REFERENCES:

COMPONENTS:	EVALUATOR:
(1) Plutonium; Pu; [7440-07-5] (2) Mercury; Hg; [7439-97-6]	C. Guminski; Z. Galus Department of Chemistry University of Warsaw Warsaw, Poland July, 1985

CRITICAL EVALUATION:

White (1) reported that the solubility of plutonium in mercury at room temperature is 1.36×10^{-2} at %. Bowersox and Leary (2,3) made more extensive measurements of the plutonium solubility; these authors reported that the solubility increases from 1.31×10^{-2} to 0.561 at % in the temperature range of 294 to 598 K. The result of (1) is in good agreement with those reported by Bowersox and Leary.

The saturated plutonium amalgam is in equilibrium with the Pu-Hg intermetallic compound, Pu_5Hg_{21} or $PuHg_3$ (2,4); however, the temperature range of stability for these compounds have not been established. The partial phase diagram has been reported by (5) and (6).

The recommended (r) and tentative values of the solubility of Pu in Hg:

T/K	Soly/at %	Reference
293	1.3×10^{-2} (r)	[1,2]
298	1.5×10^{-2}	[2]
323	2.6×10^{-2}	[2]
373	6.4×10^{-2}	[2][a]
473	2.2×10^{-1}	[2]
573	4.8×10^{-1}	[2]

[a] Interpolated value from data of (2).

References

1. White, A.G. *At. Ener. Res. Establ. Rep.*, *C/R 1468*, 1955.
2. Bowersox, D.F.; Leary, J.A. *J. Inorg. Nucl. Chem.* 1959, 9, 108.
3. Bowersox, D.F.; Leary, J.A. *U.S. At. Ener. Comm. Rep.*, *LAMS-2518*, 1961.
4. Berndt, A.F. *J. Less-Common Metals* 1966, 11, 216.
5. Schonfeld, F.W. *The Metal Plutonium*, A. S. Coffinberry, W. N. Miner, Eds., The University of Chicago Press, Chicago, 1961, p. 248.
6. Blank, H.; Brossmann, G.; Kemmerick, M. *F.R.G. At. Ener. Comm. Rep.*, *KFK-105*, 1962, p. 137.

COMPONENTS:	ORIGINAL MEASUREMENTS:
(1) Plutonium; Pu; [7440-07-5] (2) Mercury; Hg; [7439-97-6]	Bowersox, D.F.; Leary, J.A. *J. Inorg. Nucl. Chem.* <u>1959</u>, *9*, 108-112.
VARIABLES: Temperature: 20-325°C	PREPARED BY: C. Guminski; Z. Galus

EXPERIMENTAL VALUES:

Solubility of plutonium in mercury:

$t/°C$	Soly/10^2 at %
20[a]	1.61
21	1.31
24	1.61
50	2.55
100	6.25
150[a]	12.6
190	18.2
200	19.0
225	27.5
260	38.0
280	42.1
300	49.6
325[a]	56.1

[a] also reported in (1).

AUXILIARY INFORMATION

METHOD/APPARATUS/PROCEDURE:	SOURCE AND PURITY OF MATERIALS:
Mercury was outgassed in the reaction vessel at 250°C, then cooled to room temperature. The vessel was filled with helium and freshly machined plutonium turnings were added. The evacuation and filling of the vessel with helium were repeated several times. The mixture of the metals was held at 250-300°C for one day and was shaken periodically by hand. The temperature of the vessel was adjusted at desired level. The liquid phase was sampled periodically and filtered through a sintered-glass filter. Plutonium was leached from the filtrate by contacting with concentrated HCl for one day. The solution was analyzed by radio-assay for Pu content.	99.8% pure plutonium and triply-distilled mercury were used.
	ESTIMATED ERROR: Soly: accuracy ± 1%. Temp: precision ± 2%.
	REFERENCES: 1. Bowersox, D.F.; Leary, J.A. *U.S. At. Ener. Comm. Rep.*, <u>LAMS-2518</u>, <u>1961</u>.

SYSTEM INDEX

Page numbers preceded by E refer to evaluation text whereas those not preceded by E refer to compiled tables.

Actinium	E421
Aluminium	E84, E85, 86-92
Antimony	E173, E174, 175-181
Arsenic	E172
Barium	E76, E77, 78-82
Beryllium	E55, 56
Bismuth	E182, E183, 198-193
Boron	E83
Cadmium	E402-E404, 405-420
Calcium	E66, E67, 68-70
Carbon	E134
Cerium	E214, E215, 216-220
Cesium	E50, E51, 52-54
Chromium	E277, 278-281
Cobalt	E310, E311, 312, 313
Copper	E335-E337, 338-356
Dysprosium	E245, 246
Erbium	E249, 250
Europium	E234, 235-238
Francium	E421
Gadolinium	E239, 240, 241
Gallium	E93-E95, 96-102
Germanium	E135, 136-138
Gold	E139-E141, 142
Hafnium	E267
Holmium	E247, 248
Indium	E103-E105, 106-118
Iridium	E316
Iron	E301, E302, 303-306
Lanthanum	E207, E208, 209-213
Lead	E157-E159, 160-171
Lithium	E1-E3, 4-12
Lutetium	E256, 257
Magnesium	E57, E58, 59-65
Manganese	E285, E286, 287-299
Molybdenum	E282, 283
Neodymium	E225, 226-229
Nickel	E317, E318, 319-325
Niobium	E271, 272, 273
Osmium	E309
Palladium	E326, 327-329
Platinum	E330, 331-334
Plutonium	E434, 435
Polonium	E421
Potassium	E33, E34, 35-43
Praseodymium	E221, 222-224
Promethium	E421
Protactinium	E421
Radium	E421
Rhenium	E300
Rhodium	E314, 315
Rubidium	E44, E45, 46-49
Ruthenium	E307, 308
Samarium	E230, E231, 232, 233
Scandium	E206
Silicon	E134
Silver	E357-E359, 360-368
Sodium	E13-E18, 19-32
Strontium	E71, E72, 73-75
Tantalum	E274, 275, 276
Technetium	E421
Tellurium	E194, E195, 196-205
Terbium	E242, 243-244
Thallium	E119-E121, 122-133
Thorium	E422, 423, 424
Thulium	E251, 252

Tin	E139-E141, 142-156
Titanium	E258, E259, 260-262
Tungsten	E284
Uranium	E425-E427, 428-433
Vanadium	E268, 269, 270
Ytterbium	E253, 254, 255
Yttrium	E206
Zinc	E385-E387, 388-401
Zirconium	E263, E264, 265, 267

REGISTRY NUMBER INDEX

Page numbers preceded by E refer to evaluation text whereas those not preceded by E refer to compiled tables

7429-90-5	E84, E85, 86-92
7429-91-6	E245, 246
7439-88-5	E316
7439-89-6	E301, E302, E303-306
7439-91-0	E207-E208, 209-213
7439-92-1	E157-E159, 160-171
7439-93-2	E1-E3, 4-12
7439-94-3	E256, 257
7439-95-4	E57, E58, 59-65
7439-96-5	E285, E286, 287-299
7439-97-6	E1-E3, 4-12, E13-E18, 19-32, E33-E34, 35-43, E44-E45, 46-49, E50, E51, 52-54, E55, 56, E57, E58, 59-65, E66, E67, 68-70, E71, E72, 73-75, E76, E77, 78-82, E84, E85, 86-92, E93-E95, 96-102, E103-E105, 106-118, E119-E121, 122-133, E134, E135, 136-138, E139-E141, 142-156, E157-E159, 160-171, E172-E174, 175-181, E182, E183, 184-193, E194, E195, 196-205, E206-E208, 209-213, E214, E215, 216-220, E221, 222-224, E225, 226-229, E230, E231, 232, 233, E234, 235-238, E239, 240, 241, E242, 243, 244, E245, 246, E247, 248, E249, 250, E251, 252, E253, 254, 255, E256, 257, E258, E259, 260-262, E263, E264, 265, 266, E267, E268, 269, 270, E271, 272, 273, E274, 275, 276, E277, 278-281, E282, 283, E284-E286, 287-299, E300-e302, 303-306, E307, 308, E309, E310, E311, 312, 313, E314, 315, E316-E318, 319-325, E326, 327-329, E330, 331-334, E335-E337, 338-356, E357-E359, 360-368, E369-E371, 372-384, E385-E387, 388-401, E402-E404, 405-420, E421, E422, 423, 424, E425-E427, 428-433, E434, 435
7439-98-7	E282, 283
7440-00-8	E225, 226-229
7440-02-0	E317, E318, 319-325
7440-03-1	E271, 272, 273
7440-04-2	E309
7440-05-3	E326, 327-329
7440-06-4	E330, 331-334
7440-07-5	E434, 435
7440-08-6	E421
7440-09-7	E33-E34, 35-43,
7440-10-0	E221, 222-224
7440-12-2	E421
7440-13-3	E421
7440-14-4	E421
7440-15-5	E300,
7440-16-6	E314, 315
7440-17-7	E44-E45, 46-49
7440-18-8	E307, 308
7440-19-9	E230, E231, 232, 233
7440-20-2	E206
7440-21-3	E134
7440-22-4	E357-E359, 360-368
7440-23-5	E13-E18, 19-32
7440-24-6	E71, E72, 73-75
7440-25-7	E274, 275, 276
7440-26-8	E421
7440-27-9	E242, 243, 244
7440-28-0	E119-E121, 122-133
7440-29-1	E422, 423, 424

```
7440-30-4       E251, 252
7440-31-5       E139-E141, 142-156,
7440-32-6       E258, E259, 260-262
7440-33-7       E284
7440-34-8       E421

7440-36-0       E173, E174, 175-181
7440-38-2       E172
7440-39-3       E76, E77, 78-82
7440-41-7       E55
7440-42-8       E83

7440-43-9       E402-E404, 405-420
7440-44-0       E134
7440-45-1       E214, E215, 216-220
7440-46-2       E50, E51, 52-54
7440-47-3       E277, 278-281

7440-48-4       E310, E311, 312, 313
7440-50-8       E335-E337, 338-356
7440-52-0       E249, 250
7440-53-1       E234, 235-238
7440-54-2       E239, 240, 241

7440-55-3       E93-E95, 96-103
7440-56-4       E135, 136-138
7440-57-5       E369-E371, 372-384
7440-58-6       E267
7440-60-0       E247, 248

7440-61-1       E425-E427, 428-433
7440-62-2       E268, 269, 270
7440-64-4       E253, 254, 255
7440-65-5       E206
7440-66-6       E385-E387, 388-401

7440-67-7       E263, E264, 265, 266
7440-69-9       E182, E183, 184-193
7440-70-2       E66, E67, 68-70
7440-73-5       E421
7440-74-6       E103-E105, 106-118

13494-80-9      E194, E195, 196-205
```

AUTHOR INDEX

Page numbers preceded by E refer to evaluation text whereas those not preceded by E refer to compilation tables.

Adlhart, O.	E301, E302
Agasyan, P. K.	E172
Ahmann, D. H.	E425-E427, 428
Akimov, V. P.	E230, E231, E234, 236
Allibert, M.	E182, E183, 191
Amarell, G.	E93-E95
Anderko, K.	E13-E18, E57, E58, E84, E85, E139-E141
Anderson, J. T.	E369-E371, 380
Aramian, K. T.	E369-E371
Armbruster, M. H.	E33, E34, 41
Atamanova, N. M.	E402-E404
Aureggi, C.	E194, E195, 199
Aven, M.	E194, E195
Aygaraeva, M. M.	E214, E215
Babanly, M. B.	E194, E195
Babinski, J. J.	E157-E159
Babkin, G. N.	E310, E317, E318
Badavamova, G. L.	E207-E208, 213, E214, E215, 220, E221, 223, 224, E225, 229, E230, E231, E234, 238, E242, 243
Badygina, L. I.	E194, E195
Baldwin, R. R.	E425-E427, 428
Balej, J.	E13-E18, 32
Baletskaya, L. G.	E335-E337
Ball, T. R.	E33, E34
Banick, C. J.	E103-E105, 108
Baranski, A.	E317, E318, 324
Barin, I.	E13-E18
Barlow, M.	E330
Barrett, C. S.	E402-E404, 416
Bartholomay, H. W.	E93-E95, 97
Batel, R.	E317, E318, 325
Beck, H.	E214, E215
Beck, R. P.	E57, E58, 62
Benjamin, L.	E385-E387, 396
Bennett, H. C.	E1-E3, 5, E33, E34, 39, E44, E45, 48, E50, E51, 53, 54, E66, E67, E71, E72, 75, E76, E77, 79
Bennett, J. A. R.	E139-E141, 150, E385-E387, 394
Bent, H. E.	E13-E18, 28, E33, E34, 40, E93-E95
Benz, R.	E263, E264, 265
Berndt, A. F.	E214, E215
Berthelot, M.	E13-E18
Bijl, H. C.	E402-E404, 408
Biltz, W.	E44, E45, 49
Biros, J.	E13-E18
Bittner, H.	E326
Bjorklund, C. W.	E263, E264, 265
Blair, J.	E194, E195
Blanco, R. E.	E425-E427, 429
Blank, H.	E434
Bolton, W.	E274
Bonnier, E.	E139-E141, E182, E183, 191
Bottger, W.	E1-E3, 4, E13-E18, 23, E33, E34, 38, E44, E45, 46, E57, E58, 59, E71, E72, 73, E76, E77, 78, E385-E387, 389, E402-E404, 407
Bowersox, D. F.	E207, E208, 210, E214, E215, 216, E263, E264, 265, E271, 272, E274, 275, E282, 283, E301, E302, 305, E307, 308, E434, 435
Braley, S. A.	E369-E371
Brandon, J.	E263, E264, E271
Braunstein, H.	E13-E18
Braunstein, J.	E13-E18
Brebrick, R. F.	E194, E195, 200
Brewer, L.	E282

Britton, G. T.	E369-E371, 376, 377,
Bros, J. P.	E93-E95, 102
Brossmann, G.	E434
Brown, J. B.	E369-E371
Brown, O. L.	E103-E105, 112
Bruzzone, C.	E66, E67, 70, E71, E72, E73, E76, E77, 80, E207-E208, 212
Bukhman, S. P.	E13-E18, 31, E76, E77, 81, E173, E174, 175, 176, E285, E286, 294, 296, E335-E337, 348, 352, E402-E404, 420
Bulina, V. A.	E207-E208, 213, E214, 219, E221, E225, 228, E239, 241, E242, 244, E245, 246, E247, 248, E249, 250, E251, 252, E256, 257
Burger, E.	E277
Burns, C. L.	E139-E141, 152
Butler, J. N.	E1-E3, 9, E326, 328, E330, 333
Bykhanov, I. M.	E194, E195
Caillet, M.	E139-E141, 153
Cambi, L.	E57, E58, 60, E66, E67, 68
Campbell, A. N.	E139-E141, E182, E183, E285, E286, E402-E404
Campbell, J. M.	E1-E3, 8
Campenella, J. L.	E207, E208, 209
Carlson, R. G.	E194, E195
Carter, H. D.	E139-E141
Carter, M. D.	E285, E286
Chang, C. T.	E425-E427
Chao, F.	E335-E337, 346
Chevalet, J.	E317, E318, 325
Chiotti, P.	E425-E427, 433
Chiranzelli, R. V.	E103-E105, 112
Chuang, Y. D.	E425-E427
Claire, Y.	E119-E121, 133
Coffinberry, A. S.	E434
Cogley, D. R.	E1-E3, 9
Cohen, E.	E385-E387, 391
Coles, B. R.	E103-E105, 114
Costa, M.	E335-E337, 346
Crenshaw, J. L.	E33, E34, 41, E385-E387, 392
Cusack, N. E.	E103-E105, 115
D'Abramo, G.	E93-E95, 101
Dallas, H. S.	E425-E427
Daniels, F.	E119-E121, 125
Danilchenko, P. T.	E57, E58, 64
Dayananda, M. A.	E385-E387, 397
Dean, O. C.	E1-E3, E225, 226, E230, 232, E239, 240, E263, E264, E282, E307, E326, E422, 423, E425-E427, 429, 430
De Gruyter, C. J.	E84, E95, 87
De Lury, R. H.	E402-E404, 411
Delves, R. T.	E194, E195, 198
De Meester, W. A. T.	E402-E404, 415
Dergacheva, M. B.	E57, E58, 65, E172, E182, E183, 192, E357-E359, E369-E371, 384
DeRight, R.	E357-E359, 365
Desai, P. D.	E1-E3, E33, E34, E44, E45, E50, E51, E119-E121, E139-E141, E157-E159, E182, E183, E357-E359, E369-E371, E385-E387
Desre, P.	E139-E141, 153, E182, E183, 191
De Wet	E277, 278, E285, E286, 289, E301, E302, E310, E317, E318, 319
Domagala, R. F.	E422
Donche, H.	E173, E174, 181
Dortbudak, T.	E300, E307, E309, E314, 315, E316, E330
Dowgird, A.	E285, E286, 297
Dragavtseva, N. A.	E173, E174, 176, E335-E337, 348
Druzinina, E. P.	E93-E95, 99
Dubova, N. M.	E173, E174
Dzholdasova, R. M.	E207-E208, E214, E215, 217, 218, E230, E231
Dziuba, E. Z.	E194, E195, 201

Eastman, E. D.	E369-E371, 374,
Edelman, B. A.	E194, E195
Edwards, T. J.	E135
Eggers, H.	E44, E45, 49
Eggert, G. L.	E103-E105, 111
Eilert, A.	E66, E67, 69
Elliott, R. P.	E422
Enikeev, R. Sh.	E207-E208, 213, E214, E215, 218, 219, E221, 223, E225, 227, 228, E230, E234, E239, 241, E242, 246, E247, 248, E249, 250, E251, 252, E253, 255, E256, 257
Epstein, L. F.	E301, E302, 303, E317, E318
Eriksrud, E.	E335-E337, 354
Espenbetov, A. A.	E13-E18, E33, E34, E402-E404
Ettmayer, P.	E285, E286, 298, E425-E427, 432
Evans, E. J.	E369-E371
Eyring, L.	E225, E230, E231, E234, E256
Fay, H.	E157-E159
Feree, J.	E13-E18, 22, E33, E34, 37, E266, E282, E285, 295
Fielder, M.	E103-E105, 115
Filippova, L. M.	E1-E3, 11, E13-E18, 30, E33, E34, 43, E76, E77, E82, E119-E121, 132, E139-E141, 156, E157-E159, 170, E182, E183, 193, E335-E337, E385-E387, 401, E402-E404, 419
Fink, S.	E258, E259, E263, E271, E274, E277, E301, E302, 305, E310, E311, E317, E318, 323
Fisk, Z.	E103-E105, 114
Fitzer, E.	E301, E302
Flad, D.	E207-E208, E214, E221, E225
Fleitman, A. H.	E258, E259, E263, E264, E271, E274, E301, E302, E330
Fogh, I.	E84, E85, 86
Fornasini, M. L.	E214, E215, E221, E225, E230, E231, E239, E245, E253, E425-E427
Forsberg, H. C.	E425-E427
Forziati, A. E.	E13-E18
Franck, G.	E103-E105, 118
Frasunyak, V. M.	E194, E195
Frolkov, A. Z.	E207-E208, E214, E221, E225, E234, E230, 235, E242, E253, 254
Fronberg, M. G.	E182, E183, 187
Frost, B. R. T.	E425-E427
Galalenko, N. P.	E194, E195
Galus, Z.	E285, E286, 290, 297, E317, E318, 324, E326, 334
Garrod-Thomas, R. N.	E1-E3, E157-E159, E301, E302, E335-E337, 340
Gaune-Escard, M.	E93-E95, 102
Gavze, M. N.	E301, E302
Gayfullin, A. Sh.	E13-E18, 30, E119-E121, 132, E157-E159, E335-E337, E385-E387, E402-E404
Gayler, M. L. V.	E139-E141, 148
Gerasimov, Ya. I.	E230, E234
Gileadi, E.	E84, E85, 92
Gilfillan, E. S.	E33, E34, 40, E93-E95
Giorgi, A. L.	E421
Gladkikh, I. P.	E13-E18
Gladyshev, V. P.	E1-E3, 10, E33, E34, E135, 136, 138, E172, E194, E195, 204, E214, E215
Gleiser, M.	E1-E3, E33, E34, E44, E45, E50, E51, E119-E121, E139-E141, E157-E159, E182, E183, E357-E359, E369-E371, E385-E387
Goddard, J. B.	E1-E3, 8
Golonka, L.	E194, E195
Golubkov, A. V.	E194, E195
Gordin, V. L.	E173, E174
Gorley, N. P.	E194, E195

Gouy, M.	E139-E141, E157-E159, E182, E183, E301, E302, E357-E359, E369-E371, E385-E387, E402-E404
Grace, R. E.	E385-E387, 395, 397
Gradkikh, I. P.	E1-E3, E33, E34, E44, E45, E50, E51
Gramkee, B. E.	E369-E371, 378,
Griffin, R. B.	E221, 222, E225, E274
Grigoreva, M. J.	E402-E404
Grobe, G.	E1-E3, 7
Groll, W.	E326, 329
Gromakov, S. D.	E194, E195
Gronlund, F.	E335-E337, 356
Grosse, A. V.	E93-E95, 100
Gschneider, K. A.	E221, 222, E225, E274, E230, E231, E234, E256
Gudtsov, N. T.	E301, E302
Guminichenko, L. V.	E221, 223, E225, 228, E239, 241
Guminski, C.	E330, 334, E335-E337, 355
Guntz, A.	E13-E18, 22, E33, E34, 37
Guryanova, O. N.	E84, E85, E335-E337
Hajicek, O.	E385-E387
Hansen, M.	E13-E18, E57, E58, E84, E85, E139-E141
Haring, M. M.	E139-E141, 149, E157-E159, 167
Harman, T. C.	E194, E195
Harris, N. L.	E1-E3, E13-E18, E33, E34, E44, E45, E55, E57, E58, E66, E67, E76, E77, E84, E85, E103-E105, E119-E121, E134, E135, E139-E141, E157-E159, E172-E174, E182, E183, E258, E259, E263, E268, E271, E274, E277, E282, E284-E286, E301, E302, E307, E310, E314, E316-E318, E326, 327, E330, 332, E335-E337, E357-E359, E369-E371, E385-E387, E402-E404
Hatfield, M. R.	E157-E159, 167
Haul, R. A. W.	E277, 278, E285, E286, 289, E301, E302, E310, E317, E318, 319
Hawkins, D. T.	E1-E3, E33, E34, E44, E45, E50, E51, E119-E121, E139-E141, E157-E159, E182, E183, E357-E359, E369-E371, E385-E387
Henry, T. H.	E369-E371,
Herning, P. E.	E194, E195, 205
Hess, C. B.	E357-E359, 363
Heycock, C. T.	E13-E18, 20, E139-E141, 143, E157-E159, 161, E182, E183, 185, E402-E404, 406
Heyne, R.	E300
Hickling, A.	E285, E286, E335-E337
Hildebrand, J. H.	E13-E18, E369-E371, 374,
Hilpert, K.	E66, E67, E103-E105
Hinnuber, J.	E172-E174, E268, E277, E282, E284-E286, E425-E427
Hinzner, F. W.	E357-E359
Hohn, H.	E301, E302
Holden, R. B.	E55
Honda, K.	E139-E141, 147, E157-E159, 164, E402-E404, 414
Horner, L.	E1-E3, 12
Horsley, G. W.	E301, E302
Hoyt, C. S.	E157-E159, 165
Hudson, D. R.	E357-E359, 367
Hulett, G. A.	E402-E404, 411
Hultgren, R.	E1-E3, E33, E34, E44, E45, E50, E51, E119-E121, E139-E141, E157-E159, E182, E183, E357-E359, E369-E371, E385-E387
Hume-Rothery, W.	E369-E371, 383,
Humphreys, W. J.	E335-E337, 338, E357-E359, 360
Hung, H. H.	E425-E427
Hurlen, T.	E285, E286, 299, E335-E337, 354
Iandelli, A.	E214, E215, E221, E225, E230, E231, E234, E256
Ichigaki, T.	E402-E404, 414
Iggena, H.	E57, E58, 59, E71, E72, 73, E76, E77, 78, E385-E387, 389, E402-E404, 407
Ignateva, L. A.	E173, E174, E335-E337, 350, 351
Igolinskaya, I. M.	E84, E85, E335-E337

Igolinskii, V. A.	E84, E85, 91, E139-E141, E335-E337
Inoue, Y.	E13-E18, 29
Inouye, K.	E385-E387, 391
Irvin, N. M.	E258, E259, E268, E277, E282, E284-E286, E288, E301, E302, E310, E317, E318, E335-E337, 339 E425-E427
Ishigaki, T.	E139-E141, 147, E157-E159, 164
Ishimori, T.	E425-E427
Ito, H.	E103-E105, 108
Jaeger, W.	E402-E404
Janecke, E.	E13-E18, E157-E159, 163, E402-E404, 410
Jangg, G.	E66, E67, E84, E85, E103-E105, 113, E157-E159, E173, E174, 177, 179, E206, E258, E259, 260, E263, E264, E267, E268, E277, 279, E285, E286, 291, 292, 298, E300-E302, E307, E309, E310, 312, E314, 315, E316-E318, 321, E326, 329, E330, E335-E337, 344, 345, 349, E357-E359, 368, 399, E422, 424, E425-E427, 431, 432
Jayaraman, A.	E194, E195
Johnson, K. W. R.	E263, E264, 265
Joyner, R. A.	E139-E141, 146, E357-E359, 362
Kagan, I. K.	E402-E404
Kairbaeva, A. A.	E335-E337, 352, E402-E404, 420
Kakovskii, I. A.	E369-E371
Kalenberg, L.	E285, E286, 287
Kaltwasser, K.	E33, E34, 42
Kamenev, A. I.	E172
Kanda, F. A.	E119-E121
Karpinski, Z. J.	E135, 137
Kartzmark, A. A.	E402-E404
Kartzmark, E. M.	E182, E183
Kataev, G. A.	E84, E85, E139-E141, E335-E337, 350
Kazantsev, M.	E369-E371, 372,
Kelley, K. K.	E1-E3, E33, E34, E44, E45, E50, E51, E119-E121, E139-E141, E157-E159, E182, E183, E357-E359, E369-E371, E385-E387
Kells, M. C.	E55
Kelman, L. R.	E55, E263
Kemmerick, M.	E434
Kemula, W.	E285, E286, 290
Kendall, P.	E103-E105, 115
Kennedy, G. C.	E194, E195
Kerp, W.	E1-E3, 4, E13-E18, 23, E33, E34, 38, E44, E45, 46, E57, E58, 59, E71, E72, 73, E76, E77, 78 E385-E387, 389, E402-E404, 407
Kestovnikov, N. N.	E230, E234
Khanapina, K.	E230, E231, E234, 237
Khomyak, V. V.	E194, E195
Khvala, M. A.	E194, E195
Kirchmayr, H. R.	E157-E159, 168, E206, E215, E221, E225, E230, E231, E239, E242, E245, E247, E249, E253, E274, E285, E286, 292, E314, E335-E337, 345, E385-E387, 399,
Klamut, C. J.	E258, E259, E263, E264, E271, E274, E301, E302
Klement, W.	E194, E195
Klemm, W.	E84, E85, 88
Klyukas, Yu. E.	E76, E77, 82
Knacke, O.	E13-E18
Knutter, R.	E103-E105
Kobayashi, Y.	E425-E427
Kollmann, K.	E301, E302, E310, E317, E318, E335-E337, 341
Korobkina, N. P.	E33, E34, 43, E157-E159, 170, E385-E387, 401
Korshunov, V. I.	E230, E231
Korshunov, V. N.	E1-E3, E13-E18, E44, E45, E50, E51
Kotova, N. A.	E84, E85, E335-E337
Kovaleva, S. V.	E135, 138, E194, E195, 204

Kozin, L. F.	E1-E3, E33, E34, E44, E45, E50, E51, E55, E57, E58, 65, E66, E67, E71, E72, E76, E77, E83, E84 E85, E93-E95, E103-E105, 106, 110, 117, E119-E121, 129, E134, E135, E139-E141, E157-E159, E172-E174, E182, E183, 193, E194, E195, 206, E214, E215, E221, E225, E230, E234, E239, E242, E245, E247, E249, E251, E253, E256, E258, E259, E263, E267, E268, E271, E274, E277, E282, E284-E286, E300-E302, E307, E309, E310, E316-E318, E326, E330, E335-E337 E357-E359, E369-E371, 384, E385-E387, 398, E402-E404, E421, E422, E425-E427
Kozlovskii, A. J.	E214, E215
Kozlovskii, M. T.	E33, E34, E135, E173, E174, 178, E285, E286, 293, 296, E335-E337, 342
Krasnova, I. E.	E285, E286, E317, E318
Kraut, K.	E13-E18
Krebaeva, Sh. D.	E214, E215, 218, E221, E225, 227, E230, E234, E253, 255
Kremann, R.	E57, E58, E84, E85
Kresrovnikov, A. N.	E194, E195
Kristensen, B.	E335-E337, 356
Kublik, Z.	E135, 137, E335-E337, 353, E357-E359
Kuleshov, V. A.	E1-E3, 10
Kuliev, A. A.	E194, E195
Kurbanov, A. A.	E194, E195
Kurnakov, N. S.	E13-E18, 24, E33, E34, 36, E44, E45, 47, E50, E52, E119-E121
Kusma, I. B.	E206
Kuzmina, G. A.	E194, E195
Kuznetsov, F. A.	E230, E231, E234
Kuznetsova, N. K.	E1-E3, E13-E18, E33, E34, E44, E45, E50, E51
Lamoreaux, R. H.	E282
Lange, A. A.	E13-E18, 31, E76, E77, 81, E173, E174, 175 E285, E286, 294, 296, E335-E337, 352, E402-E404, 420
Latypov, Z. M.	E194, E195
Laube, E.	E206
Lavrentev, V. I.	E230, E234
Lawson, W. P.	E194, E195
Leary, J. A.	E207, E208, 210, E214, E215, 216, E263, E264, 265, E271, 272, E274, 275, E282, 283, E301, E302, 305, E307, 308, E434, 435
Lee, H. C.	E425-E427
Lee, T. S.	E425-E427, 433
Lee, Y. S.	E425-E427
Legler, E.	E173, E174, 177
Lehoczky, S. L.	E194, E195, 203
Levitskaya, S. A.	E173, E174, 180, E335-E337, 347
Levitskaya, T. D.	E194, E195
Lewis, B.	E194, E195. 198
Lewis, J. B.	E139-E141, 150, E385-E387, 394
Liao, P. -K.	E194, E195
Liebhafsky, H. A.	E84, E85, 89
Liebl, G.	E103-E105, E173, E174, E317, E318, E335-E337
Lihl, F.	E173, E174, 177, E285, E286, E302, E310, E311, E317-E318, E335-E337
Lindauer, G. C.	E93-E95
Lomov, A. L.	E230, E234
Loomis, A. G.	E57, E58, 61, E369-E371
Luborsky, E.	E301, E302
Lugscheider, E.	E258, E259, E263, E264, E335-E337, 349
Lugscheider, W.	E206, E214, E215, E221, E230, E231, E239, E242, E245, E247, E249, E253, E267
McBain, J. W.	E369-E371, 376, 377,
McCoy, H. N.	E234
Maey, E.	E1-E3, E13-E18, 19, E33, E34, E357-E359

Magel, T. T.	E425-E427
Makarova, I. A.	E76, E77, 81
Makarova, J. A.	E13-E18, 31
Makrides, A. C.	E326, 328, E330, 333
Malan, H. P.	E425-E427
Maltsev, Yu. T.	E385-E387
Mamutova, Z. A.	E214, E215
Maraman, W. J.	E263, E264, 265
Marshall, A. L.	E301, E302, 303
Mason, J. T.	E425-E427, 433
Masson, D. R.	E194, E195
Mathis, H. B.	E314
Matthes, F.	E207-E208, E214, E221, E225
Maurer, R. J.	E357-E359, 366
Maxwell, J.	E285, E286, E335-E337
Mees, G.	E369-E371, 382,
Mehl, R. F.	E402-E404, 416
Menzinger, F.	E93-E95, 101
Merlo, F.	E66, E67, 70, E71, E72, E73, E76, E77, 80, E207, E208, 212, E214, E215, E221, E225, E230, E231, E239, E245, E253, E425-E427
Merriam, M. F.	E103-E105, 114
Mertke, I.	E230
Merz, V.	E13-E18
Messing, A. F.	E225, 226, E230, 232, E239, 240, E263, E264, E282, E307, E326, E422, 423, E425-E427, 430
Minardi, A.	E301, E302, 304, E310
Miner, W. N.	E434
Mizetskaya, I. P.	E194, E195
Moers, K.	E300
Moesveld, A. L. T.	E402-E404, 415
Moffatt, W. G.	E83
Moran, W. G.	E103-E105
Morawietz, W.	E33, E34, E103-E105, 116
Morozova, V. E.	E335-E337, 350
Morrison, B. H.	E425-E427, 429
Moser, H.	E119-E121
Moshkevich, A. S.	E157-E159, 169
Mukhamedieva, Sh. M.	E214, E215, 217
Muller, R.	E57, E58, E84, E85
Mullins, L. J.	E263, E264, 265
Murphy, A. J.	E357-E359, 364
Mustafa, I.	E172
Muthman, W.	E214, E215
Mylius, F.	E84, E85
Nagaoka, H.	E301, E302, E310
Nazarov, B. F.	E317, E318
Nejedlik, J. F.	E134, E258, E259, E263, E264, E271, E274, E282, E284
Nerad, A. J.	E55, E263, E301, E302
Neville, F. H.	E13-E18, 20, E139-E141, 143, E157-E159, 161, E182, E183, 185, E402-E404, 406
Newnham, R.	E194, E195
Nielson, S.	E194, E195
Nigmatulina, A. A.	E157-E159, E181-E183, 188
Nigmetova, R. Sh.	E172, E182, E183, E230, E357-E359
Nikishin, G. D.	E207-E208, E214, E221, E225, E234, 235, E242, E253, 254
Nikushkina, N. L.	E369-E371, 384
Nizhnik, A. T.	E93-E95, 96
Noer, S.	E335-E337
North, E.	E157-E159
Norton, F. J.	E301, E302, 303
Nosek, M. V.	E139-E141, 154, E157-E159, 171, E182, E183, 189, E402-E404, 418
Novoselova, A. V.	E194, E195
Novotny, H.	E206, E326

O'Kane, D. F.	E194, E195
Oelsen, W.	E301, E302, E310, E317-E318
Ogawa, E.	E103-E105, 108
Ogg, A.	E357-E359
Olcese, G. L.	E214
Omarova, A. F.	E310, E317, E318
Omarova, N. D.	E33, E34, 43, E119-E121, 132, E335-E337, E385-E387, 401, E402-E404, 419
Omelchenko, A. V.	E194, E195, 197
Onstott, E. I.	E1-E3, 8
Osipova, G. A.	E214, E215, 218, E221, E225, 227, E230, E234, E253, 255
Ostapczuk, P.	E335-E337, 353, E357-E359
Osugi, A.	E13-E18, 29
Oteeva, G. Z.	E221, 223, E225
Pajaczkowska, A.	E194, E195, 201
Palenzona, A.	E214, E215, E221, E225, E230, E231, E234, E256
Palmaer, E.	E301, E302, E317, E318
Palman, H.	E84, E85, E173, E174, 179, E258, E259, 260, E277, 279, E285, E286, 291, E310, 312, E317, E318, 321, E335-E337, 344, E357-E359, 368, E422, 424, E425-E427, 431
Panina, L. S.	E310
Paranchich, S. Yu.	E194, E195
Parkman, M. F.	E268, 269, E277, 280, E301, E302, 306, E310, E311, E317, E318, 322
Parks, R. D.	E93-E95, 101
Parks, W. G.	E103-E105, E207, E208, 209, E422
Parravano, N.	E369-E371, 375,
Pavlovich, P.	E119-E121, 123
Peled, E.	E84, E85, 92
Pellini, G.	E194, E195, 199
Perov, E. I.	E230, E231, E234, 236
Peshkov, W.	E385-E387, 393
Petot, C.	E182, E183, 191
Petot-Ervas, G.	E139-E141, 153, E182, E183, 191
Pistorius, C. W. F. T.	E119-E121, 130, 131
Plaksin, I. N.	E330, 331, E369-E371, 379
Planting, P. J.	E330
Plate, W.	E284
Podkorytova, N. V.	E317, E318
Polyakova, T. P.	E173, E174
Popp, O.	E13-E18
Pratt, J. N.	E425-E427
Predel, B.	E93-E95, 97, E139-E141, 155, E182, E183, 190
Prener, J. S.	E194, E195
Price, E. G.	E307, E316
Prime, G. E.	E422
Prytherch, W. E.	E139-E141
Pushin, N. A.	E119-E121, 127, E139-E141, 144, E157-E159, 162, E182, E183, 186, E385-E387, 390, E402-E404, 409
Putley, E. H.	E194, E195
Raub, E.	E284
Ravdel, A. A.	E157-E159, 169
Ray, B.	E194, E195, 196
Reavis, J. G.	E263, E264, 265
Reid, R. C.	E258, E259
Reinders, W.	E357-E359, 361
Rey, J.	E119-E121, 133
Rhys, D. W.	E307, E316
Ricci, F. P.	E93-E95, 101
Richards, T. W.	E1-E3, E119-E121, 125, 126, E157-E159, E301, E302, E335-E337, 340
Richter, P. W.	E119-E121, 130, 131
Robert, J.	E103-E105
Rodot, H.	E194, E195

Rodot, M.	E194, E195
Roeder, A.	E33, E34
Rolfe, C.	E369-E371, 383,
Romano, A. J.	E258, E259, E263, E264, E271, E274, E301, E302
Roos, G. D.	E119-E121, 124
Rose, F.	E84, E85
Roslonek, H.	E330, 334
Rostocker, W.	E422
Rothacker, D.	E139-E141, 155, E182, E183, 190
Rotner, Yu. M.	E194, E195
Royce, H. D.	E285, E286, 287
Ruban, L. M.	E1-E3, 10, E33, E34
Russell, A. S.	E258, E259, E268, E277, E282, E284-E286, 288, E301, E302, E310, E317, E318, E335-E337, 339, E425-E427
Sagadieva, K. Zh.	E207-E208, 211, 213, E214, E215, 217, 218, 220, E221, 223, 224, E225, 229, E230, E231, 233, E234, 237, 238, E242, 243, E285, E286, 293, E335-E337, 342
Saito, A.	E425-E427
Sarieva, L. S.	E135, 138, E194, E195, 204
Sasim, D.	E335-E337, 355
Sayun, M. G.	E239
Schadler, H. W.	E385-E387, 395
Schenck, H.	E182, E183, 187
Schlosser, W.	E369-E371
Schmidt, W.	E84, E85, 90, E335-E337
Schmitt, R. E.	E1-E3, 12
Schneider, R. F.	E369-E371
Schonfeld, F. W.	E434
Schuhmann, H.	E33, E34, 42
Schuller, A.	E13-E18, 25
Schulz, L. G.	E119-E121, 128
Schulze, A.	E402-E404, 413
Schulze, R. C.	E277
Schupp, O. E.	E335-E337, 343
Schurmann, H. K.	E93-E95, 101
Schweitzer, D. G.	E425-E427
Segre, G. J.	E425-E427
Semibratova, N. M.	E139-E141, 154, E157-E159, 171, E402-E404, 418
Serebrennikov, V. V.	E230, E231, E234, 236
Shalaevskaya, V. N.	E84, E85, 91, E139-E141, E335-E337
Shalamov, A. E.	E157-E159, 171
Shirinskikh, A. V.	E402-E404
Shklar, R. Sh.	E369-E371
Shunk, F.	E173, E174
Shvedov, V. P.	E207-E208, E214, E221, E225, E234, 235, E242, E253, 254
Siede, B.	E119-E121
Smaaberg, R.	E285, E286, 299
Smith, A. R.	E301, E302
Smith, D. L.	E182, E183
Smith, F. E.	E402-E404, 412
Smith, G. McP.	E1-E3, 5, E33, E34, 39, E44, E45, 48, E50, E51, 53, 54, E66, E67, E71, E72, 75, E76, E77, 79
Smits, A.	E57, E58, E84, E85
Smurigina, T. V.	E76, E77
Smyth, C. P.	E119-E121, 126
Soshmikov, V. I.	E194, E195, 197
Spencer, J. F.	E119-E121, E157-E159, E194, E195, E335-E337, E402-E404
Spengler, H.	E103-E105, E173, E174, 196, E317, E318, E335-E337
Speranskaya, E. F.	E310
Spereni, G.	E57, E58, 60, E66, E67, 68
Spicer, W. M.	E93-E95, 97, E103-E105, 107
Spiegler, P.	E119-E121, 128
Srudka, M.	E335-E337, 355
Staurset, A.	E335-E337, 354

Author Index

Stegman, G.	E157-E159, 165
Steininger, J.	E194, E195, 202
Steinmetz, E.	E182, E183, 187
Stepanova, O. S.	E93-E95, E135
Steppan, F.	E285, E286, E317, E318, E330, E422, E425-E427
Stevenson, D. A.	E357-E359
Stormont, R. W.	E83
Strachan, J. F.	E1-E3, E13-E18, E33, E34, E44, E45, E55, E57, E58, E66, E67, E76, E77, E84, E85, E103-E105, E119-E121, E134, E139-E141, E157-E159, E172-E174, E182, E183, E258, E259, E263, E268, E271, E274, E277, E282, E284-E286, E301, E302, E307, E310, E314, E316-E318, E326, 327, E330, 332, E335-E337, E357-E359, E369-E371, E385-E387, E402-E404
Strauss, A. J.	E194, E195, 200
Straussfurth, T.	E335-E337, E357-E359
Strickland-Constable, R. F.	E385-E387, 396
Sturch, E.	E425-E427, 429
Su, C.-H.	E194, E195
Sucheni, A.	E119-E121
Sudakov, V. A.	E103-E105, 117
Sunden, N.	E103-E105, 109
Sunier, A. A.	E357-E359, 363, E369-E371, 378, 381,
Suranov, A. V.	E194, E195
Suvorovskaya, N. A.	E330, 331
Swift, E.	E13-E18, 28
Syroeshkina, T. V.	E135, 138
Szofran, F. R.	E194, E195, 203
Tammann, G.	E13-E18, 21, E33, E34, 35, E119-E121, E139-E141, 142, E157-E159, 160, E172, E182, E183, 184, E268, E277, E282, E284-E286, E301, E302, E310, E317, E318, E335-E337, 341, E357-E359, E369-E371, 373, E385-E387, 388, E402-E404, 405, E425-E427
Tannanaeva, N. N.	E103-E105, 106
Taylor, D. F.	E139-E141, 152
Teeter, C. E.	E402-E404
Tember, G. A.	E135, 136
Temmerman, E.	E173, E174, 181
Thibault, M.	E103-E105
Thompson, E. S.	E301, E302
Thompson, H. E.	E157-E159, 166
Timofeyev, G.	E157-E159
Toibaev, B. K.	E173, E174
Tomashik, V. N.	E194, E195
Toner, D. F.	E317, E318, 320
Trebukhov, A. A.	E357-E359
Tung, T.	E194, E195
Ugai, Ya. A.	E173, E174
Upthegrove, W. R.	E385-E387, 400, E402-E404, 417
Usachev, P. V.	E194, E195
Usenova, K. A.	E214, E215, 218, E221, E225, 227, E230, E234, E245, E253, 255
Van Heteran	E139-E141, 145
Van Lent, P. H.	E139-E141, 151
Vanstone, E.	E13-E18, 26, 27
Vanyarko, V. G.	E194, E195
Vanyukov, A. V.	E194, E195
Vargo, E. J.	E134, E271, E274, E277, E282, E284
Vengel, P. F.	E194, E195
Verbeek, F.	E173, E174, 181
Verpiaetse, H.	E173, E174, 181
Vier, D. T.	E421
Vokhrysheva, L. E.	E239
Vol, A. E.	E402-E404
Volkov, A. G.	E1-E3, E33, E34, E44, E45, E50, E51

Volkova, V. N.	E173, E174
Volosatova, N. S.	E194, E195
Vorona, Yu. V.	E194, E195
Wald, F.	E83
Walker, R. A.	E425-E427
Walls, H. A.	E385-E387, 400, E402-E404, 417
Wang, J. Y. N.	E55, 56, E258, E259, 261, E263, E264, E301, E302, E335-E337
Wang, L. C.	E425-E427
Watters, J. I.	E335-E337, 343
Weeks, J. R.	E258, E259, 262, E263, 265, E268, 270, E271, 273, E274, 276, E277, 281, E301, E302, 303, E310, E311, 313, E317, E318, 323, E330, E425-E427
Weibke, F.	E44, E45, 49
Weihs, G.	E66, E67
Weiner, L. G.	E369-E371, 381,
Weiss, P.	E84, E85, 88
Weith, W.	E13-E18
Whaley, D. K.	E268, E277, E310, E311, E317, E318
White, A. G.	E434
White, C. M.	E369-E371, 378,
White, J. C.	E139-E141, 149
Whitman, C. I.	E55
Wilkinson, W. D.	E55, E263
Williams, D. J.	E194, E195
Williams, E. J.	E57, E58, 63
Williams, T. C.	E369-E371
Wilson, A. S.	E425-E427, 428
Winkler, J.	E134
Winter, H.	E1-E3, 4, E13-E18, 23, E33, E34, 38, E44, E45, 46
Winterhagen, H.	E369-E371
Witteman, G. W.	E421
Wolf, W.	E1-E3, 7
Woolley, J. C.	E194, E195
Wymer, R. G.	E425-E427
Yaggee, F. L.	E55, E263
Yan-Sho-Syan, G. V.	E139-E141, 154, E157-E159, 171, E182, E183, 189, E402-E404, 418
Yanagaze, T.	E103-E105, 108
Yatsenko, S. P.	E93-E95, 99
Yoshida, Z.	E330
Youness, T.	E335-E337, 343
Young, A. S.	E194, E195
Yu, T. L.	E425-E427
Zaichko, L. F.	E173, E174
Zakharov, M. S.	E93-E95, E135, E173, E174, E335-E337
Zakharov, Z. A.	E335-E337
Zakharova, E. A.	E173, E174, E335-E337, 351
Zapponi, P. P.	E157-E159, 167
Zaripova, L. G.	E194, E195
Zebreva, A. I.	E1-E3, 11, E13-E18, 30, E33, E34, 43, E76, E77, 82, E119-E121, 132, E135, E139-E141, 156, E157-E159, 170, E173, E174, 178, 180, E181-E183, 188, 193, E207, E208, 211, 213, E214, E215, 217-220, E221, 223, 224, E225, 228, 229, E230, E231, 233, E234, 237, 238, E239, 241, E242, 243, E245, 246, E247, 248, E249, 250, E251, 252, E256, 257, E285, E286, 295, E335-E337, 347, E385-E387, 401, E402-E404, 419
ldybaeva, B.	E207-E208, 211
akanov, V. Z.	E1-E3, 11, E33, E34, E76, E77, 82, E139-E141, 156, E182, E183, 193
el, G.	E84, E85, 92
hov, V. P.	E194, E195
atskaya, I. V.	E194, E195
ker, D.	E55
ovsky, G. J.	E1-E3, 6, E44, E45, 47, E50, E51, 52
ic, V.	E317, E318, 325
golskaya, E. V.	E93-E95, 96
jagincev, O. E.	E13-E18

Sci Ref QD 543 .S6629 v.25
1986
Metals in mercury
JAN 14 1987